Lecture Notes in Physics

The Editorial Policy for Proceedings

The series Lecture Notes in Physics reports new developments in physical research and teaching – quickly, informally, and at a high level. The proceedings to be considered for publication in this series should be limited to only a few areas of research, and these should be closely related to each other. The contributions should be of a high standard and should avoid lengthy redraftings of papers already published or about to be published elsewhere. As a whole, the proceedings should aim for a balanced presentation of the theme of the conference including a description of the techniques used and enough motivation for a broad readership. It should not be assumed that the published proceedings must reflect the conference in its entirety. (A listing or abstracts of papers presented at the meeting but not included in the proceedings could be added as an appendix.)
When applying for publication in the series Lecture Notes in Physics the volume's editor(s) should submit sufficient material to enable the series editors and their referees to make a fairly accurate evaluation (e.g. a complete list of speakers and titles of papers to be presented and abstracts). If, based on this information, the proceedings are (tentatively) accepted, the volume's editor(s), whose name(s) will appear on the title pages, should select the papers suitable for publication and have them refereed (as for a journal) when appropriate. As a rule discussions will not be accepted. The series editors and Springer-Verlag will normally not interfere with the detailed editing except in fairly obvious cases or on technical matters.
Final acceptance is expressed by the series editor in charge, in consultation with Springer-Verlag only after receiving the complete manuscript. It might help to send a copy of the authors' manuscripts in advance to the editor in charge to discuss possible revisions with him. As a general rule, the series editor will confirm his tentative acceptance if the final manuscript corresponds to the original concept discussed, if the quality of the contribution meets the requirements of the series, and if the final size of the manuscript does not greatly exceed the number of pages originally agreed upon. The manuscript should be forwarded to Springer-Verlag shortly after the meeting. In cases of extreme delay (more than six months after the conference) the series editors will check once more the timeliness of the papers. Therefore, the volume's editor(s) should establish strict deadlines, or collect the articles during the conference and have them revised on the spot. If a delay is unavoidable, one should encourage the authors to update their contributions if appropriate. The editors of proceedings are strongly advised to inform contributors about these points at an early stage.
The final manuscript should contain a table of contents and an informative introduction accessible also to readers not particularly familiar with the topic of the conference. The contributions should be in English. The volume's editor(s) should check the contributions for the correct use of language. At Springer-Verlag only the prefaces will be checked by a copy-editor for language and style. Grave linguistic or technical shortcomings may lead to the rejection of contributions by the series editors. A conference report should not exceed a total of 500 pages. Keeping the size within this bound should be achieved by a stricter selection of articles and not by imposing an upper limit to the length of the individual papers. Editors receive jointly 30 complimentary copies of their book. They are entitled to purchase further copies of their book at a reduced rate. As a rule no reprints of individual contributions can be supplied. No royalty is paid on Lecture Notes in Physics volumes. Commitment to publish is made by letter of interest rather than by signing a formal contract. Springer-Verlag secures the copyright for each volume.

The Production Process

The books are hardbound, and the publisher will select quality paper appropriate to the needs of the author(s). Publication time is about ten weeks. More than twenty years of experience guarantee authors the best possible service. To reach the goal of rapid publication at a low price the technique of photographic reproduction from a camera-ready manuscript was chosen. This process shifts the main responsibility for the technical quality considerably from the publisher to the authors. We therefore urge all authors and editors of proceedings to observe very carefully the essentials for the preparation of camera-ready manuscripts, which we will supply on request. This applies especially to the quality of figures and halftones submitted for publication. In addition, it might be useful to look at some of the volumes already published. As a special service, we offer free of charge LATEX and TEX macro packages to format the text according to Springer-Verlag's quality requirements. We strongly recommend that you make use of this offer, since the result will be a book of considerably improved technical quality. To avoid mistakes and time-consuming correspondence during the production period the conference editors should request special instructions from the publisher well before the beginning of the conference. Manuscripts not meeting the technical standard of the series will have to be returned for improvement.

For further information please contact Springer-Verlag, Physics Editorial Department II, Tiergartenstrasse 17, D-69121 Heidelberg, FRG

J. Ehlers H. Friedrich (Eds.)

Canonical Gravity: From Classical to Quantum

Proceedings of the 117th WE Heraeus Seminar
Held at Bad Honnef, Germany,
13-17 September 1993

Springer-Verlag Berlin Heidelberg GmbH

Editors

J. Ehlers
H. Friedrich
MPI für Astrophysik
Karl-Schwarzschild-Strasse 1
D-85748 Garching bei München

ISBN 978-3-662-13954-7 ISBN 978-3-540-48665-7 (eBook)
DOI 10.1007/978-3-540-48665-7

CIP data applied for

Softcover reprint of the hardcover 1st edition 1994

Typesetting: Camera ready by author/editor
SPIN: 10127032 55/3140-543210 - Printed on acid-free paper

Preface

When asking a worker in the field about the progress in quantum general relativity in the last decade, one shouldn't be surprised to hear: "We understand the problems better". If it referred to a lesser task, such an answer would sound ironic. But the search for quantum gravity, a theory expected to combine in a suitable way the principles of classical general relativity and quantum theory, has been going on now for more than half a century and in spite of a number of ingenious proposals, a satisfactory theory is still lacking.

In view of the many approaches to quantum gravity it is difficult (not only) for a beginner to enter the field. The 117th WE Heraeus Seminar, 13 - 17 September 1993, on the "Canonical Formalism in Classical and Quantum Relativity" was organized to offer an introduction to an area of quantum gravity which on the one hand is still close to the original ideas of classical relativity and which on the other hand has seen a lot of activity and some significant progress recently. The present volume is a collection of the main talks given at that seminar.

Though we consider the present volume as a useful introduction to the problems of quantum gravity, it cannot replace a textbook, of course. The articles are different in style; they show, as is in the nature of the problem, different degrees of rigor, their representations vary from the mathematically well defined to the speculative, and, in spite of the restriction to a certain area of quantum gravity, the articles deal with a wide variety of questions, ranging from problems of pure mathematics and physics to ones of a conceptual nature. Furthermore, since the lectures rely on background material from two quite different theories, general relativity and quantum theory, the reader will invariably have to fill in some of this material herself.

However, since the field is still far from being closed, we consider it indeed an advantage to see related problems discussed from different perspectives. Moreover, in planning the seminar it was agreed with the speakers to make an effort to present the field of canonical general relativity in a coherent way, from the classical to the quantum aspects, so as to make the subject accessible also to non-specialists. The speakers were quite successful in doing so and we are grateful for their willingness to write up their talks. In particular we want to thank those authors who were prepared to report on material which is not part of their main research interests or which at present is not at the center of the discussion, but which is needed to complete the picture.

Garching, 1994

Jürgen Ehlers
Helmut Friedrich

Contents

Contributors

Abhay Ashtekar
Center for Gravitational Physics and Geometry
Pennsylvania State University
University Park, PA 16802-6300, USA

Robert Beig
Institut für Theoretische Physik
Universität Wien, Boltzmanngasse 5
1090 Wien, Austria

Bernd Brügmann
Max-Planck-Institut für Physik
Werner-Heisenberg-Institut, Foehringer Ring 6
80805 München, Germany

Domenico Giulini
Fakultät für Physik,
Universität Freiburg
Hermann-Herder-Str. 3
79104 Freiburg, Germany

Petr Hájiček
Inst. für Theoretische Physik
Universität Bern
Sidlerstr. 5
CH-3012 Bern, Switzerland

Chris J. Isham
Blacket Laboratory
Imperial College of Science and Technology
South Kensington
London SW7 2BZ, U.K.

Claus Kiefer
Fakultät für Physik
Universität Freiburg
Hermann-Herder-Str. 3
79104 Freiburg, Germany

Renate Loll
Center for Gravitational Physics and Geometry
Pennsylvania State University
University Park, PA 16802-6300, USA

Alan D. Rendall
Max-Planck-Institut für Astrophysik
Karl-Schwarzschild-Strasse 1, Postfach 1523
85740 Garching, Germany

Thomas Thiemann
Center for Gravitational Physics and Geometry
Pennsylvania State University
University Park, PA 16802-6300, USA

Andreas Wipf
Institut für Theoretische Physik
ETH - Hoenggerberg
CH - 8093 Zürich, Switzerland

Prima Facie Questions in Quantum Gravity *

C.J. Isham

1 Introduction

1.1 Preliminary Remarks

The many ideas and suggestions that have become attached to the study of quantum gravity form a substantial web whose subtle interconnections often cause considerable confusion, particularly amongst those approaching the subject for the first time. Therefore, when seeking to assess any particular scheme (such as canonical quantisation), it is helpful to begin by looking at the subject in the broadest possible terms with the aim of unravelling this web a little.

The present essay is intended to serve this need via an examination of certain *prima facie* questions that can be used to clarify the different structural and conceptual frameworks adopted by the various approaches to quantum gravity. The paper starts with a brief discussion of the four general ways whereby a quantum theory of gravity might be constructed. This is followed by some motivation for studying the subject in the first place and an explanation of what is meant by a '*prima facie*' question. Next there is a short sketch of the major current research programmes in quantum gravity: this ensures that the subsequent discussion does not take place in a complete technical vacuum. Then we move to three questions that are of exceptional importance in quantum gravity: (i) the role of background structure; (ii) the role of the spacetime diffeomorphism group; and (iii) the 'problem of time'. Finally, several of the current approaches to quantum gravity are used to illustrate some of the different ways in which these fundamental issues can be addressed.

What follows is a pedagogical exposition of some basic ideas in quantum gravity. It is *not* a full review of the field and, for this reason, references to original work are limited in number. Comprehensive reference lists can be found in recent genuine reviews of quantum gravity: for example [Alv89], [Kuc92] and [Ish92, Ish93].

* Based on a lecture given at the "Seminar in Memory of David Bohm", London, May 1993; and lectures given at the UK Institute for Particle Physics, St. Andrews, Scotland, September 1993.

1.2 What is Quantum Gravity?

Research in quantum gravity could perhaps be defined as any attempt to construct a theoretical scheme in which ideas from general relativity and quantum theory appear together in some way. A fundamental property of any such scheme is the existence of units with *dimensions* formed from Newton's constant G, Planck's constant $\hbar$, and the ubiquitous speed of light c. For example

- the Planck length $L_P := (G\hbar/c^3)^{1/2} \simeq 10^{-33}cm$;
- the Planck time $T_P := L_P/c \simeq 10^{-42}s$;
- the Planck mass $M_P := \hbar/cL_P \simeq 10^{-5}gms$;
- the Planck energy $E_P = M_P c^2 \simeq 10^{18}GeV$;

A major question in any approach to quantum gravity is what role is played by these fundamental units.

This definition of quantum gravity is very broad and includes, for example, studies of a quantum field propagating in a spacetime manifold equipped with a fixed background Lorentzian metric. However, in practice, references to 'quantum gravity' usually include the idea that a quantum interaction of the gravitational field with itself is involved in some way; quantum field theory in a fixed background is then better regarded as a way of probing certain aspects of quantum gravity proper.

Understood in this more limited sense, attempts to construct a quantum theory of gravity can be divided into four broad categories that I shall refer to as type I, type II, type III, and type IV.

I. The *quantisation* of general relativity.

The idea is to start with the classical theory of general relativity and then to apply some type of quantisation algorithm. This is intended to be analogous to the way in which the classical theory of an atom bound by the Coulomb potential is 'quantised' by replacing certain classical observables with self-adjoint operators on a Hilbert space. Of course, this is essentially also the approach used in developing important elementary-particle physics ideas like the Salam-Weinberg electro-weak theory and the quantum chromodynamics description of the strong nuclear force.

Approaches to quantum gravity of this type have been studied extensively and divide into two main categories: (i) 'canonical' schemes that start with a pre-quantum division of four-dimensional spacetime into a three-dimensional space plus time; and (ii) 'covariant' schemes that try to apply quantum ideas in a full spacetime context. The perturbative non-renormalisability of covariant quantum gravity was proved in the early 1970s, and most activity in this area ceased thereafter. On the other hand, the major advance inaugurated by Ashtekar's work has led to the canonical approach becoming one of the most active branches of quantum gravity research [Ash91].

II. *General-relativise* quantum theory.

Schemes of this type are much rarer than those of type I. The main idea is to begin with some prior idea of quantum theory and then to force it to be compatible with general relativity. The biggest programme of this type is due to Haag and his collaborators [FH87].

III. General relativity appears as a *low-energy limit* of a theory that is formed using conventional quantum ideas but which does *not* involve a type-I quantisation of the classical theory of relativity.

The dimensional nature of the basic Planck units lends credence to the idea of a theory that could reproduce standard general relativity in regimes whose scales are well away from that of the Planck time, length, energy *etc*. Superstring theory is the most successful scheme of this type and has been the subject of intense study during the last decade.

IV. Both general relativity *and* standard quantum theory appear only in certain limiting situations in the context of a theory that starts from radically new perspectives.

Very little is known about potential schemes of this type or, indeed, if it is necessary to adopt such an iconoclastic position in order to solve the problem of quantum gravity. However, the recurring interest in such a possibility is based on the frequently espoused view that the basic ideas behind general relativity and quantum theory are fundamentally incompatible and that any complete reconciliation will necessitate a total rethinking of the central categories of space, time and matter.

1.3 Why Should We Study Quantum Gravity?

Notwithstanding many decades of intense work, we are still far from having a complete quantum theory of gravity. The problem is compounded by the total lack of any empirical data (either observational or experimental) that is manifestly relevant to the problem. Under these circumstances, some motivation is necessary to explain why we should bother with the subject at all.

- *We must say something*. The value of the Planck length suggests that quantum gravity should be quite irrelevant to, for example, atomic physics. However, the non-renormalisability of the perturbative theory means it is impossible to actually compute these corrections, even if physical intuition suggests they will be minute. Furthermore, no consistent theory is known in which the gravitational field is left completely classical. Hence we are obliged to say *something* about quantum gravity, even if the final results will be negligible in all normal physical domains.
- *Gravitational singularities*. The classical theory of general relativity is notorious for the existence of unavoidable spacetime singularities. It has long been suggested that a quantum theory of gravity might cure this disease by some sort of 'quantum smearing'.
- *Quantum cosmology*. A particularly interesting singularity is that at the beginning of a cosmological model described by, say, a Robertson-Walker metric. Classical physics breaks down here, but one of the aims of quantum gravity has always been to describe the 'origin' of the universe as some type of quantum event.

- *The end state of the Hawking radiation process.* One of the most striking results involving general relativity and quantum theory is undoubtedly Hawking's famous discovery of the quantum thermal radiation produced by a black hole. Very little is known of the final fate of such a system, and this is often taken to be another task for a quantum theory of gravity.
- *The unification of fundamental forces.* The weak and electromagnetic forces are neatly unified in the Salam-Weinberg model, and there has also been a partial unification with the strong force. It is an attractive idea that a consistent quantum theory of gravity *must* include a unification of all the fundamental forces.
- *The possibility of a radical change in basic physics.* The deep incompatibilities between the basic structures of general relativity and of quantum theory have led many people to feel that the construction of a consistent theory of quantum gravity requires a profound revision of the most fundamental ideas of modern physics. The hope of securing such a paradigm shift has always been a major reason for studying the subject.

1.4 What Are Prima Facie Questions?

By a 'prima facie' question I mean the type of question that can be asked of almost any approach to quantum gravity and which is concerned with the most basic issues in the subject. The following general classes are of particular importance:

1. *General issues concerning the relation between classical and quantum physics.* The minimal requirement of any quantum theory of gravity is that it should reproduce classical general relativity and standard quantum theory in the appropriate physical domains. However, it is difficult to make general categorical statements about either (i) what is meant by the 'classical limit' of a given quantum system; or (ii) how to construct a quantum analogue of any given classical system: in practice, the matter is usually decided on an *ad hominum* basis. This produces significant problems in attempts to quantise gravity using, for example, a type-I canonical formalism.

2. *Specific issues in quantum gravity.* Three especially important issues of direct relevance to quantum gravity are as follows:

- How relevant are the *spacetime* concepts associated with the classical theory of general relativity? Do the Planck length and time signify the scales at which all normal ideas of space and time break down?
- To what extent is it appropriate to construct a quantum theory of gravity using the technical and conceptual apparatus of *standard* quantum theory? For example, the traditional 'Copenhagen' interpretation of quantum theory is often asserted to be quite inappropriate for a theory of quantum cosmology. What should take its place?
- Does a consistent quantum theory of gravity necessarily involve the *unification* of the fundamental forces of physics or is it possible to construct a

theory that involves the gravitational field alone? This questions signals one of the key differences between superstring theory and the Ashtekar version of canonical quantum gravity: the latter asserts that a quantisation of pure gravity *is* possible whereas one of the main claims of superstring theory is to provide a scheme that encompasses all the forces.

Most readers will probably agree that questions like the above are likely to be relevant in any approach to constructing a quantum theory of gravity. But there is a hidden danger that should not be underestimated. As theoretical physicists, we are inclined towards a simple realist philosophy that sees our professional activities in terms of using an appropriate conceptual scheme to link what is 'out there' (the world of 'actual facts') to some mathematical model. However, one of the important lessons from the philosophy of science is that facts and theories cannot be so neatly separated: what we call a 'fact' does not exist without some theoretical schema for organising experimental and experiential data; and, conversely, in constructing a theory we inevitably impose some prior idea of what we mean by a fact.

In most branches of physics no real problem arises when handling this interconnected triad of facts, mathematical model and bridging conceptual framework: it has simply become part of the standard methodology of science. However, the situation in quantum gravity is rather different since there are no known ways of directly probing the Planck regime. This lack of hard empirical data means that research in the subject has tended to focus on the construction of abstract theoretical schemes that are (i) internally consistent (in a mathematical sense), and (ii) are compatible with some preconceived set of concepts. This rather introspective situation helps to fuel the recurrent debate about whether the construction of a comprehensive theory of quantum gravity requires a preliminary fundamental reappraisal of our standard concepts of space, time and matter, or whether it is better to try first to construct an internally consistent mathematical model and only then to worry about what it 'means'.

Of course, a sensible pragmatist will strive to maintain a proper balance between these two positions, but this ever-present tension between conceptual framework and mathematical model does lend a peculiar flavour to much research in the field. In particular, the wide range of views on how to approach the subject has generated a variety of different research programmes whose practitioners not infrequently have difficulty in understanding what members of rival schools are trying to do. This is one reason why it is important to uncover as many as possible of the (possibly hidden) assumptions that lie behind each approach: one person's 'deep' problem may seem irrelevant to another simply because the starting positions are so different. This situation also shows how important it is to try to find some area of physics where the theory can be tested directly. A particularly important question in this context is whether there are genuine quantum gravity effects at scales well below the Planck energy. Needless to say, the answer to a question like this is itself likely to be strongly theory dependent.

1.5 Current Research Programmes in Quantum Gravity

At this stage it might be helpful to give a brief account of some of the major current research programmes in quantum gravity. I find the following scheme particularly useful, although the plethora of topics studied could certainly be organised in many other ways.

A. *Quantum Gravity Proper*

The two major current programmes that attempt to construct a full-blown theory of quantum gravity are the Ashtekar version of canonical quantum gravity, and superstring theory.

Canonical quantum gravity has a long history and has been used extensively to discuss a variety of conceptual issues including the problem of time and the possible meaning of a quantum state of the entire universe. However, the great difficulties that arise when trying to make proper mathematical sense of the crucial equations (in particular, the Wheeler-DeWitt equation) that arise in this type-I approach eventually impose insurmountable obstacles. One of the reasons why the Ashtekar programme is so interesting is that many of these issues can now be reopened within the context of a mathematical framework that is much better behaved.

The technical and conceptual framework within which superstring theory is currently discussed is very different from that of canonical quantum gravity and owes far more to its origins in elementary particle physics that it does to the classical theory of general relativity, which arises only as a low-energy limit of the theory. As a consequence, questions concerning the status of space and time take on a quite different form from that in the canonical formalism. This is to be expected of any type-III approach to quantum gravity.

Another significant type-I scheme is the 'euclidean programme' whose basic ingredient is a functional integral over metrics with a Riemannian signature [GH93]. This view of quantum gravity cannot be called a 'full-blown' approach to quantum gravity since, as yet, there is no known way of making mathematical sense of such integrals; in practice, most work involves a saddle-point, semi-classical approximation. On the other hand, the heuristic functional integral can easily be extended to include a sum over different manifolds, and hence the scheme is a natural one in which to discuss topology change; something that is rather difficult in the canonical approach.

B. *Quantum Cosmology*

One of the major reasons for studying quantum gravity is to understand the Planck era of the very early universe. Most discussions of quantum cosmology have employed 'minisuperspace' models (sometimes in the context of the euclidean programme) in which only a finite number of the gravitational modes are quantised. Too much weight should not be attached to the results of such crude approximations, especially those (the great bulk) that use the ill-defined equations of standard canonical quantum gravity. However, models of this type can be valuable tools for exploring the many conceptual problems that arise in the typical quantum-cosmology situation where one aspires to describe the quantum state of the entire universe. For example, there has been much discussion of the problem of time and the, not unrelated, inapplicability of the normal Copenhagen interpretation of quantum theory. One of the currently most active ways of tackling these issues involves the 'consistent histories' approach to quantum theory [Har93].

C. *Model Systems*

In addition to minisuperspace techniques many other model systems have been studied with the aim of probing specific aspects of the full theory. For example:

- *Quantum gravity theories in three or less spacetime dimensions.* Low-dimensional quantum gravity has been widely studied. In particular, the three-dimensional theory has been solved completely using a variety of different methods and provides valuable information on how these might be related in general.
- *Quantum field theory on a spacetime with a fixed Lorentzian geometry.* There has been a substantial interest in this subject since Hawking's discovery of black-hole radiation. It throws useful light on certain aspects of the full theory and might also have direct astrophysical significance.
- *Semi-classical quantisation.* There has been much activity in recent years devoted to the WKB approximation to the Wheeler-DeWitt equation. This approach is unlikely to reveal much about the Planck regime proper but, nevertheless, there have been a number of interesting results, especially suggestions that the techniques may yield genuine results away from the scales set by the Planck length and time.
- *Regge calculus.* The idea of approximating spacetime by a simplicial complex has been of interest in both classical and quantum gravity for a long time. The quantum results to date are rather modest but could increase in the future with the rapid increase in the power of computing systems.

D. *Spacetime Structure at the Planck Length and Time*

A major motivation for many who elect to study quantum gravity is the belief that something really fundamental happens to the structure of space and time at the Planck scale. This has inspired a number of fragmented attempts that start *ab initio* with a theoretical framework in which standard spacetime concepts are radically altered. Schemes of this type include twistor theory, various approaches to a discrete models of physics, non-commutative geometry, quantum topology or set theory, quantum causal sets, and the like. The main difficulty is that the starting point of programmes of this type is so far from conventional physics that it is difficult to get back to the mundane world of Einstein field equations in a continuum spacetime.

2 Prima Facie Questions in Quantum Gravity

I wish now to consider in more detail three exceptionally important questions that can be asked of any approach to quantum gravity. These are: (i) what background structure is assumed?; (ii) what role is played by the spacetime diffeomorphism group?; (iii) how is the concept of 'time' to be understood?

2.1 Background Structure

The phrase 'background structure' can mean several things. It can refer to a specific choice of, say, a manifold or Lorentzian metric that is fixed once and for all and which therefore is not itself subject to quantum effects. For example, the early attempts to construct a quantum theory of gravity using ideas drawn from particle physics involved writing the spacetime metric $g_{\alpha\beta}(X)$ as a sum $\eta_{\alpha\beta} + h_{\alpha\beta}(X)$ where $\eta_{\alpha\beta}$ is the Minkowski metric diag$(-1,1,1,1)$. The idea was to quantise $h_{\alpha\beta}(X)$ using standard relativistic quantum field theory (see section 3.1). Thus, in this type-I approach, the topology and differential structure of spacetime are fixed as that of the vector space $\mathbb{R}^4$, while the background Minkowski metric plays a key role in the quantum field theory via its associated Poincaré group of isometries.

Analysis of such background structure is a useful aid in classifying and distinguishing the various approaches to quantum gravity. However, 'background' can be used in another way that, if anything, is even more important but which is frequently not articulated in such a concrete way. I mean the entire *conceptual* and *structural* framework within whose language any particular approach is couched. Different approaches to quantum gravity differ significantly in the frameworks they adopt, which causes no harm—indeed the selection of such a framework is an essential pre-requisite for theoretical research—provided the choice is made consciously. The problems arise when practioners from one particular school become so accustomed to a specific structure that it becomes, for them, an almost *a priori* set of truths, and then they find it impossible to understand how any other position could ever be valid. Unfortunately, a fair number of such misunderstandings have occurred during the history of the study of quantum gravity.

Bearing all this in mind let us begin now to examine some of the specific issues that concern the choice of such technical or conceptual background structure.

A. *The Use of Standard Quantum Theory*
One of the common features of superstring theory and the Ashtekar programme is their use of standard quantum theory. True, the normal formalism has to be adapted to handle the constraints that appear in both approaches, but most of the familiar apparatus is present: linear vector space, linear operators, inner products *etc.* However, there have been recurrent suggestions that a significant change in the formalism is necessary before embarking on a full quantum gravity programme. For example:

1. General relativity may induce an essential *non-linearity* into quantum theory [KFL86, Pen86, Pen87]. In particular, it may be possible to regard the infamous 'reduction of the state vector' as a genuine dynamical process induced by interactions involving the gravitational field. The major approaches to quantum gravity would change radically if it was necessary to start *ab initio*

with a non-linear theory rather than, say, deriving the non-linear effects as some type of higher-order correction.

2. How valid are the *continuum* concepts employed in quantum theory, in particular the use of real and complex numbers? The idea here is roughly as follows. One reply to the question "why do we use real numbers in quantum theory?" is that we want the eigenvalues of self-adjoint operators to be real numbers because they represent the possible results of physical measurements. And why should the results of measurements be represented by real numbers? Because all measurements can ultimately be reduced to the positions of a pointer in space, and space is modelled using real numbers. In other words, in using real or complex numbers in quantum theory we are arguably making a prior assumption about the continuum nature of space. However, it has often been suggested that the Planck length and time signal the scale at which standard spacetime concepts break down, and that a more accurate picture might be a discrete structure that looks like a familiar differentiable manifold only in some coarse-grained sense. But we will not be able to construct such a theory if we start with a quantum framework in which the continuum picture has been assumed *a priori*.

   This argument is not water-tight, but it does illustrate quite well how potentially unwarranted assumptions can enter speculative theoretical physics and thereby undermine the enterprise.

3. The Hawking radiation from a black hole is associated with a loss of information through the event horizon and corresponds to what an external observer would regard as a transition from a pure state to a mixed state. If the idea of a potential loss of information has to be imposed on the formalism *ab initio* it would enforce a significant change in the current quantum gravity programmes.

These three possible objections to standard quantum theory are all concerned with the mathematical structure of the subject. However, another serious problem arises with the *interpretation* of quantum theory, especially in the context of quantum cosmology. This particular objection has been taken very seriously in recent years and by now there is a fairly widespread agreement that the familiar 'Copenhagen' view is not appropriate. In particular, there is a strong desire to find an alternative interpretation whose fundamental ingredients do not include the notion of a measurement by an external observer.

The consistent-histories approach may provide such a scheme although, even here, there is a problem in so far as most discussions of this subject presuppose the existence of a fixed spacetime. In conventional quantum theory there is certainly a strong case for arguing that this is necessary, both for the mathematical foundations and for the conceptual interpretation of the theory. This raises what turns out to be one of the most interesting *prima facie* questions in quantum gravity: how much of the standard spacetime structure must be imposed as part of the fixed background?

B. *How Much Spacetime Structure Must Be Fixed?*
The mathematical model of spacetime used in classical general relativity is a differentiable manifold equipped with a Lorentzian metric. Some of the more important pieces of substructure underlying this picture are illustrated in Figure 1.

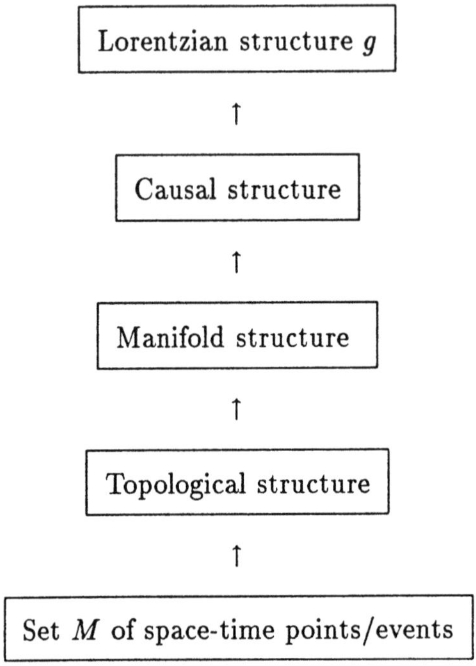

Figure 1. The spacetime structure of classical general relativity

The bottom level is a set M whose elements are to be identified with spacetime 'points' or 'events'. This set is formless with its only general mathematical property being the cardinal number. In particular, there are no relations between the elements of M and no special way of labelling any such element.

The next step is to impose a topology on M so that each point acquires a family of neighbourhoods. It now becomes possible to talk about relationships between points, albeit in a rather non-physical way. This defect is overcome by adding the key ingredient of all standard views of spacetime: the topology of M must be compatible with that of a differentiable manifold. A point can then be labelled uniquely in M (at least, locally) by giving the values of four real numbers. Such a coordinate system also provides a more specific way of describing relationships between points of M, albeit not intrinsically in so far as these depend on which coordinate systems are chosen to cover M.

In the final step a Lorentzian metric g is placed on M, thereby introducing the ideas of the lengths of a path joining two spacetime points, parallel transport with respect to a Riemannian connection, causal relations between pairs of points *etc.* There are also a variety of possible intermediate steps between the manifold and Lorentzian pictures; for example, as signified in Figure 1, the idea of a causal structure is more primitive than that of a Lorentzian metric.

The key question is how much of this classical structure is to be held fixed in the quantum theory. In the context of the Copenhagen interpretation the answer is arguably "all of it". I suspect that Bohr would have identified the spacetime Lorentzian structure as an intrinsic part of that classical world which he felt was such an essential epistemological prerequisite for the discussion of quantum objects. It seems probable therefore that he would not have approved at all of the subject of quantum gravity!

However, since we wish to assert that some sort of quantum spacetime structure *is* meaningful, the key question for any particular approach to quantum gravity is how *much* of the hierarchy illustrated in Figure 1 must be kept fixed. For example, in most type-I approaches to 'quantising' the gravitational field, the set of spacetime points, topology and differential structure are all fixed, and only the Lorentzian metric g is subject to quantum fluctuations. But one can envisage more interesting type-I possibilities in which, for example, the set M and its topology are fixed, but quantum fluctuations are permitted over all those manifold structures that are compatible with this particular topology. Moving back a step, one can envisage the exotic idea of fixing only the set-theoretic structure of M and allowing quantum fluctuations over topologies that can be placed on this set, including perhaps many that are not compatible with a manifold structure at all. Finally, one might even imagine 'quantising' the point set M itself: presumably by allowing quantum fluctuations in its cardinal number.

The ideas sketched above are all examples of what might be called 'horizontal' quantisation in which the quantum fluctuations take place only within the category of objects specified by the classical theory. However, it is also possible to contemplate 'vertical' quantisation in which fluctuations take place in a wider category. A simple example, and one that arises in several type-I schemes, is to permit quantum fluctuations in the metric to include fields that are degenerate, or have signature other than $(-1,1,1,1)$. A more exotic possibility would be to allow fluctuations in manifold structure that include non-commutative manifolds.

The ideas of 'horizontal' and 'vertical' quantisation arise most naturally in the context of type-I approaches in which a given classical system is 'quantised' in some way. However, the question of how much of the classical structure of spacetime remains at different levels of the full theory can be asked meaningfully in all the four general approaches to quantum gravity. This is related to the question of the role of the Planck length in such theories. A common expectation is that the standard picture of space and time is applicable only at scales well above the Planck regime, and that the Planck length, time, energy *etc* signal the point at which *phase transitions* take place.

The notion of different phases is attractive but it also suggests that a complete theory of quantum gravity should assume no prior spacetime structure at all. Of course, this does not forbid the construction of partial theories that describe the theory in a particular phase; indeed, this may be a necessary first stage in the construction of the full structure.

C. *Background Causal Structure*

A piece of potential background of particular importance is causal structure. For example, consider the problem of constructing the quantum theory of a scalar field ϕ propagating on a spacetime manifold M equipped with a fixed Lorentzian metric g. In such a theory, a key role is played by the microcausal condition

$$[\hat{\phi}(X), \hat{\phi}(Y)] = 0 \tag{1}$$

for all spacetime points X and Y that are spacelike separated with respect to g. More rigorously, if $\mathcal{A}(O)$ denotes the C^*-algebra of local observables associated with the spacetime region $O \subset M$, then we require $[\mathcal{A}(O_1), \mathcal{A}(O_2)] = 0$ for any disjoint regions O_1 and O_2 that are spacelike separated.

Now consider what happens in a type-I, 'covariant' approach to quantum gravity. If the Lorentzian metric g becomes quantised then the light cone associated with any spacetime point is no longer fixed and it is not meaningful to impose a microcausal relation like (1): any pair of spacetime points are 'potentially' null or time-like separated, and hence spacetime quantum fields can never commute.

This collapse of one of the bedrocks of conventional quantum field theory is probably the single greatest reason why spacetime approaches to quantum gravity have not got as far as might have been hoped. In the original, particle-physics based schemes the problem was circumscribed by introducing a background Minkowskian metric η and then quantising the graviton using the microcausal structure associated with η. Such a background is also necessary for the idea of 'short-distance' behaviour (which plays a key role in discussing renormalisability) to have any meaning. However, the use of a fixed causal structure is an anathema to most general relativists and therefore, even if this approach to quantum gravity had worked (which it did not), there would still have been a strong compunction to reconstruct the theory in a way that does not employ any such background.

This very non-trivial problem is one of the reasons why the canonical approach to quantum gravity has been so popular. However, a causal problem arises here too. For example, in the Wheeler-DeWitt approach, the configuration variable of the system is the Riemannian metric $q_{ab}(x)$ on a three-manifold Σ, and the canonical commutation relations invariably include the set

$$[\hat{q}_{ab}(x), \hat{q}_{cd}(x')] = 0 \tag{2}$$

for all points x and x' in Σ. In normal canonical quantum field theory such a relation arises because Σ is a space-like subset of spacetime, and hence the fields at x and x' should be simultaneously measurable. But how can such a relation be justified in a theory that has no fixed causal structure? This problem is rarely mentioned but it means that, in this respect, the canonical approach to quantum gravity is no better than the covariant one. It is another aspect of the 'problem of time' to which my second lecture is devoted

2.2 The Role of the Spacetime Diffeomorphism Group Diff(M)

The group Diff(M) of spacetime diffeomorphisms plays a key role in the classical theory of general relativity and so the question of its status in quantum gravity is of considerable *prima facie* interest. We shall restrict our attention to diffeomorphisms with compact support, by which I mean those that are equal to the unit map outside some closed and bounded region of M. Thus, for example, a Poincaré-group transformation of Minkowski spacetime is not deemed to belong to Diff(M). This restriction is imposed because the role of transformations with a non-trivial action in the asymptotic regions of M is quite different from those that act trivially.

The role of Diff(M) in quantum gravity depends strongly on the approach taken to the subject. For example, in a type-III or type-IV scheme the structure of classical relativity is expected to appear only in a low-energy limit and so there is no strong reason to suppose that Diff(M) will play any fundamental role in the quantum theory. A type-II scheme is quite different since the group of spacetime diffeomorphisms is likely to be a key ingredient in forcing a quantum theory to comply with the demands of general relativity. On the other hand, the situation for type-I approaches is less clear. Any scheme based on a prior canonical decomposition into space plus time is bound to obscure the role of spacetime diffeomorphisms, and even in the covariant approaches quantisation may affect enough of the classical theory to detract from the significance of such transformations.

Some insight can be gained by looking at certain aspects of spacetime diffeomorphisms in classical general relativity. It is helpful here to distinguish between the pseudo-group of local coordinate transformations and the genuine group Diff(M) of global diffeomorphisms of M. Compatibility with the former can be taken to imply that the theory should be written using tensorial objects on M. On the other hand, as Einstein often emphasised, Diff(M) appears as an active group of transformations of M, and invariance under this group implies that the points in M have no direct physical significance. Of course, this is also true in special relativity but it is mitigated there by the existence of inertial reference frames that can be transformed into each other by the Poincaré group of isometries of the Minkowski metric.

Put somewhat differently, the action of Diff(M) on M induces an action on the space $\mathcal{F}$ of spacetime fields, and the only thing that has immediate physical meaning is the quotient space $\mathcal{F}/\mathrm{Diff}(M)$ of orbits, *i.e.*, two field configurations are regarded as physically equivalent if they are connected by a Diff(M) transformation. Technically, this is analogous to the situation in electromagnetism whereby a vector potential A_μ is equivalent to $A_\mu + \partial_\mu f$ for all functions f. However, there is an important difference between the electromagnetism and general-relativity. Electromagnetic gauge transformations occur at a fixed spacetime point X, and the physical configurations can be identified with the values of the electromagnetic field $F_{\mu\nu}(X)$, which depends *locally* on points of M. On the other hand, Diff(M) maps one spacetime point into another, and therefore the obvious way of constructing a Diff(M)-invariant object is to take some scalar function of spacetime fields and integrate it over the whole of M, which gives something that is very *non*-local. The idea that 'physical observables' are naturally non-local is an important ingredient in many approaches to quantum gravity.

2.3 The Problem of Time

One of the major issues in quantum gravity is the so-called 'problem of time'. This arises from the very different roles played by the concept of time in quantum theory and in general relativity. Let us start by considering standard quantum theory.

1. Time is not a physical observable in the normal sense since it is not represented by an operator. Rather, it is treated as a *background* parameter which, as in classical physics, is used to mark the evolution of the system; in this sense it can be regarded as part of Bohr's background classical structure. In particular, it provides the parameter t in the time-dependent Schrödinger equation
$$i\hbar\frac{d\psi_t}{dt} = \hat{H}\psi_t. \tag{3}$$
This is why the meaning assigned to the time-energy uncertainty relation $\delta t\,\delta E \geq \frac{1}{2}\hbar$ is quite different from that associated with, for example, the position and the momentum of a particle.
2. The idea of events happening at a single time plays a crucial role in the technical and conceptual foundations of quantum theory:
   - The notion of a *measurement* made at a particular time is a fundamental ingredient in the conventional Copenhagen interpretation. In particular, an *observable* is something whose value can be measured at a fixed time.
   - One of the central requirements of the *scalar product* on the Hilbert space of states is that it is conserved under the time evolution (3). This is closely connected to the unitarity requirement that probabilities always sum to one.
   - More generally, a key ingredient in the construction of the Hilbert space for a quantum system is the selection of a complete set of observables that are required to *commute* at a fixed value of time.
3. These ideas can be extended to systems that are compatible with special relativity: the unique time system of Newtonian physics is simply replaced with the set of relativistic inertial reference frames. The quantum theory can be made independent of a choice of frame if it carries a unitary representation of the Poincaré group. In the case of a relativistic quantum field theory, this is closely related to the microcausality requirement, which—as emphasised earlier—becomes meaningless if the light cone is itself the subject of quantum fluctuations.

The background Newtonian time appears explicitly in the time-dependent Schrödinger equation (3), but it is pertinent to note that such a time is truly an abstraction in the sense that no *physical* clock can provide a precise measure of it [UW89]: there is always a small probability that a real clock will sometimes run backwards with respect to Newtonian time.

When we come to a Diff(M)-invariant theory like classical general relativity the role of time is very different. If M is equipped with a Lorentzian metric g, and if its topology is appropriate, it can be foliated in many ways as a one-parameter family of space-like surfaces, and each such parameter might be regarded as a possible definition of time. However several problems arise with this way of looking at things:

- There are many such foliations, and there is no way of selecting a particular one, or special family of such, that is 'natural' within the context of the theory alone.
- Such a definition of time is rather non-physical since it provides no hint as to how it might be measured or registered.
- The possibility of defining time in this way is closely linked to a fixed choice of the metric g. It becomes untenable if g is subject to some type of quantum fluctuation.

The last problem is crucial in any type-I approach to quantum gravity and raises a number of important questions. In particular:

- How *is* the notion of time to be incorporated in a quantum theory of gravity?
- Does it play a fundamental role in the construction of the theory or is it a 'phenomenological' concept that applies, for example, only in some coarse-grained, semi-classical sense?
- In the latter case, how reliable is the use at a basic level of techniques drawn from standard quantum theory?

The three main ways that have been suggested for solving the problem of time are as follows.

1. Fix some *background* causal structure and use that to determine temporal concepts in the quantum theory. Such a background might arise from two possible sources.
   - It might come from a *contingent* feature of the actual universe; for example, the 3^0K thermal radiation. However, structure of this type is approximate and therefore works only if fine details are ignored. Also, there is a general matter of principle: do we expect a quantum theory of gravity to work for 'all possible' universes (whatever that might mean), or only for the actual one in which we happen to live?
   - An asymptotic causal structure could be associated with a spacetime manifold that is spatially non-compact and asymptotically flat. However, this would not help in the typical cosmological situation, and it is by no means obvious that 'time' defined in this way can be measured in any physically meaningful way.
2. Attempt to locate events both spatially and temporally with specific functionals of the gravitational and other fields. This important idea is based on the observation that, for example, if ϕ is a scalar field then, as emphasised

earlier, the value $\phi(X)$ of ϕ at a particular $X \in M$ has no physical meaning. On the other hand, the value of ϕ where something 'is' *does* have a physical meaning in the sense that 'ϕ(thing)' is Diff(M) invariant.
The hope is that an 'internal time' of this type can be introduced in such a way that the normal dynamical equations of the classical theory are reproduced *precisely*. Then one would try to apply a similar technique to the quantum case. Ideas of this type have played a major role in the development of canonical quantum gravity.

3. The third approach starts by constructing some sort of quantum theory but with no reference to time at all. Physical time is then introduced as a reading on 'real clocks' but it is accepted that such a scheme will never *exactly* reproduce the standard notion of time and that all physical clocks will at best work in some semi-classical limit. Approaches of this type are truly 'timeless' and raise the key issue of whether a meaningful quantum theory can indeed be created in a way that contains no fundamental reference to time. That this is not a trivial matter is exemplified by the remarks made earlier about the crucial role of time in conventional quantum theory.

3 Approaches to Quantum Gravity

As explained earlier, there are four general ways of trying to construct a quantum theory of gravity: I., quantize general relativity; II., general-relativise quantum theory; III., schemes constructed using standard quantum theory in which general relativity emerges only in some low-energy limit; and IV., schemes in which both general relativity and quantum theory emerge in some appropriate domain in the context of a theory that contains radical new ideas. Our task now is to see how the *prima facie* issues discussed in the previous section are addressed in some of these schemes.

3.1 Quantize General Relativity

A. *The Particle-Physics Approach*
The early particle-physics based approaches to quantum gravity illustrate quite well a number of the issues discussed above. The starting point is to fix the background topology and differential structure of spacetime M to be that of Minkowski space, and then to write the Lorentzian metric g on M as

$$g_{\alpha\beta}(X) = \eta_{\alpha\beta} + h_{\alpha\beta}(X) \tag{4}$$

where h measures the departure of g from flat spacetime η.

The background metric η provides a fixed causal structure with the usual family of Lorentzian inertial frames. Thus, at this level, there is no problem of time. The causal structure also allows a notion of microcausality, thereby permitting a conventional type of relativistic quantum field theory to be applied to the field $h_{\alpha\beta}$. In particular, the quanta of this field (defined as usual using representations of the Poincaré group of isometries of η) are massless spin-2 particles. A typical task would then be to compute perturbative scattering-matrix elements for these gravitons, both with each other and with the quanta of various matter fields. Note that there is no immediate problem of interpretation: the existence of a background spacetime manifold and causal structure fits in well with the standard Copenhagen view of quantum theory.

The action of Diff(M) is usually studied infinitesimally and is reflected in the quantum theory via a set of Ward identities that must be satisfied by the n-point functions of the theory. Thus the role of spacetime diffeomorphisms is also relatively straightforward.

It is clear that many of the *prima facie* issues discussed earlier are resolved in an approach of this type by virtue of its heavy use of background structure. However, many classical relativists object violently to an expansion like (4), not least because the background [2] causal structure cannot generally be identified with the physical one. Also, one is restricted to a specific background topology, and so a scheme of this type is not well adapted for addressing many of the most interesting questions in quantum gravity: black-hole phenomena, quantum cosmology, phase changes *etc*. Nevertheless, if the scheme above had worked it would have been a major result and would undoubtedly have triggered a substantial effort to construct a covariant type-I theory in a non-perturbative way; a good analogue is the great increase in studies of lattice gauge theory that followed the proof by t'Hooft that Yang-Mills theory is perturbatively renormalisable.

Of course, this did not happen in the gravitational case because the ultraviolet divergences are sufficiently violent to render the theory perturbatively non-renormalisable. One reaction has been to regard this pathology as a result of using the expansion (4); an expansion that is, anyway, unpleasant when viewed from the canons of the classical theory. Several attempts have been made to construct a non-perturbative, covariant scheme, but none is particularly successful and it was only when Ashtekar made his important discoveries in the context of the *canonical* theory that the idea of non-perturbative quantisation really began to bear fruit.

The majority of particle physicists followed a different line and tried to enlarge the classical theory of general relativity with carefully chosen matter fields with the hope that the ultraviolet divergences would cancel, leaving a theory that is perturbatively well-behaved. The cancellation of a divergence associated with a loop of bosonic particles (like the graviton) can be achieved only by the introduction of *fermions*, and hence supergravity was born. However, supersymmetry requires very special types of matter, which supports the idea that a successful theory of quantum gravity must lead to a unified theory; *i.e.*, the extra fields needed to cancel the graviton infinities might be precisely those associated with some grand unified scheme of the fundamental forces.

Early expectations were high following successful low-order results but it is now generally accepted that if higher-loop calculations could be performed (they are very complex) intractable divergences would appear once more. However, this

---

[2] The scheme can be extended to use an arbitrary background metric, but this does not change the force of the objection.

line of thinking is far from dead and the torch is currently carried by perturbative superstring theory.

Superstring theory has the great advantage over the simple covariant approaches that the individual terms in the appropriate perturbation expansion *can* be finite and, furthermore, the particle content of theories of this type could well be such as to relate the fundamental forces in a unified way.

The low-energy limit of these theories is a form of supergravity but, nevertheless, standard spacetime ideas do not play a very significant role. This is reflected by the graviton being only one of an infinite number of particles in the theory; similarly, the spacetime diffeomorphism group appears only as part of a much bigger structure. This down-playing of classical general relativity is typical of a type-III approach.

Notwithstanding the successes of superstring theory, some of the earlier objections to perturbative schemes still hold and, in addition, the superstring perturbation series is highly divergent. Hence much current attention is being devoted to the challenge of constructing a non-perturbative version of the theory. Most of the suggestions made so far work within the context of standard quantum theory and, in this sense, they are still of type III. However, the possibility also arises of finding a genuine type-IV structure whose low-energy limits would include standard quantum theory as well as supergravity.

B. *The Canonical Approach*

The response to our *prima facie* questions given by the canonical approach to quantum gravity differs significantly from that of the particle-physics based schemes. Since my second lecture is devoted to canonical quantisation, I will merely sketch here some of the most important features.

1. *Use of standard quantum theory.* The basic technical ideas of standard quantum theory are employed, albeit adapted to handle the non-linear constraints satisfied by the canonical variables. On the other hand the traditional, Copenhagen type of interpretation of the theory is certainly *not* applicable in quantum cosmology, which is one of the most important potential uses of canonical quantum gravity.

2. *Background manifold stucture.* The canonical theory of classical relativity assumes *ab initio* that the spacetime manifold M is diffeomorphic to $\Sigma \times \mathbb{R}$ where Σ is some three-manifold. This three-manifold becomes part of the fixed background in the quantum theory and so, for example, there is no immediate possibility of discussing quantum topology.

3. *Background metric structure.* One of the main aspirations of the canonical approach to quantum gravity has always been to build a formalism with no background spatial, or spacetime, metric. This is particularly important in the context of quantum cosmology.

4. *The spacetime diffeomorphism group.* In the canonical form of general relativity the spacetime diffeomorphism group $\mathrm{Diff}(M)$ is replaced by a more complex entity (the 'Dirac algebra') which contains $\mathrm{Diff}(\Sigma)$ as a subgroup but which is not itself a genuine group. Invariance under $\mathrm{Diff}(\Sigma)$ means that the functionals

of the canonical variables that correspond to physical variables are non-local with respect to Σ. The role of the full Dirac algebra is more subtle and varies according to which canonical scheme is followed.

5. *The problem of time.* In the absence of any background metric, this becomes a major issue. It is closely connected with the role of the Dirac algebra and the general question of what is meant by an 'observable'. None of the several suggested ways for handling this problem works fully and it seems plausible that the standard concept of time can be recovered only in some semi-classical sense.

6. *The unification of the fundamental forces.* Perturbation theory in simple canonical quantum gravity is as badly defined mathematically as is its particle-physics based cousin. However, developments in the Ashtekar programme imply that it may be possible to construct a non-perturbative theory that is finite and that involves just the gravitational field alone. In this sense, canonical quantum gravity does *not* suggest that a unification of the forces is a necessary ingredient of a technically successful theory.

3.2 General-Relativise Quantum Theory

I cannot say much about the idea 'general-relativising' quantum theory (*i.e.*, type-II schemes) in relation to our *prima facie* issues because little research has been done in this area. A key role would probably be played by the spacetime diffeomorphism group Diff(M): indeed, a type-II scheme might be *defined* as any attempt to force standard quantum theory to be compatible with Diff(M); of course the structure of M itself would then necessarily be part of the fixed background.

Important questions that arise in an approach of this type include:

- What form of quantum theory should be used? In particular, does it require a prior notion of 'time'?
- What is the role of the field equations of classical general relativity? Do they also need to be imposed as part of the structural background, or is the fundamental input the spacetime diffeomorphism group alone?
- Is there a 'canonical' version in which standard quantum theory is forced to be compatible with the Dirac algebra rather than with Diff(M)?
- There is a long-standing and extensive research programme to construct quantum field theories (usually linear) in a background spacetime manifold equipped with a fixed Lorentzian metric. Does this work throw any light on the idea of general-relativising quantum theory?

3.3 The Use of Radical New Concepts

In the classification we have been using, a type-IV scheme is any approach to quantum gravity that starts with a view of quantum theory and spacetime physics that is radically different from that of conventional theories, and with the expectation that these standard ideas will emerge only in some limited domain. Almost by definition, schemes of this type dispense with much of the background structure of other approaches to quantum gravity, including, possibly, standard quantum theory as well as many normal spacetime concepts. Unfortunately, such schemes tend to be individualistic in form, and their manner of dealing with our *prima facie* issues has to be treated on a strictly *ad hominum* basis.

4 Conclusions

The two major current approaches to quantum gravity proper—the Ashtekar programme, and superstring theory—differ so much in their starting positions and lines of development that it is hard to say much in conclusion other than that the problem of quantum gravity is still wide open. In particular, and *pace* the discussion above, the jury is still out on the all-important question of whether a consistent theory of quantum gravity can be achieved within the framework of our existing understanding of physics, or whether some radical change is needed before any real headway can be made.

The problem of time is crucial in this respect, and its resolution is still very unclear. However, one reading of the current situation is that normal notions of time and space are applicable only at scales well above the Planck regime. If true, such a position throws great doubt on the use of *any* standard quantum ideas as a basic ingredient in the theory; indeed, a more plausible scenario is that standard quantum theory becomes applicable at precisely the same point in the formalism as does the normal notion of time. Such a situation is exciting for those who, like myself, enjoy indulging in speculative metaphysics/theoretical-physics, but it is also most frustrating in the absence of any clear empirical data that could point us in the right direction. The problem of quantum gravity continues to be a challenge for the next century!

References

[Alv89] Alvarez, E.: Quantum gravity: an introduction to some recent results, *Rev. Mod. Phys.* **61** (1989) 561–604

[Ash91] Ashtekar, A.: Lectures on Non-Perturbative Canonical Gravity, World Scientific Press, Singapore (1991)

[FH87] Fredenhagen, K. & Haag, R.: Generally covariant quantum field theory and scaling limits, *Comm. Math. Phys* **108** (1987) 91–115

[GH93] Gibbons, G.W. & Hawking, S.W.: Euclidean Quantum Gravity, World Scientific Press, Singapore (1993)

[Har93] Hartle, J.: Spacetime quantum mechanics and the quantum mechanics of spacetime, *in* 'Proceedings on the 1992 Les Houches School, Gravitation and Quantisation' (1993)

[Ish92] Isham, C.: Conceptual and geometrical problems in quantum gravity, *in* H. Mitter & H. Gausterer, eds, 'Recent Aspects of Quantum Fields', Springer-Verlag, Berlin (1991) 123–230

[Ish93] Isham, C.: Canonical quantum gravity and the problem of time, *in* 'Integrable Systems, Quantum Groups, and Quantum Field Theories', Kluwer Academic Publishers, London (1993) 157–288

[KFL86] Károlyházy, F., Frenkel, A. & Lukács, B.: On the possible role of gravity in the reduction of the state vector, *in* R. Penrose & C. Isham, eds, 'Quantum Concepts in Space and Time', Clarendon Press, Oxford (1986) 109–128

[Kuc92] Kuchař, K.: Time and interpretations of quantum gravity, *in* 'Proceedings of the 4th Canadian Conference on General Relativity and Relativistic Astrophysics', World Scientific, Singapore (1992) 211–314

[Pen86] Penrose, R.: Gravity and state vector reduction, *in* R. Penrose & C. Isham, eds, 'Quantum Concepts in Space and Time', Clarendon Press, Oxford (1986) 129–146

[Pen87] Penrose, R.: Newton, quantum theory and reality, *in* S. Hawking & W. Israel, eds, 'Three Hundred Years of Gravitation', Cambridge University Press, Cambridge (1987) 17–49

[UW89] Unruh, W. & Wald, R.: Time and the interpretation of quantum gravity, *Phys. Rev.* **D40**, (1989) 2598–2614

Hamilton's Formalism for Systems with Constraints

A. Wipf

1. Introduction

All fundamental field theories in physics are invariant with respect to some group of *local symmetry transformations*. For Yang-Mills theories these are the gauge transformations, for string theory and gravity space-time diffeomorphisms and for supersymmetric theories local supersymmetry transformations. In such theories which are called *gauge theories* or more generally *singular systems*, the local symmetry relates different solutions stemming from the same initial conditions and the general solution of the equations of motion contains arbitrary time-dependent functions. Hence there is a continuous set of accelerations which belong to the same initial position and velocity and we expect that all accelerations correspond only to a subset of initial conditions. This subset is defined by the Lagrangean constraints[1]. Hence all gauge theories are systems with constraints. In the Hamiltonian formalism this means that there are conditions on the allowed initial momenta and positions. These conditions must then be conserved by the time evolution, and this requirement may lead to further constraints.

One should distinguish between gauge theories with internal symmetries and those which are *generally covariant.*
For the former all constraints are linear in the momenta and the Hamiltonian does not vanish. The local symmetry transformations are generated by the first class constraints.
For generally covariant theories at least one constraint is quadratic in the momenta[2] and there are canonical variables for which the Hamiltonian H itself is a constraint, usually called *superhamiltonian.* This leads to the question whether H generates the dynamical time-evolution or kinematical local symmetries as the other first class constraints. We will hear more about this problem of interpreting time in the lectures of R. Beig and P. Hajicek.

Attempts to handle constrained systems date back more than forty years. In his classical works Dirac set up a formalism to treat such systems self-consistently (Dirac 1950-1964). Later Bergmann et.al. (Anderson and Bergmann 1951, Bergmann et.al

[1] this statement is made more precise by the generalized Bianchi identities
[2] topological field theories are exceptional in this respect

1956, Bergmann 1961) in a series of papers investigated the connection between constraints and invariances. After the introduction of Grassmann variables to describe fermions (Berezin 1966), the formalism has been extended to include fields with half-integer spins (Gervais and Sakita 1971, Casalbuoni 1976, Berezin and Marinov 1977). The development culminated with the advent of the elegant and powerful BRST formalism (Becchi, Rouet and Stora 1974-1976, Tuytin 1975). These and other classical results have been a prerequisite for the quantization of gauge theories both in the path integral formalism (Fadeev 1969, Batalin and Vilkovisky 1977) and in the functional Schrödinger picture (see e.g. Symanzik 1983, Jackiw 1988, Kiefer and Wipf 1993).

Besides the classical lectures of Dirac (Dirac 1964) there are several excellent reviews on the treatment of constrained systems. Some focus more on systems with a finite number of degrees of freedom (Sudarshan and Mukunda 1974), others on field theories (Hanson, Regge and Teitelboim 1976) and some on both (Sundermeyer 1982, Govaerts 1991, Henneaux and Teitelboim 1992). For generally covariant theories you may consult (Gitman and Tuytin 1990).

In these hopefully self-contained lectures I applied the developed formalism at various stages to the abelian Chern-Simons model with sources (Hagen 1984, Polyakov 1988, Witten 1989). In this way the reader may become acquainted with the constrained dynamics by way of example. There are several aspects I could not cover in these lectures, especially the inclusion of fermions which leads to graded Poisson-structures (Gervais and Sakita 1971, Casalbuoni 1976) and the popular BRST formalism (Becchi et.al 1974-1976). Generally covariant theories and in particular gravity are covered by the lectures of R. Beig and N. Giulini on classical gravity and the other contributions to the proceedings.

In the first part of these lectures I followed the conventional discussion of constrained systems. The second chapter is devoted to *singular Lagrangean systems*. I discuss the generalized Bianchi identities and show that all gauge theories are constrained systems. In chapter three some important facts about *constrained Hamiltonian systems* are reviewed and discussed. In particular primary/secondary and first/second class constraints, the generalized Legendre transformation and the Dirac-Bergmann algorithm are introduced. The general formalism is applied to the *abelian Chern-Simons theory* in chapter four. Then we introduce the reduced phase space for first and second class systems. Here the important Dirac brackets for second class (SC) systems, the concept of observables and gauge transformations for first class (FC) systems and the first order formalism for mixed SC and FC systems are discussed. Again I apply the general results to the abelian Chern-Simons theory and show that the only observables are the Wilson-loops. In the last chapter I investigate the relation between Lagrangean symmetries and the Hamiltonian gauge transformations generated by the first class constraint (Mukhanov and Wipf 1994). We shall see that the latter must be supplemented by transformations which vanish on-shell in order to recover the Lagrangean symmetries. Also, I will discuss for which theories the equations of motion follow from the local symmetries. Some of the results in this last chapter are new and have not previously been published. We feel that the results offered are somewhat novel.

I am indebted to V. Mukhanov for sharing with me many insights and contributing to these notes and to J. Ehlers for many helpful remarks when I prepared these notes.

2. Singular Lagrangean Systems

2.1 Singular Lagrangeans

In these lectures I consider systems whose dynamics can be derived from Hamilton's variational principle. I assume that all Lagrangeans depend at most on first derivatives, up to divergence terms. For higher derivative theories, and in particular for higher derivative gravity, see (Gitman and Tuytin 1990). I use local coordinates, unless I am forced to address global questions, e.g. the Gribov problem or the role of topologically nontrivial field configurations.

With these assumptions the classical trajectories of a system with N degrees of freedom make the action

$$S = \int_{t_1}^{t_2} L(q^i, \dot{q}^i) dt \quad , \quad i = 1, \ldots, N \tag{2.1}$$

stationary under variations $\delta q(t)$ which vanish at the endpoints. The q and $\dot{q}$ are local coordinates on the velocity phase space TQ. The necessary and sufficient conditions for S to be stationary are the *Euler-Lagrange equations*

$$L_i \equiv -\frac{d}{dt}\left(\frac{\partial L}{\partial \dot{q}^i}\right) + \frac{\partial L}{\partial q^i} = 0 \tag{2.2}$$

which can be rewritten as

$$L_i = -\frac{\partial^2 L}{\partial \dot{q}^i \partial \dot{q}^j}\ddot{q}^j - \frac{\partial^2 L}{\partial \dot{q}^i \partial q^j}\dot{q}^j + \frac{\partial L}{\partial q^i} \equiv -W_{ij}(q, \dot{q})\ddot{q}^j + V_i = 0. \tag{2.3}$$

We see that the accelerations at a given time are uniquely determined by $(q, \dot{q})$ at that time only if the Hessian (W_{ij}) can be inverted. Such systems are called *regular.*

If, on the other hand, $\det W = 0$, the accelerations and thus the time evolution will not be uniquely fixed by the $(q, \dot{q})$. Such systems are called *singular* (Anderson and Bergmann 1951). For singular systems different time evolutions will stem from the same initial conditions.

The rank R of W, which we assume for simplicity to be constant on TQ, being $R<N$ implies the existence of $M=N-R$ null-eigenvectors

$$Y^i_m(q, \dot{q}) W_{ij}(q, \dot{q}) = 0 \quad , \quad m = 1, \ldots, M. \tag{2.4}$$

Contracting the E-L equations (2.3) with the Y_m we get

$$\phi_m(q, \dot{q}) \equiv Y^i_m V_i = 0 \quad , \quad m = 1, \ldots, M, \tag{2.5}$$

as necessary and sufficient conditions in order that eqs. (2.3), considered as algebraic eqs. for the unknown $\ddot{q}^j$, have a solution. These equations do not contain accelerations. Assume that $M' \leq M$ relations

$$\phi_{m'} = 0 \quad , \quad m' = 1, \ldots, M', \tag{2.6}$$

are functionally independent of the others, and the remaining ones are either dependent or identically fulfilled. The independent $\phi_{m'}$ are the so called *Lagrange constraints.*

For *field theories* the dynamics is described by functions $\varphi^a(x)$ on spacetime with values in a certain target space. The index a may belong to an internal symmetry, it may be a spacetime index or both internal and spacetime index as in nonabelian gauge theories. When going from point mechanics to field theory one may think of replacing the discrete label i by a continuous one (a, x):

$$q^i(t) = q(t,i) \longrightarrow q(t,a,x) = \varphi^a(t,x) = \varphi^a(x). \tag{2.7}$$

Summations include integrals, e.g.

$$\sum_i \dot{q}^i \dot{q}^i \longrightarrow \sum_a \int dx\, \dot{\varphi}^a(x) \dot{\varphi}^a(x)$$

and functions of $(q, \dot{q})$ become functionals of φ and $\dot{\varphi}$. Also, derivatives with respect to q^i or $\dot{q}^i$ become functional derivatives, e.g.

$$\frac{\partial L}{\partial \dot{q}^i} \longrightarrow \frac{\delta L}{\delta \dot{\varphi}^a(x)}$$

The suitable velocity phase space TQ is chosen so that the Lagrange-functional L is continuous and sufficiently often differentiable. If the target space is linear one may choose a Banach space (typically a Sobolov space), otherwise one tries to model the theory on a C^k-Banach manifold (Chernoff and Marsden 1974) since the implicit function theorem still applies then. Banach manifolds are modeled on Banach spaces and are straightforward generalizations of finite-dimensional manifolds.

A functional on a Banach space X is called continuous if

$$\lim_{n\to\infty} F[\varphi_n] = F[\varphi] \quad \text{for} \quad X \ni \varphi_n \to \varphi.$$

F is called Frechet-differentiable at φ if there exists a linear functional F'_φ such that

$$|F[\varphi + \delta\varphi] - F[\varphi] - F'_\varphi[\delta\varphi]| = o(\| \delta\varphi \|) \quad \text{as} \quad \| \delta\varphi \| \to 0.$$

For local theories the Lagrange-an has the form

$$L[\varphi, \dot{\varphi}] = \int dx\, \mathcal{L}(\varphi, \partial_i\varphi, \dot{\varphi}) \tag{2.8}$$

with a Lagrangean density $\mathcal{L}$ depending only on the field and its derivatives at the same point. For such theories the Euler-Lagrange equations are

$$L_a \equiv -\frac{\partial}{\partial t}\frac{\delta L}{\delta \dot{\varphi}^a} + \frac{\delta L}{\delta \varphi^a} = -\partial_\mu \frac{\partial \mathcal{L}}{\partial(\partial_\mu \varphi^a)} + \frac{\partial \mathcal{L}}{\partial \varphi^a} = 0, \tag{2.9}$$

where I adopted the common notation

$$F'_\varphi[\delta\varphi] \equiv \int \frac{\delta F}{\delta\varphi(x)} \delta\varphi(x)\, dx.$$

Note that L_a does not change if $\mathcal{L}$ is replaced by $\mathcal{L} - \partial_\nu K^\nu$, where $K^\nu(\varphi, \partial\varphi)$ is an arbitrary tensor density. Rewriting the field equations as

$$\begin{aligned} L_a &= -\frac{\partial^2 \mathcal{L}}{\partial(\partial_\mu \varphi^a)\partial(\partial_\nu \varphi^b)}\partial_\mu\partial_\nu\varphi^b - \frac{\partial^2 \mathcal{L}}{\partial(\partial_\mu \varphi^a)\partial\varphi^b}\partial_\mu\varphi^b + \frac{\partial \mathcal{L}}{\partial \varphi^a} \\ &\equiv -W^{\mu\nu}_{ab}\partial_\mu\partial_\nu\varphi^b + V_a = -W^{00}_{ab}\ddot{\varphi}^b + \tilde{V}_a = 0 \end{aligned} \tag{2.10}$$

we can see that theories with x^0 taken as evolution parameter are regular if W^{00}_{ab} is invertible and singular if it is not. For singular systems there exist (for each x) $M = N - R$ null-vectors

$$Y^a_m(\varphi, \partial\varphi) W^{00}_{ab}(\varphi, \partial\varphi) = 0 \quad , \quad m = 1, \ldots, M \tag{2.11}$$

which lead to nontrivial and independent relations

$$\phi_{m'}(\varphi, \partial\varphi) \equiv Y^a_{m'}\tilde{V}_a = 0 \quad , \quad m' = 1, \ldots, M' \leq M, \tag{2.12}$$

involving only the fields and their first derivatives. These are the Lagrangean constraints.

How one proceeds for singular systems is neatly explained in (Shanmugadhasan 1963, Sundermeyer 1982). There are two points which have to be considered. Firstly the rank of the Hessian may decrease if one takes the independent constraints (2.6,12) into account. This may lead to new independent constraints. Again the rank may decrease leading to further constraints, etc. This process terminates as soon as the rank does not change any more.

Secondly one needs to check whether the constraints one has found after the above algebraic process has terminated are respected by the *time evolution*. This may lead to new constraints. Again and again differentiate newly emerging constraints until no new ones arise. Add those relations involving accelerations to those already present. Consistency of the old relations with the new ones may lead to further constraints. After all that one needs again to check whether the rank of the Hessian has changed. If this is the case one needs to start from the beginning etc.

2.2 Generalized Bianchi Identities

If a theory possesses a *local* gauge invariance we may map solutions into solutions without affecting the initial conditions. Thus we expect that *gauge theories are singular systems.* Actually this follows from the generalized Bianchi identity (Utiyama 1959, Trautman 1964) which we derive next.

The point transformations

$$\begin{aligned} x' &= x'(x) \sim x + \delta x \\ \varphi'(x') &= \varphi'(\varphi(x), x) \sim \varphi(x) + \delta\varphi \end{aligned} \tag{2.13}$$

which satisfy

$$\int d^d x' \, \mathcal{L}(\varphi', \partial'\varphi') = \int d^d x \left(\mathcal{L}(\varphi, \partial\varphi) - \partial_\nu K^\nu\right) \tag{2.14}$$

and thus leave the L_a invariant, called symmetries, form a group which we assume to be continuous. For infinitesimal transformations, $d^d x' = d^d x(1 + \partial_\mu \delta x^\mu)$, and the invariance (2.14) is equivalent to

$$\bar{\delta}\mathcal{L} + \partial_\nu(\mathcal{L}\delta x^\nu) = \partial_\mu \lambda^\mu \tag{2.15}$$

with some λ. Using

$$\bar{\delta}\mathcal{L} = \frac{\partial \mathcal{L}}{\partial \varphi^a}\bar{\delta}\varphi^a + \frac{\partial \mathcal{L}}{\partial(\partial_\nu \varphi^a)}\bar{\delta}(\partial_\nu \varphi^a)$$

it follows at once that

$$\bar{\delta}\mathcal{L} + \partial_\mu(\mathcal{L}\delta x^\mu) = \partial_\mu(\mathcal{L}\delta x^\mu) + \partial_\mu\Big(\frac{\partial \mathcal{L}}{\partial(\partial_\mu \varphi^a)}\bar{\delta}\varphi^a\Big) + L_a \bar{\delta}\varphi^a, \tag{2.16a}$$

where the Euler derivatives L_a have been defined in (2.9) and

$$\bar{\delta}\varphi^a = \delta\varphi^a - \partial_\mu \varphi^a \delta x^\mu \sim \varphi^{a\prime}(x) - \varphi^a(x) \tag{2.16b}$$

is the infinitesimal difference of the old and the transformed fields at the same point. We used that $[\bar{\delta}, \partial_\mu] = 0$. Thus the gauge invariance implies the Bessel-Hagen identity

$$\partial_\mu\Big(\mathcal{L}\delta x^\mu + \frac{\partial \mathcal{L}}{\partial(\partial_\mu \varphi^a)}\bar{\delta}\varphi^a - \lambda^\mu\Big) + L_a \bar{\delta}\varphi^a = 0. \tag{2.17}$$

Nowhere did we use the equation of motion and thus (2.17) are *off-shell identities.*

First assume that S is invariant under global transformations forming a n-dimensional Lie-group. Then

$$\lambda^\mu = \epsilon_\alpha \lambda^{\alpha\mu}, \quad \delta_\epsilon x^\mu = \epsilon_\alpha A^{\alpha\mu}, \quad \delta_\epsilon \varphi^a = \epsilon_\alpha B^{\alpha a}, \tag{2.18}$$

where the $\epsilon_\alpha, \alpha = 1, \dots, n$ are the constant parameters of the infinitesimal transformations. Inserting this into (2.17) and going on shell, $L_a = 0$, we conclude

$$\partial_\mu j^{\alpha\mu} = 0, \quad \text{where} \quad j^{\alpha\mu} = \frac{\partial \mathcal{L}}{\partial(\partial_\mu \varphi^a)}(B^{\alpha a} - A^{\alpha\nu}\partial_\nu \varphi^a) + \mathcal{L}A^{\alpha\mu} - \lambda^{\alpha\mu}, \tag{2.19}$$

which *is Noether's first theorem.* When deriving (2.19) we imposed the equations of motion so that the currents are *conserved only on-shell.*

Let us now assume that the symmetry transformations are local. In that case the parameters become space-time dependent. Then (2.18) generalizes to

$$\delta_\epsilon x^\mu = \epsilon_\alpha A^{\alpha\mu} \quad , \quad \delta_\epsilon \varphi^a = \epsilon_\alpha B^{\alpha a} + (\partial_\mu \epsilon_\alpha)C^{\alpha a\mu}, \tag{2.20}$$

where the 'descriptors' $\epsilon_\alpha(x)$ parametrize the infinitesimal local gauge transformations and B and C in general depend on the fields and their derivatives (Anderson and Bergmann 1951). I assumed that no second or higher derivatives of ϵ enter because this covers most interesting examples. With

$$\bar{\delta}_\epsilon \varphi^a = \epsilon_\alpha\big(B^{\alpha a} - (\partial_\mu \varphi^a)A^{\alpha\mu}\big) + (\partial_\mu \epsilon_\alpha)C^{\alpha a\mu} \tag{2.21}$$

the integrated form of (2.17), after a partial integration, reads

$$0 = \int \epsilon_\alpha\big[L_a(B^{\alpha a} - (\partial_\mu \varphi^a)A^{\alpha\mu}) - \partial_\mu(L_a C^{\alpha a\mu})\big].$$

Since it must hold for arbitrary functions ϵ_α this implies that the expression between the square brackets must vanish. Thus, the Euler-Lagrange derivatives L_a satisfy identically ('off shell') a first order system of linear partial differential equations, the *generalized Bianchi identities.* This fact is called *Noether's second theorem.* Inserting L_a from (2.10) we end up with

$$\begin{aligned}0 = L_a\big(B^{\alpha a} - \partial_\mu \varphi^a A^{\alpha\mu} - \partial_\mu C^{\alpha a\mu}\big) - C^{\alpha a\mu}\partial_\mu V_a \\ + C^{\alpha a\mu}\big(\partial_\mu W^{\rho\sigma\rho}_{ab}\partial_\rho\partial_\sigma\varphi^b + W^{\rho\sigma}_{ab}\partial_\mu\partial_\rho\partial_\sigma\varphi^b\big).\end{aligned} \tag{2.22}$$

Since these are off-shell identities we conclude

$$C^{\alpha a(\mu}W^{\rho\sigma)}_{ab} = 0,$$

where the brackets around the indices mean symmetrization. In particular, the $C^{\alpha a0}$ which are not identically zero are null-eigenvectors of the Hessian,

$$C^{\alpha a0}W^{00}_{ab} = 0 \tag{2.23}$$

and render the system singular. If all $C^{\alpha a\mu}$ vanish, then (2.22) reduces to

$$0 = \big(B^{\alpha a} - \partial_\rho\varphi^a A^{\alpha\rho}\big)L_a \Longrightarrow \big(B^{\alpha a} - A^{\alpha\rho}\partial_\rho\varphi^a\big)W^{(\mu\nu)}_{ab} = 0. \tag{2.24}$$

Recalling (2.21) we see that if $C \equiv 0$ then the $B^{\alpha a} - A^{\alpha\rho}\partial_\rho\varphi^a$ cannot all vanish, else $\bar\delta_\epsilon\varphi^a$ would be zero or the field would be unaffected by all infinitesimal symmetry transformations. Since this never happens, we again conclude that the system must be singular. So we have the important result that *gauge theories are necessarily singular.* However, the converse is not true. Not all conceivable singular systems are gauge theories.

3. Hamilton's Formalism for Constrained Systems

3.1 Primary Constraints

The departing point for the Hamiltonian formalism is to define the canonical momenta (momentum densities) by

$$p_i = \frac{\partial L}{\partial \dot q^i}(q,\dot q) \quad \text{resp.} \quad \pi_a(x) = \frac{\delta L}{\delta\dot\varphi^a(x)}, \tag{3.1}$$

where we assume that $L \in C^2(TQ)$. Only if

$$W_{ij} = \frac{\partial p_i}{\partial \dot q^j} \quad \text{resp.} \quad W^{00}_{ab}(x,y) = \frac{\delta\pi_a(x)}{\delta\dot\varphi^b(y)}$$

is invertible can this relation be solved for all velocities in terms of the phase space variables, $\dot q = \dot q(q,p)$ resp. $\dot\varphi = \dot\varphi(\varphi,\pi)$ [3]. In the other case not all momenta (3.1) are independent, but there are some relations

$$\phi_m(q,p) = 0 \quad \text{resp.} \quad \phi_m(\varphi,\pi) = 0 \quad , \quad m = 1,\dots,M \tag{3.2}$$

that follow from the definition (3.1) of the momenta. In these lectures I assume that the constraints (3.2) are independent.

---

[3] for field theories we assume TQ to be a Banach manifold so that the inverse function theorem applies

In the following we restrict ourselves to finite dimensional systems and only comment on the related results for field theories. The corresponding field theoretical formulae, if they apply, are obtained if one uses deWitt's condensed notation (DeWitt 1965) in which i becomes a composite index as explained above and in chapter 7.

The conditions (3.2) are the $M=N-R$ *primary constraints*. They define the $2N-M$-dimensional *primary constraint surface*, denoted by Γ_p. The equations of motion have not been used to derive them and they imply no restriction on the $(q,\dot{q})$. (3.1) maps the $2N$-dimensional velocity phase space TQ to the lower-dimensional submanifold Γ_p in the momentum phase space Γ. Hence the inverse images of a given point in Γ_p form a manifold of dimension M.

To pass to the Hamiltonian formalism we impose some *regularity conditions* on the primary constraints. They can be alternatively formulated as:

i) the independent functions $\phi_m, m=1,\ldots,M$ can locally be taken as the first M coordinates of a new, regular, coordinate system in the vicinity of Γ_p.
ii) The gradients $d\phi_1,\ldots,d\phi_M$ are linearly independent on Γ_p; i.e., $d\phi_1\wedge\ldots\wedge d\phi_M\neq 0$ on Γ_p.

For example, if ϕ is an admissible constraint, ϕ^2 is not, since $d(\phi^2)=2\phi d\phi=0$ on Γ_p. If the constraints are regular the following properties hold.

Theorem 1 *If a smooth function $F(q,p)$ vanishes on Γ_p, then in a neighborhood of Γ_p $F=f^m\phi_m$ for some functions f^m.*

Theorem 2 *If $\lambda_i\delta q^i+\mu^i\delta p_i=0$ for arbitrary variations $\delta q^i,\delta p_i$ tangent to the constraint surface, then*

$$\lambda_i=u^m\frac{\partial\phi_m}{\partial q^i}\quad\textit{and}\quad\mu^i=u^m\frac{\partial\phi_m}{\partial p_i}\qquad\textit{on}\quad\Gamma_p$$

for some u^m.

Before proving these two important theorems it is useful to distinguish between weak and strong equations. A function $F(q,p)$ defined in a neighborhood of Γ_p is called *weakly zero* if

$$F|_{\Gamma_p}=0,\quad\text{abbreviated as}\quad F\approx 0 \tag{3.3a}$$

and *strongly zero* if

$$F|_{\Gamma_p}=0\quad\text{and}\quad\left(\frac{\partial F}{\partial q^i},\frac{\partial F}{\partial p_i}\right)|_{\Gamma_p=0},\quad\text{written}\quad F\simeq 0. \tag{3.3b}$$

These definitions are useful since the equations of motion contain gradients of functions on Γ_p. The primary constraint surface can itself be defined by weak equations. We have

$$\phi_m\approx 0\quad\text{but}\quad\phi_m\not\simeq 0$$

because of our regularity conditions on the constraints.
Since $\nabla_x(f^m\phi_m)\approx f^m\nabla_x\phi_m$, where $x=(q,p)$ denotes the phase space coordinates, the first theorem implies

Lemma 1 *$F \approx 0 \Longrightarrow F - f^m\phi_m \simeq 0$ and $\nabla_x F \approx f^m \nabla_x \phi_m$ for some functions f^m.*

To prove the first theorem we choose the independent constraints ϕ_m as first coordinates of a regular coordinate system $x = (\phi, \tilde{x})$ in a neighborhood of a point of Γ_p. Since $F(0, \tilde{x}) = 0$ we have

$$F(\phi, \tilde{x}) = \int_0^1 \frac{d}{d\tau} F(\tau\phi, \tilde{x}) d\tau = \phi_m \int_0^1 F_{,m}(\tau\phi, \tilde{x}) d\tau$$

and thus

$$F = f^m \phi_m \quad \text{with} \quad f^m = \int_0^1 F_{,m}(\tau\phi, \tilde{x}) d\tau.$$

This proves theorem 1 in some neighborhood U of any point on Γ_p. We cover a neighborhood of Γ_p by open sets U_i, on each of which theorem 1 applies. Together with the open sets V_k on which $\phi_k \neq 0$, the U_i cover the whole phase space. On V_k we can set $F = (F/\phi_k)\phi_k$, and theorem 1 holds there. Finally, to guarantee that the f^m are the same on the overlap of U_i and $U_{i'}$ one uses a locally finite partition of unity.

Theorem 2 follows immediately from the regularity condition which implies that at a given point x on Γ_p a basis of $T_x\Gamma_p$ (the vectors tangent to Γ_p at x), together with the gradients $\nabla_x \phi_m$ form a basis of $T_x\Gamma$. The assumption in theorem 2 means that (λ, μ) is orthogonal to $T_x\Gamma_p$. Thus it must be a linear combination of the gradients $\nabla_x \phi_m$.

Similarly, for *field theories* one finds for a Frechet-differentiable functional

$$F[\phi, \tilde{x}] \approx 0 \Rightarrow F = \int f^m \phi_m \quad , \quad f^m(x) = \int d\tau \frac{\delta F}{\delta \phi_m(x)} [\tau\phi, \tilde{x}]$$

and concludes that a weakly vanishing functional is a linear combination of smeared constraints. The test functions should lie in the space dual to the space of the constraints (Benguria et.al 1977).

3.2 Generalized Legendre Transformations

The canonical Hamiltonian

$$H = \dot{q}^i p_i - L \quad \text{resp.} \quad H = \int dx\, \pi_a(x) \dot{\varphi}^a(x) - L = \int dx\, \mathcal{H} \tag{3.4}$$

has the remarkable property that $\dot{q}$ enters H only through the combination $p(q, \dot{q})$. This follows from

$$\begin{aligned} \delta H &= \dot{q}^i \delta p_i + \delta \dot{q}^i p_i - \delta \dot{q}^i \frac{\partial L}{\partial \dot{q}^i} - \delta q^i \frac{\partial L}{\partial q^i} \\ &= \dot{q}^i \delta p_i - \delta q^i \frac{\partial L}{\partial q^i} \end{aligned} \tag{3.5}$$

which shows that H is a function of p and q only. Here δp is to be regarded as linear combination of δq and $\delta \dot{q}$ so that $\delta q, \delta p$ are tangent to Γ_p. H is only defined on Γ_p

since we used the constraints. We would like to extend H to the whole phase space Γ in order to obtain canonical equations of motion.

The equation (3.5) can be rewritten, if H henceforth denotes any smooth extension of the original H to a neighborhood of Γ_c in Γ, as

$$(\frac{\partial H}{\partial q^i} + \frac{\partial L}{\partial q^i})\delta q^i + (\frac{\partial H}{\partial p_i} - \dot{q}^i)\delta p_i = 0. \tag{3.6}$$

Applying theorem 2 we conclude that

$$\begin{aligned} \dot{q}^i &\approx \frac{\partial H}{\partial p_i} + u^m \frac{\partial \phi_m}{\partial p_i} \\ -\frac{\partial L}{\partial q^i} &\approx \frac{\partial H}{\partial q^i} + u^m \frac{\partial \phi_m}{\partial q^i}. \end{aligned} \tag{3.7}$$

These relations give *equations* on Γ_p only. We extend them to Γ (or, at least, to a neighborhood of Γ_p in Γ) by setting

$$\begin{aligned} \dot{q}^i &\approx \frac{\partial H}{\partial p_i}|_{\Gamma_p} + u^m \frac{\partial \phi_m}{\partial p_i}|_{\Gamma_p}, \\ -\frac{\partial L}{\partial q^i} &\approx \frac{\partial H}{\partial q^i}|_{\Gamma_p} + u^m \frac{\partial \phi_m}{\partial q^i}|_{\Gamma_p}. \end{aligned} \tag{3.8a}$$

and letting, at each point of Γ_p, the u^m vary freely in a neighborhood of its value at that point. Then $(\tilde{x}, u)$ are local coordinates on Γ near Γ_p,if $\tilde{x}$ are local coordinates on Γ_p. Therefore, the first set of eqs. (3.8a), $\dot{q}^i = \dot{q}^i(\tilde{x}, u) = \dot{q}^i(q, p)$, defines an invertible map from Γ to TQ. The inverse map is given by the equations

$$p_i = \frac{\partial L}{\partial \dot{q}^i}(q, \dot{q}) \quad \text{and} \quad u^m = u^m(q, \dot{q}), \tag{3.8b}$$

the last M of which arise by solving the equations $\dot{q}^i = \dot{q}^i(\tilde{x}, u)$ for the u^m.

We had to extend the Hamiltonian, which was originally defined only on Γ_p, to a neighborhood of Γ_p. According to theorem 1 two possible extensions differ by a term $c^m \phi_m$. Thus the formalism should be unchanged by the replacement

$$\tilde{H} \longrightarrow \tilde{H} + c^m(q, p)\phi_m.$$

Indeed, making this transformation in (3.7) just shifts the u to $u + c$.

Finally, the relations (3.7) allow us to rewrite the equation of motion (2.2) in the equivalent Hamiltonian form

$$\dot{q}^i \approx \frac{\partial H}{\partial p_i} + u^m \frac{\partial \phi_m}{\partial p_i} \quad \text{and} \quad \dot{p}_i \approx -\frac{\partial H}{\partial q^i} - u^m \frac{\partial \phi_m}{\partial q^i}, \tag{3.9}$$

where we dropped the tilde atop H. The Lagrangean equations of motion (2.2) are equivalent to (3.9). A phase space function

$$H_p \equiv H + u^m \phi_m \tag{3.10}$$

is called a *primary Hamiltonian.* The passage from $q, \dot{q}, L(q, \dot{q})$ to $q, p, H_p(q, p)$ is called a *generalized Legendre transformation.*

Introducing the Poisson bracket of two phase space functions

$$\{F,G\} \equiv \frac{\partial F}{\partial q^i}\frac{\partial G}{\partial p_i} - \frac{\partial F}{\partial p_i}\frac{\partial G}{\partial q^i} \qquad \text{resp.}$$
$$\{F,G\} \equiv \int dx \Big(\frac{\delta F}{\delta \varphi^a(x)}\frac{\delta G}{\delta \pi_a(x)} - \frac{\delta F}{\delta \pi_a(x)}\frac{\delta G}{\delta \varphi^a(x)} \Big) \tag{3.11}$$

and using $u^m \nabla_x \phi_m \approx \nabla_x(u^m \phi_m)$, the Hamiltonian equations of motion can be rewritten as

$$\dot{q}^i \approx \{q^i, H_p\} \approx \{q^i, H\} + \{q^i, \phi_m\}u^m$$
$$\dot{p}_i \approx \{p_i, H_p\} \approx \{p_i, H\} + \{p_i, \phi_m\}u^m. \tag{3.12a}$$

Besides there are still the equations defining Γ_p:

$$\phi_m(q,p) = 0. \tag{3.12b}$$

For any phase-space function $F(q,p)$ the *time evolution* follows then from

$$\dot{F} \approx \{F, H_p\} \approx \{F, H\} + u^m\{F, \phi_m\}. \tag{3.13}$$

3.3 Dirac-Bergmann Algorithm

As in the Lagrangean formalism the constraints must be consistent with the time evolution. If initially (q,p) is on Γ_p it should remain there at later times. This means that the equations of motion should preserve the constraints and this gives rise to the *consistency conditions* (Dirac 1950, Anderson and Bergmann 1951)

$$\dot{\phi}_m \approx \{\phi_m, H\} + \{\phi_m, \phi_n\}u^n \equiv h_m + C_{mn}u^n \approx 0. \tag{3.14}$$

For non-admissible Lagrangeans these relations will be inconsistent. As an example take $L = \dot{q} - q$ which leads to $H = q$ and $\phi = p - 1$ so that (3.14) reads $1 \approx 0$. For such inconsistent models the action has no stationary points.

To discuss the consistency relations (3.14) we distinguish the two following cases:

i) $\det C \not\approx 0$:
In this case u is weakly uniquely fixed by (3.14) to be $u^n \approx -C^{nm}h_m$, where C^{nm} is the inverse of C_{nm}. The time evolution (3.13) of a phase space function becomes

$$\dot{F} \approx \{F, H\} - \{F, \phi_m\}C^{mn}\{\phi_n, H\}. \tag{3.15}$$

No additional conditions appear. For any initial data (q,p) on Γ_p the time evolution stemming from (3.15) is unambiguous and stays on Γ_p.

ii) $\det C \approx 0$:
In this case u is not fixed and (3.14) is only solvable if $h_m w_a^m \approx 0$ for all left null-eigenvectors w_a of C. Either these equations are fulfilled or they lead to a certain number K_1 of new constraints

$$\phi_k \approx 0 \quad , \quad k = M+1, \ldots, M+K_1 \equiv J_1,$$

called *secondary constraints.* The primary and secondary constraints $\phi_j \approx 0$, $j = 1, \ldots, J_1$ define a submanifold $\Gamma_1 \subset \Gamma_p$.
Now one has to check consistency for the primary and newly generated secondary constraints on Γ_1,

$$\dot{\phi}_j = \{\phi_j, H\} + \{\phi_j, \phi_n\}u^n \equiv h_j + C_{jn}u^n = 0 \quad \text{on} \quad \Gamma_1 \tag{3.16}$$

with the rectangular $J_1 \times M$ matrix C. The left null-eigenvectors w_a^j of C_{jn} imply further conditions $w_a^j h_j = 0$ on Γ_1 and may lead to further, so called *tertiary constraints* which, together with the primary and secondary constraints, define a hypersurface $\Gamma_2 \subset \Gamma_1$, etc.

This procedure either leads to an inconsistency, or to a discrete set of phase space or the following situation: There is a submanifold $\Gamma_c \subset \Gamma$ defined by

$$\phi_j \approx 0 \quad , \quad j = 1, \dots, M + K \equiv J. \tag{3.17}$$

For every left null-eigenvector w_a^j of the rectangular matrix $C_{jm} = \{\phi_j, \phi_m\}$ the conditions $w_a^j\{\phi_j, H\} \approx 0$ are fulfilled. Thus, there exist multiplier fields u^m satisfying the equations

$$\{\phi_j, H\} + \{\phi_j, \phi_m\}u^m \approx 0, \tag{3.18}$$

where $\approx$ now means equality on Γ_c. We note that the primary constraints are merely consequences of the definition of the momenta, whereas we used the equations of motion to arrive at the secondary constraints[4].

We make the same regularity assumptions on the full set of constraints ϕ_j defining Γ_c as we made on the primary constraints ϕ_m defining Γ_p. Also, we assume that the rank of C is constant on Γ_c.

3.4 First and Second Class Constraints

The distinction between primary and secondary constraints will be of minor importance in the final form of the Hamiltonian theory. A different classification of constraints, namely into first and second class (Dirac 1950,1964), will play a central part. Let v_a be a basis of the 'right'-kernel of C,

$$\{\phi_j, \phi_m\}v_a^m \approx 0 \quad , \quad a = 1, \dots, \dim \mathrm{Ker}\, C = M - \mathrm{rank}\, C. \tag{3.19}$$

The general solution for the multipliers u in (3.18) has then the form

$$u = \tilde{u} + \mu^a v_a, \tag{3.20}$$

where $\tilde{u}$ is a particular solution. We have separated the part of u that remains undetermined by the consistency conditions. This part contains $M - \mathrm{rank}\, C$ free functions μ^a.

The combinations of primary constraints

$$\phi_a = v_a^m \phi_m \tag{3.21}$$

weakly commute with all other constraints,

$$\{\phi_a, \phi_j\} \approx 0 \quad , \quad j = 1, \dots, J. \tag{3.22}$$

Moreover, since the v^a form a base of $\mathrm{Ker}\, C$, the ϕ_a are a complete set of primary constraints with this property. This leads to the concept of *first class functions* and

---

[4] in the sequel I call all non-primary constraints secondary

in particular *first class constraints* (FCC). A function $F(q,p)$ is said to be first class if its Poisson brackets with *all constraints* vanish weakly (on Γ_c),

$$\{F, \phi_j\} \approx 0 \quad , \quad j = 1, \dots, J. \tag{3.23}$$

The set of first class functions is closed under Poisson brackets (Dirac 1958). This is proved as follows: if F, G are first class, then according to theorem 1

$$\{F, \phi\} = \phi' \quad , \quad \{G, \phi\} = \phi''$$

for any constraint ϕ, where ϕ', ϕ'' are some linear combinations of the constraints. Using the Jacobi identity we have

$$\{\{F,G\},\phi\} = \{F,\{G,\phi\}\} - \{G,\{F,\phi\}\} = \{F,\phi''\} - \{G,\phi'\} \approx 0.$$

In particular the constraints ϕ_a are a complete set of first class primary constraints. Also, as a result of the Dirac-Bergmann algorithm H_p is first class.

A function that is not first class is called *second class.* I use a notation adapted to this new classification. A basis of the set of primary and secondary FCC is denoted by $\{\gamma_a\}$. The remaining constraints are called second class constraints (SCC) and I denote a basis for them by $\{\chi_\alpha\}$.

The first property we need is that the matrix of SCC

$$\Delta_{\alpha\beta} = \{\chi_\alpha, \chi_\beta\} \tag{3.24}$$

is *non-singular.* Indeed, if it was singular, then there would exist a null vector $r^\alpha \Delta_{\alpha\beta} \approx \{r^\alpha \chi_\alpha, \chi_\beta\} \approx 0$. Since $r^\alpha \chi_\alpha$ also commutes weakly with the FCC (by their first class property) it would weakly commute with all constraints and would be first class which contradicts our assumption. For counting degrees of freedom it is important to note that the number of SCC must be even. Otherwise the antisymmetric Δ would be singular.

Now consider the consistency conditions (3.18). They are identically fulfilled for the γ_a. For the SCC we have

$$\{\chi_\alpha, H\} + \Delta_{\alpha\beta} u^\beta \approx 0,$$

where $u^\beta = 0$ if χ_β is a secondary SCC. Solving for the multipliers we obtain

$$\Delta^{\beta\alpha}\{\chi_\alpha, H\} = \begin{cases} u^\beta, & \chi_\beta \ \text{primary} \\ 0, & \chi_\beta \ \text{secondary} \end{cases} ,$$

where $\Delta^{\alpha\beta}\Delta_{\beta\gamma} = \delta^\alpha_\gamma$. Thus all multipliers belonging to the primary SCC are determined by the consistency conditions and we remain with the undetermined multipliers μ^a in (3.20). We have the important result that *the number of undetermined multipliers is equal to the number of independent primary FCC.*

Inserting that into the equations of motion (3.13) we end up with

$$\dot{F} \approx \{F, H\} + \{F, \phi_a\}\mu^a - \{F, \chi_\alpha\}\Delta^{\alpha\beta}\{\chi_\beta, H\}, \tag{3.25}$$

where the ϕ_a are the primary FCC. One can easily check that all constraints are preserved in time.

4. A Simple Model: Abelian Chern-Simons Theory with Sources

To see how the general formalism works in an explicit example I consider the abelian Chern-Simons model (Hagen 1984, Polychronakos 1990, Iengo and Lechner 1992, Fröhlich and Studer 1993). This is a field theory for a gauge potential A_a in 3 space-time dimensions with coordinates $x = (x^0, x^1, x^2) \equiv (t, \boldsymbol{x})$ with first order Lagrangean density

$$\begin{aligned} \mathcal{L} &= \frac{\kappa}{4} A^a \epsilon_{abc} F^{cb} + A^a J_a, \quad \text{where} \\ F_{ab} &= \partial_a A_b - \partial_b A_a \quad , \quad \partial^a J_a = 0. \end{aligned} \tag{4.1}$$

Indices are lowered with the metric $\eta_{ab} = \mathrm{diag}(1,-1,-1)$ and ϵ_{abc} is the Levi-Civita symbol, $\epsilon_{012} = 1$. We enclose the system in a finite box $[0, L] \times [0, L]$. The quantum theory is sensitive to the value of the coupling constant κ. For rational $2\pi\kappa$ and vanishing external current J the Hilbert space becomes finite-dimensional (Polychronakos 1990).

For arbitrary periodic currents the action is invariant under $U(1)$-gauge transformations

$$A^a \to A^a + \partial^a \lambda \quad , \quad S \to S + \oint n^a \left(\frac{\kappa}{4} \lambda \epsilon_{abc} F^{cb} + \lambda J_a\right) \tag{4.2}$$

provided λ vanishes at the initial and final times and λ, F_{01}, F_{02} are periodic in x^1, x^2 with period L. So we shall *assume* that λ and F_{ab} are both periodic.

Since $\mathcal{L}$ is linear in the first derivatives the Hessian vanishes identically and the model is singular. Thus we expect 3 independent primary constraints (per space point). More explicitly, the canonical momentum densities are

$$\pi_a(\boldsymbol{x}) = \frac{\delta L}{\delta \dot{A}^a(\boldsymbol{x})} = \frac{\kappa}{2} A^b(\boldsymbol{x}) \epsilon_{ba0} \tag{4.3}$$

and immediately lead to the *primary constraints*

$$\{\phi_m\} = \{\pi_0, \pi_1 + \frac{\kappa}{2} A^2, \pi_2 - \frac{\kappa}{2} A^1\} \quad , \quad m = 1, 2, 3. \tag{4.4}$$

The canonical Hamiltonian becomes

$$\begin{aligned} H &= \int d\boldsymbol{x} \Big(\pi_a(\boldsymbol{x}) \dot{A}^a(\boldsymbol{x}) - \mathcal{L}\Big) = \int d\boldsymbol{x}\, \mathcal{H} \\ &= - \int d\boldsymbol{x} \Big(\frac{\kappa}{2} A^a \epsilon_{abi} \partial^i A^b + A^a J_a\Big) \quad , \quad i = 1, 2. \end{aligned} \tag{4.5a}$$

and the time evolution is determined by (3.13) with primary Hamiltonian

$$H_p = \int d\boldsymbol{x}\, \mathcal{H}_p \quad , \quad \mathcal{H}_p = \mathcal{H} + u^m \phi_m, \tag{4.5b}$$

and fundamental Poisson bracket

$$\{A^a(\boldsymbol{x}), \pi_b(\boldsymbol{y})\} = \delta^a_b \delta(\boldsymbol{x} - \boldsymbol{y}). \tag{4.6}$$

Let us now see whether secondary constraints arise from the consistency conditions $\dot{\phi}_m \approx 0$. One computes

$$\begin{aligned}\dot{\phi}_1(x) &= \int dy\ \{\pi_0(x), \mathcal{H}_p(y)\} \\ &= -\int dy\{\pi_0(x), \kappa A^0(y)\epsilon_{0ji}\partial^i A^j(y) + J_0(y)A^0(y)\} \\ &= \int \Big(\kappa\epsilon_{0ji}\partial^i A^j(y) + J_0(y)\Big)\delta(x-y) \\ &= \epsilon_{0ji}\kappa\partial^i A^j(x) + J_0(x)\end{aligned}$$

leading to the *secondary constraint*

$$\phi_4(x) = \kappa F_{12} - J_0(x). \tag{4.7}$$

Of course there is a quicker way to arrive at this conclusion, since $\dot{\phi}_1 = -\partial\mathcal{H}_p/\partial A^0$.

The time derivatives of the other two primary constraints are

$$\begin{aligned}\dot{\phi}_2(x) &= \kappa\big(u^3 - \partial^2 A^0(x)\big) + J_1(x) \\ \dot{\phi}_3(x) &= \kappa\big(-u^2 + \partial^1 A^0(x)\big) + J_2(x)\end{aligned} \tag{4.8}$$

and putting them weakly to zero fixes the multipliers u^2, u^3. Finally, we must have

$$\dot{\phi}_4(x) = \kappa\partial^2 u^2(x) - \kappa\partial^1 u^3(x) + \partial_0 J_0(x) \approx 0.$$

Inserting u^2, u^3 from (4.8) this becomes $\partial_a j^a = 0$ and yields no further condition. Thus the Dirac-Bergmann algorithm leads to 3 primary and 1 secondary constraint.

Obviously $\phi_1 \equiv \gamma_1$ is first class. Also the combination

$$\gamma_2 = \partial_1\phi_2 + \partial_2\phi_3 + \phi_4 = \partial_i\pi_i + \frac{\kappa}{2}F_{12} - J_0 \tag{4.9}$$

is first class and is the analog of the Gauss constraint in electrodynamics. As SCC we may take

$$\chi_1 = \phi_2 \quad \text{and} \quad \chi_2 = \phi_3 \Longrightarrow \Delta_{\alpha\beta} = \kappa\begin{pmatrix} 0 & 1 \\ -1 & 0\end{pmatrix}\delta(x-y). \tag{4.10}$$

Δ has inverse $\Delta^{-1} = -\Delta/\kappa$ and (3.25) reads

$$\dot{F} \approx \{F, H\} + \{F, \gamma_1\}u^1 + \frac{1}{\kappa}\int dy\{F, \chi_i(y)\}\epsilon_{ij}\{\chi_j(y), H\}. \tag{4.11}$$

Since the FCC commute with all constraints they generate infinitesimal transformations on Γ_c, i.e. if (A, π) is on Γ_c then

$$\begin{aligned}\delta_{\mathcal{N}} A(x) &= \int dy\{A(x), \gamma_a(y)\}\mathcal{N}^a(y) \\ \delta_{\mathcal{N}}\pi(x) &= \int dy\{\pi(x), \gamma_a(y)\}\mathcal{N}^a(y)\end{aligned} \tag{4.12}$$

are variations tangent to it. This follows from $\delta_{\mathcal{N}}\phi_j = \int\{\phi_j, \gamma_a\}\mathcal{N}^a \approx 0$. Also, since these transformations commute with H_p, one expects that they are related to infinitesimal gauge transformations. Indeed, defining

$$G = \int dy \Big(\partial_0\lambda(y)\gamma_1(y) + \lambda(y)\gamma_2(y)\Big) \tag{4.13}$$

one finds

$$\delta_\lambda A^a(x) = \partial^a\lambda(x). \tag{4.14}$$

Only a particular combination of the 2 FCC generates the gauge transformations (4.2).

Contrary to the gauge generator G the primary Hamiltonian contains only the primary FCC. A symmetric treatment of all FCC is achieved in the extended formalism (see Dirac 1964) discussed below. We will come back to the important connection between FCC and gauge transformations in chap.7.

5. The Reduced Phase Space

First we shall consider SC systems, for which no multipliers remain in the time evolution (3.25). There is no ambiguity in the dynamics. The term in (3.25) containing the inverse of Δ forces the system to stay on Γ_c. This surface will be the *reduced phase space* for SC systems.

In the second subsection we consider FC systems. These are the most important systems since all gauge theories are of this type. Gauge related points should be identified and this leads us to the problem of gauge invariant functions and/or the gauge fixing problem. The FCC together with a complete set of gauge fixing conditions form then a SC system. Thus for FC systems the gauge fixing define a subset $\Gamma_r \in \Gamma_c$ and this set is the reduced phase space.

5.1 Second Class Constraints and Dirac Bracket

Motivated by (3.25) one introduces the *Dirac bracket* (see Dirac 1964) for two phase space functions as

$$\{F, G\}^* \equiv \{F, G\} - \{F, \chi_\alpha\}\Delta^{\alpha\beta}\{\chi_\beta, G\}, \tag{5.1}$$

in terms of which

$$\dot{F} \approx \{F, H\}^* \tag{5.2}$$

for SC systems. This bracket posses the same properties as the Poisson bracket, i.e. it is antisymmetric, bilinear and obeys the Jacobi identity and product rule. In addition we have

$$\{F, \chi_\alpha\}^* = 0 \quad , \quad \{F, G\}^* \approx \{F, G\} \quad , \quad \{F, \{G, K\}^*\}^* \approx \{F, \{G, K\}\} \tag{5.3}$$

for arbitrary F and first class G, K. These properties follow easily from the definition (5.1) and the property that first class functions have vanishing Poisson brackets with all constraints, e.g.

$$\{F, \chi_\alpha\}^* = \{F, \chi_\alpha\} - \{F, \chi_\beta\}\Delta^{\beta\gamma}\{\chi_\gamma, \chi_\alpha\} = 0. \tag{5.4}$$

Let us draw an immediate consequence of (5.4). According to theorem 1 any function can be replaced by its restriction to Γ_c, up to a linear combination of the constraints. Thus when calculating the Dirac bracket (5.1) between two functions we may replace them by their restriction to Γ_c since the other brackets vanish on account of (5.4). It follows that the SCC can be set equal to zero either before or after evaluating the Dirac bracket.

To understand the *geometric meaning* of SCC we recall some facts from symplectic geometry (see e.g. Arnold 1978):

In most cases the phase space Γ is the cotangental bundle T^*Q over the configuration space Q and hence is equipped with a natural symplectic structure (a non-degenerate closed two-form)

$$\omega = \omega_{\mu\nu} dx^\mu \wedge dx^\nu \tag{5.5a}$$

which, according to Darboux, can locally be written as

$$\omega = dq^i \wedge dp_i. \tag{5.5b}$$

Given ω, we can assign to a functions its corresponding Hamiltonian vector field:

$$F \Longrightarrow X_F \quad \text{by} \quad i_{X_F}\omega = dF, \tag{5.6}$$

where i_X and d are the interior and exterior derivatives, respectively. In local coordinates we find

$$\omega_{\mu\nu} X_F^\mu Y^\nu = \omega(X_F, Y) = i_{X_F}\omega(Y) = dF(Y) = \partial_\nu F\, Y^\nu$$

for any vector field Y, or

$$X_F^\mu = -\omega^{\mu\nu}\partial_\nu F, \quad \text{where} \quad \omega^{\mu\nu}\omega_{\nu\rho} = \delta^\mu{}_\rho. \tag{5.7}$$

The Poisson bracket of two functions is

$$\{F, G\} = -\partial_\rho F \omega^{\rho\sigma}\partial_\sigma G = \omega_{\mu\nu}\omega^{\mu\rho}\partial_\rho F \omega^{\nu\sigma}\partial_\sigma G = \omega(X_F, X_G). \tag{5.8}$$

In particular, the change of F under the Hamiltonian flow generated by G can be written as

$$F' \equiv \{F, G\} = \omega(X_F, X_G) = i_{X_F}\omega(X_G) = dF(X_G) = X_G^\mu \partial_\mu F. \tag{5.9}$$

In other words, the flow generated by G is just the motion along the Hamiltonian vector field X_G. For $G{=}H$ (5.9) are the Hamiltonian equations of motion.

Finally there is an important relation between Poisson and Lie bracket,

$$[X_F, X_G] = -X_{\{F,G\}}, \quad \text{where} \quad [X, Y]^\mu = X^\alpha \partial_\alpha Y^\mu - Y^\alpha \partial_\alpha Y^\mu \tag{5.10}$$

are the Lie bracket. (5.10) is a consequence of the Jacobi identity.

Let us now return to the SC systems. The inclusion map $j : \Gamma_c \longrightarrow \Gamma$ induces a two-form on Γ_c, namely the pull back of the symplectic form ω on Γ, $\omega_c = j^*\omega$. ω_c is closed since ω has this property, but it may be degenerate. In this case it is called *pre-symplectic*. However, for SCC it is indeed symplectic, as follows from

Theorem 3 *All constraints are second class if and only if $\omega_c = j^*\omega$ is non-degenerate.*

Actually, we shall see that the Dirac bracket belongs to ω_c. Most properties of the Dirac bracket, and in particular the Jacobi identity follow then at once from the corresponding properties of ω.

To prove this theorem we must show that ω is non-degenerate on the vectors tangent to Γ_c iff all constraints are SCC. A vector field Y is tangent to Γ_c if $Y^\mu \partial_\mu \chi_\alpha$ vanishes for all constraints χ_α. With (5.9) this is equivalent to

$$\omega(X_\alpha, Y) \approx 0 \quad \text{for all} \quad X_\alpha \equiv X_{\chi_\alpha}. \tag{5.11}$$

On the other hand, from (5.8) it follows that if χ_β is a SCC, there is an α with

$$\omega(X_\alpha, X_\beta) = \{\chi_\alpha, \chi_\beta\} \equiv \Delta_{\alpha\beta} \not\approx 0 \tag{5.12}$$

so that the Hamiltonian vector fields generated by the SCC are not tangent to Γ_c. Let us now determine the vectors X which obey

$$\omega(X, Y) \approx 0 \quad \text{for all tangent} \quad Y. \tag{5.13}$$

Since ω is non-degenerate, (5.13) has $\dim\Gamma$-$\dim\Gamma_c$ linearly independent independent solutions X. But because of our regularity conditions on the constraints the $\dim\Gamma$-$\dim\Gamma_c$ Hamiltonian vector fields X_α which are not tangent, are independent solutions. Thus any X which obeys (5.13) is a combination of the X_α. Hence there cannot be a tangent X obeying (5.13) which proves that $j^*\omega$ is non-degenerate.

Note that we used the SC nature of the constraints and in particular that the *flows generated by the SCC lead off the constraint surface.*

Now it is easy to prove that the Dirac bracket furnishes an explicit representation for the induced Poisson bracket. For that consider on Γ_c

$$\{F, G\}^* = \omega(X_F, X_G) - \omega(X_F, X_\alpha)\Delta^{\alpha\beta}\omega(X_\beta, X_G) \equiv \omega^*(X_F, X_G). \tag{5.14}$$

It is easy to see that $\omega^*(X_F + X_\chi, X_G) = \omega^*(X_F, X_G)$ for any Hamiltonian vector field belonging to the constraints. Thus ω^* depends only on the tangent components of X_F, X_G. But for tangent X_F we have $\omega(X_F, X_\alpha) \approx 0$ (see 5.11) and ω^* can be replaced by ω without changing the value of (5.14). This proves that ω^* *is just the pull-back of* ω *to* Γ_c.

5.2 First Class Constraints and Gauge Transformations

For purely FC systems the time evolution is governed by

$$\dot{F} \approx \{F, H\} + \{F, \phi_a\}\mu^a, \tag{5.15}$$

where the ϕ_a are the primary FCC. For the same initial conditions we get different evolutions, depending on the multipliers μ^a. The presence of arbitrary functions μ^a in the primary Hamiltonian tells us that not all $x = (q, p)$ are observable, i.e. there are several x representing a given physical state. Assume that the initial value $x(0)$ is given and represents a certain state. Then the equation of motion should fully determine the physical state at later times. So if $x'(t) \neq x(t)$ stem from the same physical state $x(0)$ they should be identified.

Consider two infinitesimal time evolutions of $F = F(0)$ given by H_p with different values of the multipliers,

$$F_i(t) = F(0) + t\{F, H\} + t\{F, \phi_a\}\mu_i^a \quad \text{i=1,2} \quad . \tag{5.16}$$

The difference $\delta F = F_2(t) - F_1(t)$ between the values is then

$$\delta_\mu F = \{F, \mu^a \phi_a\}, \quad , \quad \mu = t(\mu^2 - \mu^1). \tag{5.17}$$

Such a transformation does not alter the the physical state at time t, and hence is called a *infinitesimal gauge transformation*. Now we calculate

$$[\delta_\mu, \delta_\nu] F = \{\{\mu^a \phi_a, \nu^b \phi_b\}, F\}$$

and conclude that the commutator of any two primary FCC also generates gauge transformations. Also, performing a gauge transformation at $t=0$ with multipliers ν and then time evolve with multipliers μ should lead to the same state as doing these transformations in the reverse order. We find

$$[\delta_{t,\mu}, \delta_\nu] F = -t\{\{\nu^a \phi_a, H\}, F\} - t\{\{\nu^a \phi_a, \mu^b \phi_b\}, F\}$$

and conclude that the commutators $\{\phi_a, H\}$ also generates gauge transformations.

We have seen that the set of first class functions is closed with respect to the Poisson bracket and thus the $\{\phi_a, \phi_b\}$ and $\{\phi_a, H\}$ are linear combinations of the FCC. However, in general there will appear secondary FCC in these combinations. Also, if we compared the higher order terms in the time evolutions (5.16) we would find that time derivatives of $\{\phi_a, H_p\}$ generate gauge transformations. This way secondary FCC show up as gauge transformations in all relevant systems. However, it is not possible to show that every FCC is of the form $\{\phi_a, \phi_b\}$ or $\{\phi_a, H\}$ for some first class primary constraints (ϕ_a, ϕ_b) and thus is a gauge generator.Dirac conjectured that this might be the case, that is that *all* FCC γ_a show up this way and generate gauge transformations. We shall assume this conjecture to hold in what follows (although there are some exotic counterexamples [5] (Allcock 1975, Cawley 1979)).

Note, however, that if the structure coefficients in

$$\{\gamma_a, \gamma_b\} = t_{ab}^c \gamma_c \tag{5.18}$$

depend on the canonical variables, then $[\delta_\mu, \delta_\nu]F$ is a gauge transformation only on the constraint surface. Also, above we made the hidden assumption that time (or the space-time coordinates in field theory) is not affected by the transformation. Else we would have to take $F + \delta_\mu F$ at the transformed time $t + \delta_\mu t$ before calculating the second variation δ_ν. We come back to this point when discussing generally covariant theories in chap.7.

We conclude that the most general physically permissible motion should allow for an arbitrary gauge transformation to be performed during the time evolution. But H_p contains only the primary FCC. We thus have to add to H_p the secondary FCC multiplied by arbitrary functions. This led Dirac to introduce the *extended Hamiltonian*

---

[5] see (Henneaux and Teitelboim 1992, p. 82) for a discussion of Diracs conjecture.

$$H_p \longrightarrow H_e = H + \mathcal{N}^a \gamma_a \tag{5.19}$$

which contains *all* FCC (Dirac 1964). H_e accounts for all the gauge freedom.

Clearly, H_p and H_e should imply the same time evolution for the classical observables. *Observables* are gauge invariant functions on Γ_c, that is phase space functions that weakly commute with the gauge generators,

$$F \quad \text{observable} \quad \Longleftrightarrow \{F, \gamma_a\} \approx 0 \quad \text{for all} \quad a. \tag{5.20}$$

Since $H_e - H_p$ is a combination of the secondary FCC, we have

$$\dot{F} \approx \{F, H_p\} \approx \{F, H_e\} \tag{5.21}$$

for any observable F, as required. In the extended formalism one makes no distinction between primary and secondary FCC since they are treated symmetrically. The introduction of H_e is a new feature of the Hamiltonian scheme. It does not follow from the Lagrangean formalism.

Let us now investigate the *geometric meaning* of FC systems. As a preparation we show:

The induced 2-form $j^*\omega$ has rank $\geq 2(N-M)$, where M is the number of independent first or second class constraints.

Let us assume that the tangent vectors X_p, $p=1,\dots,P$ form a basis of the kernel of $j^*\omega$, i.e. $j^*\omega(X_p, Y) = \omega(X_p, Y) = 0$ for all tangents Y. Let us now see how big P can be. All vectors $a^p X_p$ obey

$$a^p \omega(X_p, Z_q) = 0, \quad \text{where the} \quad Z_q, \quad q = 1, \dots, M$$

are the Hamiltonian vectors associated with the constraints which, together with the tangent vectors, form a basis of $T\Gamma$ at the point under consideration. These are M equations for P unknowns. So, if $P \geq M$ then there would always exist a solution $X = a^p X_p$ with $\omega(X, Z_q) = 0$ for all Z_q. Being also a null-eigenvector of $j^*\omega$ we would conclude that $\omega(X, Z) = 0$ for all vectors $Z \in T\Gamma$ or that ω is degenerate. This then proves the statement above. Now we have the following

Theorem 4 *For a FC system the induced two-form $j^*\omega$ is maximally degenerate. Its kernel is spanned by the Hamiltonian vector fields belonging to the FCC.*

First, if $X^\mu \partial_\mu \gamma_b \approx 0$ for all constraints, then X is tangent. But since $X^\mu_a \partial_\mu \gamma_a \approx \{\gamma_a, \gamma_b\} \approx 0$, all Hamiltonian vectorfields belonging to the constraints are tangent. Second, for an arbitrary tangent vector X we have $\omega(X_a, X) = i_{X_a}(X) = \partial_\mu \chi_a X^\mu \approx 0$. Thus the M X_a's are null-eigenvectors of the induced two-form and the rank of $j^*\omega$ equals $2N-2M$, i.e. it is maximally degenerate.

Thus we have the following situation: The M FCC generate *flows which stay on the constraint surface* and which we identified with gauge transformations. The Hamiltonian vector fields belonging to the constraints are the null-directions of the induced pre-symplectic 2-form. That these null-vector fields generate gauge orbits follows from

Theorem 5 *On Γ_c the vectors X_a generate M-dimensional manifolds.*

The proof uses the Frobenius integrability condition, which says that M linearly independent vector fields define an integrable distribution (i.e. through each point in Γ_c there is a surface, the gauge orbit, to which the X_a are tangent) iff all Lie brackets $[X_a, X_b]$ are linear combinations of $(X_1, \ldots, X_M)$. Indeed,

$$[X_a, X_b] = -X_{\{\gamma_a, \gamma_b\}} = -t_{ab}^{\ c} X_c + \gamma_c \omega^{\mu\nu} \partial_\nu t_{ab}^{\ c} \approx -t_{ab}^{\ c} X_c, \tag{5.22}$$

where we used (5.18). Note that for (q,p)-dependent structure coefficients (as in gravity) the null vector fields are integrable only on the constraint surface.

In a next step one wants to eliminate the gauge degrees of freedom, that is identify points on the same gauge orbit. This can in principle be achieved by introducing gauge invariant variables, e.g. the transverse potential or holonomies in electrodynamics, or alternatively by fixing the gauge. A *gauge fixing* must obey two conditions: first it must be attainable and second it should fix the gauge uniquely [6]. We can fix the gauge by imposing the independent conditions

$$F_a(q,p) = 0, \quad \text{a=1,\ldots,M}. \tag{5.23}$$

The surface defined by these conditions should intersect every gauge orbit in exactly one point. A necessary condition is that at least one gauge fixing function F_b should change in the direction of all null-vectors X_a. In other words, there is at least one F_b so that

$$\lambda^a (X_a, \nabla F_b) = \lambda^a \{\gamma_a, F_b\} \neq 0$$

for all λ. This implies that

$$\det\{\gamma_a, F_b\} \equiv \det F_{ab} \neq 0. \tag{5.24}$$

In particular, if we could choose gauge fixings canonically conjugate to the constraints, $\{\gamma_a, F_b\} = \delta_{ab}$, then any gauge orbit and the gauge fixing surface both are Lagrangean submanifolds of Γ, conjugate to each other, like the surfaces q =const. and p =const., and in this case $\det F_{ab} = 1$. The determinant of F plays an important part in the quantization of gauge systems and is the well known *Fadeev-Popov determinant* (Fadeev and Popov 1967).

Because of (5.24) the FCC together with the gauge fixings form a SC system and we can take over the result from the previous subsection. The reduced phase space Γ_r consists of the points fulfilling the constraints and gauge fixings. Denoting the γ_a and F_a collectively as Ω_p, $p = 1, \ldots, 2M$, we find for the Hamiltonian equation of motion for any phase space function

$$\dot{F} = \{F, H\} - \{F, \Omega_p\} G^{pq} \{\Omega_q, H\}, \tag{5.25}$$

where G^{pq} is the inverse matrix of the Poisson bracket matrix of the constraints and gauge fixings, $G_{pq} = \{\Omega_p, \Omega_q\}$.

[6] There may be obstructions to fulfilling these requirements as has been demonstrated by Gribov and Singer (Gribov 1978, Singer 1978).

5.3 Systems with Second and First Class Constraints

Before gauge fixing the evolution is governed by the 'first class partner of the extended Hamiltonian'

$$H_e^* = H + \gamma_a \mathcal{N}^a - \chi_\alpha \Delta^{\alpha\beta}\{\chi_\beta, H\} \tag{5.26}$$

since for observables F,

$$\dot{F} = \{F, H_e^*\} \approx \{F, H\} + \{F, \gamma_a\}\mathcal{N}^a - \{F, \chi_\alpha\}\Delta^{\alpha\beta}\{\chi_\beta, H\}. \tag{5.27}$$

For a discussion of the *starred variables* see (Bergmann and Komar 1960). After gauge fixing one can again introduce starred variables with respect to the SC system $\psi_I = (\chi_\alpha, \gamma_a, F_b)$. Denoting the Poisson bracket matrix of all these constraints ψ_I by $\tilde{\Delta}$, we have

$$\dot{F} = \{F, H^*\}, \quad \text{where} \quad H^* = H - \psi_I \tilde{\Delta}^{IJ}\{\psi_J, H\}. \tag{5.28}$$

5.4 Gauge Fixing of Chern-Simons Theory

In section 4 we have seen that the Chern-Simons Lagrangean (4.1) leads to two SCC and two FCC. Now we supplement those by two gauge fixing conditions, namely

$$F_1 = A_0 \quad \text{and} \quad F_2 = \partial_i A^i. \tag{5.29}$$

Altogether the conditions $(\chi_1, \chi_2, \gamma_1, F_1, \gamma_2, F_2)$ form a SC system and define Γ_r. The surface defined by χ_i, γ_1 and F_1 can be parametrized by the spatial components of the gauge potential which can be decomposed as

$$A_i = \epsilon_{ij}\partial_j \varphi + \partial_i \lambda + \frac{1}{L} q_i \tag{5.30a}$$

with constant q_i [7]. Imposing further γ_2 and F_2 we see that

$$A_i = -\frac{1}{\kappa}\epsilon_{ij}\partial_j \frac{1}{\Delta} J^0 + \frac{1}{L} q_i \tag{5.30b}$$

so that $\Gamma_r = \{q_1, q_2\}$ is finite-dimensional. Furthermore, $\gamma_2 = 0$ and the periodicity of the A_i imply that the total charge $Q = \int dx J^0$ must vanish.

The inverse Poisson bracket 'matrix' reads

$$(\tilde{\Delta}^{IJ})(x,y) = \frac{1}{\kappa}\begin{pmatrix} 0 & -1 & 0 & 0 & \frac{1}{\Delta}\partial_2 & 0 \\ 1 & 0 & 0 & 0 & -\frac{1}{\Delta}\partial_1 & 0 \\ 0 & 0 & 0 & \kappa & 0 & 0 \\ 0 & 0 & -\kappa & 0 & 0 & 0 \\ \frac{1}{\Delta}\partial_2 & -\frac{1}{\Delta}\partial_1 & 0 & 0 & 0 & -\kappa\frac{1}{\Delta} \\ 0 & 0 & 0 & 0 & \kappa\frac{1}{\Delta} & 0 \end{pmatrix} \delta(x-y)$$

and one finds the following Dirac bracket for the coordinates on Γ_r

[7] The $U(1)$-bundle over the torus is non-trivial and A must be periodic only up to non-trivial gauge transformations (see e.g. Sachs and Wipf 1992). For simplicity we assume here that A is periodic and hence $\int F_{12} = 0$

$$\{q_i, q_j\} = -\delta_{ij}.$$

Calculating the starred Hamiltonian, one needs to remember that for a periodic function one has $\Delta^{-1}\Delta f = f - V^{-1}\int f$. After some algebra one finds

$$H^* = -\int dx \Big\{ A^0 (\kappa F_{12} - J_0) + \frac{1}{\kappa} J_0 \frac{1}{\Delta} \epsilon_{ij} \partial_i J_j \Big\} - \frac{1}{2} j_i q^i - \frac{1}{\kappa} \epsilon_{ij} j_i p_j,$$

where we have introduced the mean 'fluxes'

$$j_i \equiv \frac{1}{L} \int dx\, J_i \quad , \quad q^i \equiv \frac{1}{L} \int dx\, A^i \quad \text{and} \quad p_i \equiv \frac{1}{L} \int dx\, \pi_i. \tag{5.31}$$

After imposing the constraints χ_i, γ_2 and F_1 the non-trivial equations of motion take the simple form

$$\dot{q}_i = -\frac{1}{\kappa} \epsilon_{ij} j_j. \tag{5.32}$$

Of course, the evolution belonging to H^* stays on Γ_r. Since Γ_r is 2-dimensional the (topological) Chern-Simons theory is effectively a simple mechanical system. This was expected from the beginning since there are 6 constraints and gauge fixings for 6 degrees of freedom (per x).

To see the meaning of this result more clearly, let us see what the observables are. As coordinates on Γ_c we may take λ and q_i in (5.30a), so that we are considering functionals $F[A^0, \lambda, q_i, J_a]$. Such F commute with the FCC if they are independent of λ and A^0. Hence observables have the form

$$F = F[J_a, q_i] \tag{5.33}$$

and depend only on the zero-modes of the A_i.

Let us finally remark that for a pure CS theory ($J = 0$) the Lagrangean density is invariant, up to a total time derivative, under global gauge transformations for which only $e^{i\lambda}$ must be periodic. This introduces global gauge transformations with windings around the handles of the torus (the box with opposite points identified). Hence we must identify gauge potentials which are related by such global gauge transformations

$$A_i \sim A_i + \frac{2\pi}{L} n_i \quad \text{or} \quad q_i \sim q_i + 2\pi n_i.$$

Gauge invariant functionals must be invariant under such transformations. Thus they depend only on $\exp(i \sum m_i q_i)$. For pure Chern-Simons theories we have

$$e^{i \sum m_i q_i} = W(C, A) = \exp\left\{ i \oint_C A \right\}$$

on the constraint surface ($F_{12} = 0$) if the loop C winds m_i-times around the torus in the direction i. For a contractible loop $W(C, A)$ vanishes. Thus, observables have the form

$$F(A) = F\left(e^{i \oint_C A} \right). \tag{5.34b}$$

Let C, D be 2 loops which wind m_i, n_i-times around the torus in the direction i. We parametrize them by $x(\tau), y(s)$. We compute

$$\oint_C \oint_D \{A(x(\tau)), A(y(s))\} = -\oint_C \oint_D \epsilon_{ij}\dot{x}^i(\tau)\dot{y}^j(s)d\tau ds = -(n_1m_2 - n_2m_1).$$

Upon deformation of the curves the commutator is invariant and therefore is a *topological invariant.* This can be understood by noting that for $J=0$ the Chern-Simons model (4.1) is invariant under space-time diffeomorphisms. In particular the *spatial* ones are generated by

$$G_{diff} = \int dx\epsilon^i A_i \gamma_2$$

and hence observables must be invariant under spatial diffeomorphisms.

Finally note, that for the pure Chern-Simons theory the phase space variables q_i lie in $[0, 2\pi]$, that is Γ_r is compact and as a consequence the Hilbert space becomes finite dimensional.

5.5 First Order Action Principles

The solutions to the primary Hamiltonian equations of motion (3.12) extremize the primary (or total) first order action,

$$\delta S_p = \delta \int_{t_1}^{t_2} \Big(\dot{q}^i p_i - H - \sum_{primary} u^m \phi_m\Big) dt = 0 \tag{5.35}$$

with respect to variations $\delta q, \delta p, \delta u$ subject only to the restriction $\delta q(t_1) = \delta q(t_2) = 0$. The variables u^m which have been introduced to make the Legendre transformation invertible appear now as Lagrange multipliers enforcing the *primary constraints.* It is clear that the theory is invariant under $H \to H + c^m\phi_m$ since such a change can be absorbed into the Lagrange multipliers.

The variational principle (5.35) is equivalent to

$$\delta \int \Big(\dot{q}^i p_i - H\Big) dt = 0 \quad \text{subject to} \quad \phi_m = 0. \tag{5.36}$$

There is yet another variational principle which for gauge invariant observables leads to the same time evolution, namely the *extended action principle.* The equations of motion for the extended formalism follow from

$$\delta S_e = \delta \int \Big(\dot{q}^i p_i - H - \sum_{all\,constr.} u^j \phi_j\Big) dt = 0, \tag{5.37}$$

where the sum extends over *all constraints.*

Take the case of purely SCC and let $y^i \to x^\mu(y^p)$ be the imbedding $\Gamma_r \to \Gamma$. The Lagrange multiplier method guarantees that the implementation of the constraints χ_α, either directly or via the Lagrange multipliers, are equivalent. Now let us solve the constraints directly in (5.37). Recall that a symplectic 2-form can locally be written as $\omega = d\theta$. The pull-back of the one-form potential θ is

$$j^*(\theta) = j^*(p_i dq^i) \equiv j^*(a_\mu dx^\mu) = a_\mu(x(y))\frac{\partial x^\mu}{\partial y^p} dy^p.$$

Inserting this into the extended action one finds

$$S_e[y] = \int \{\theta - H(x(y))dt\} = \int \left(a_p \dot{y}^p - h\right) dt. \tag{5.38}$$

The corresponding variational principle yields then the equation of motion for SC systems:

$$\delta S_e[y] = 0 \Longleftrightarrow \dot{y}^p = \{y^p, h(y)\}^*. \tag{5.39}$$

This can be checked directly by using

$$j^*\omega = j^* d\theta = d\, j^*\theta = d\, a_p(y) dy^p. \tag{5.40}$$

The fact that (5.37) yields (5.38) is of practical use when calculating Dirac brackets. One solves for the constraints inside the action and from the new kinetic term one reads off the induced potential form $j^*\theta$ on Γ_c. From (5.40) one computes the induced symplectic form and thus the Dirac bracket.

For FC systems it is also legitimate to solve the FCC inside the action

$$S_e = \int \left(\dot{q}^i p_i - H - \mathcal{N}^a \gamma_a\right) dt. \tag{5.41}$$

However, since the induced 2 form is degenerate the equations of motion on Γ_c are not canonical. To get Hamiltonian equations one needs to go to Γ_r by imposing additional gauge conditions. Then one may write down the corresponding action for the SC system as discussed above.

Let us demonstrate how this works for the *Chern-Simons theory.* Inserting the fields on the reduced phase space (5.30b) into the first order action results in

$$S = \int dt \int \left(\frac{\kappa}{2} q_i \epsilon_{ij} \dot{q}_j - q_i j_i\right) + \frac{1}{\kappa} \int dx J^0 \frac{1}{\Delta} \epsilon_{ij} \partial_i J_j \tag{5.42}$$

which of course reproduces the correct equations of motion (5.32).

6. Yang-Mills Theories

In this chapter I consider YM theories (see e.g. Abers and Lee 1973) without coupling to matter and emphasize the role of the constraints (Hanson et.al. 1976, Sundermeyer 1982). Pure *non-abelian* YM theories are interesting in their own right and they are highly non-trivial.

The YM action for the gauge fields is

$$S = -\frac{1}{4} \int \mathrm{tr}\left[F_{\mu\nu} F^{\mu\nu}\right] d^3x dt, \tag{6.1}$$

where the field strength is Lie-algebra valued,

$$F_{\mu\nu} = \partial_\mu A_\nu - \partial_\nu A_\mu - i[A_\mu, A_\nu] \quad , \quad A_\mu = A_\mu^a T_a \quad , \quad [T_a, T_b] = i f_{ab}^c T_c. \tag{6.2}$$

Here $a, b, \dots$ denote internal indices and $\mu, \nu \dots$ space-time indices. The T_a are hermitean generators and the structure constants f_{ab}^c are totally antisymmetric. The basis $\{T_a\}$ can be chosen orthogonal with respect to the Killing metric $k_{ab} = f_{ad}^c f_{cb}^d$

on the Lie algebra. Then the positions of the internal indices a, b does not matter. The trace in (6.1) refers to those indices.
The action is invariant under local gauge transformations

$$A_\mu \longrightarrow e^{-i\theta} A_\mu e^{i\theta} + i e^{-i\theta} \partial_\mu e^{i\theta} \tag{6.3}$$

with $\theta = \theta^a T_a$. The functions $\theta^a = \theta^a(x,t)$ are arbitrary functions on space-time. The infinitesimal form of these gauge transformations is

$$\delta_\theta A^a_\mu = -(\partial_\mu \theta^a + f^a_{bc} A^b_\mu \theta^c) = -(D_\mu \theta)^a. \tag{6.4}$$

The local gauge invariance implies generalized Bianchi identities

$$D_\mu D_\nu F^{\mu\nu} = 0$$

and renders the system singular. Among the field equations $D_\mu F^{\mu\nu} = 0$ there are some containing second time-derivatives of A,

$$D_\mu F^{\mu i} = 0 \quad , \quad i = 1,2,3 \tag{6.5}$$

and which therefore are dynamical equations of motion. The others

$$D_\mu F^{\mu 0} = D_i F^{i0} \quad \text{or} \quad \phi_m(A, \dot{A}) = \partial_i F^{i0}_m + f_{ab}{}^m A^a_i F^{i0}_b = 0, \tag{6.6}$$

where $m = 1, \dots, N$=dim(Gauge Group), are Lagrangean constraints. No further constraints appear since the time derivatives of the ϕ_m vanish on account of the field equations and the constraints themselves.

The canonical momenta conjugate to the A's are

$$\pi^\mu_a = -F^{0\mu}_a \quad , \quad \{A^a_\mu(x), \pi^\nu_b(y)\} = \delta^a_b \delta^\nu_\mu \delta(x-y). \tag{6.7}$$

Since the field strength tensor is antisymmetric we obtain N *primary constraints*

$$\phi_m(A,\pi) = \pi^0_m \approx 0.$$

After a partial integration the canonical Hamiltonian is found to be

$$H = \int dx \Big(\frac{1}{2} \pi^i_a \pi^i_a + \frac{1}{4} F^a_{ij} F^a_{ij} - A^a_0 D_i \pi^i_a \Big). \tag{6.8}$$

It determines the time evolution

$$\dot{F} = \{F, H_p\} \quad , \quad H_p = H + \int dx \, u^m \phi_m. \tag{6.9}$$

We need to check the consistency of the primary constraints:

$$\dot{\phi}_m = \{\phi_m, H_p\} = 0 \Longrightarrow \tilde{\phi}_m = (D_i \pi^i)_m \approx 0. \tag{6.10}$$

These N secondary constraints are the generalizations of the *Gauss constraint* in electrodynamics.

The only non-trivial Poisson brackets of the algebra of constraints are

$$\{\tilde{\phi}_m(x), \tilde{\phi}_n(y)\} = f^p_{mn} \tilde{\phi}_p(x) \delta(x-y). \tag{6.11}$$

The algebra is closed and therefore the $2N$ constraints $\gamma_a = (\phi_m, \tilde{\phi}_n)$ form a FC system. Their Poisson brackets with H are computed to be

$$\{\phi_m, H\} = \tilde{\phi}_m \approx 0 \quad , \quad \{\tilde{\phi}_m, H\} = -f_{mn}^{\ \ p} A_0^n \tilde{\phi}_p \approx 0. \tag{6.12}$$

Let us now investigate the relation between the Hamiltonian gauge symmetries generated by the FCC and the Lagrangean gauge transformations (6.4). A general combination of the FCC $\phi = \int(\epsilon_1^m \phi_m + \epsilon_2^m \tilde{\phi}_m)$ generates the canonical symmetries

$$\begin{aligned} \delta A_\mu^a &= \{\phi, A_\mu^a\} = \delta_\mu^0 \epsilon_1^a - \delta_\mu^i D_i \epsilon_2^a \\ \delta \pi_a^\mu &= \{\phi, \pi_a^\mu\} = \delta_i^\mu f_{ab}^{\ \ c} \epsilon_2^b \pi_c^i + \int \phi_m \{\epsilon_1^m, \pi_a^m\}, \end{aligned} \tag{6.13}$$

where we have already anticipated that ϵ_1 depends on A_0. ¿From (6.4) we read off how the ϵ's must be chosen to correspond to Lagrangean gauge transformations. We find that the particular combination

$$G = D_0 \theta^m \phi_m - \theta^m \tilde{\phi}_m \tag{6.14}$$

generates those transformations. Both primary and secondary FCC enter the Lagrangean gauge transformations similarly as for the CS theory.

Alternatively we can introduce gauge invariant variables, e.g. the Wilson loops (Wu and Yang 1975, lecture of R. Loll) or fix the gauge. To fix the gauge freedom we need $2N$ gauge fixing conditions on the phase space variables (A, π). Contrary to the situation in electrodynamics the gauge fixing in YM theories is rather subtle due to the Gribov problem (Gribov 1978). Let $F_a(A_\mu)$ be local gauge fixings (which we assume not to depend on the momenta). Then the following problem may arise: *There are several fields $A_\mu^{(j)}$ which are related by finite gauge transformations and all of them obey the gauge fixing.*

This happens for example for QED_2 on the Euclidean torus, if one decomposes a gauge field as in (5.30a). The local condition $\partial^\mu A_\mu$ eliminates the gauge function λ but does not constrain the q_i. But 2π and q_i+2π are gauge equivalent configurations and this freedom cannot be fixed by local gauge conditions. This is an example to a more general situation which has been proven by Singer (Singer 1978): For compactified YM-theories no global continuous gauge choice of the (local) form $F_a(A)=0$ exists which completely specifies the gauge.

Rather then dwelling on the various gauge fixings, their merits and drawbacks, which is of course important in a functional quantization, let me make some remarks about the variational problem.

The primary FCC ϕ_m are sort of uninteresting, since they can easily be eliminated in the extended formalism. We eliminate them by setting $\pi^0 = A_0 = 0$ and find the first order action

$$S = \int \Big[\pi_a^i \dot{A}_i^a - \mathcal{N}^a \gamma_a - \frac{1}{2}(\pi_a^i \pi_i^a + B_a^i B_i^a)\Big] dt \quad , \quad \gamma_a = (D_i \pi^i)^a, \tag{6.15}$$

with multiplier fields $\mathcal{N}_a$. This form of the action is the one which is usually met in the literature (for example, in gravity one does not keep the momenta conjugate to the lapse and shift functions in the first order action). After having eliminated one pair of canonical variables one may wonder how one can reconstruct the full set of Lagrangean gauge transformations (6.4). This is indeed possible for all relevant systems in physics, e.g. YM-theories, the relativistic particle, string and gravity (Mukhanov and Wipf 1994). In particular for YM-theories one sees that the first order action is invariant under the transformations

$$\begin{aligned}\delta A_i^a &= \{A_i^a, \int \theta^b \gamma_b\} = -(\partial_i \theta^a + f^a bc A_i^b \lambda^c \\ \delta \pi_a^i &= \{\pi_a^i, \int \theta^b \gamma_b\} = -f_{bc}^a \mathcal{N}^b \theta^c \\ \delta \mathcal{N}^a &= \dot{\theta}^a - f_{bc}^a \mathcal{N}^b \theta^c\end{aligned} \tag{6.16}$$

which are just the transformations generated by G in (6.14) if we identify the Lagrangean mulipliers $\mathcal{N}^a$ with the A_0^a. This implies then that the transformations (6.16) correspond to Lagrangean gauge transformations. Actually the transformations (6.16) *can be exponentiated in phase space* and the exponentiated ones are just the transformation (6.3) with A_0 replaced by $\mathcal{N}$ and supplemented by

$$\pi^i \longrightarrow e^{-i\lambda} \pi^i e^{i\theta}, \quad \text{where} \quad \pi^i = \pi_a^i T^a. \tag{6.17}$$

For generally covariant theories it is not clear what are the exponentiated transformations in phase space (belonging to Lagrangean symmetries) which leave the first order action invariant. We will come back to this point in the next chapter, and in particular sect. 7.2.

7. Lagrangean Symmetries of First-Class Hamiltonian Systems

7.1 The Relativistic Particle

We describe the relativistic particle in 4-dimensional Minkowski spacetime by 4 scalar fields $\phi^\mu(t)$, $\mu = 0,1,2,3$, in 1-dimensional 'spacetime'. To the latter we assign a metric $d\tau^2 = -g(t)dt^2$ (τ =proper time), $g < 0$; t denotes an arbitrary time parameter. (g is not related to the Minkowski metric $\eta_{\mu\nu}$ whose components are used here only as a set of constants.) We consider $g(t)$, along with the $\phi^\mu(t)$, as our 'field' variables and take as action

$$S = \frac{1}{2} \int \left(\frac{d\phi^\mu}{d\tau} \frac{d\phi_\mu}{d\tau} - m^2\right) d\tau$$

rewritten as

$$S = -\frac{1}{2} \int \sqrt{-g} [g^{00} \dot{\phi}^\mu \dot{\phi}_\mu + m^2] dt \tag{7.1}$$

where the overdot denotes differentiation with respect to time t and $\phi^\mu \phi_\mu = -(\phi^0)^2 + \sum_1^3 (\phi^i)^2$. The m^2-term may be viewed as 'cosmological constant' in 1-dimensional 'spacetime'.

S is invariant with respect to general coordinate transformations (reparametrization invariance). The infinitesimal form of these transformations reads

$$t \to t - \xi, \qquad g \to g + \mathcal{L}_\xi g, \qquad \phi^\mu \to \phi^\mu + \mathcal{L}_\xi \phi^\mu, \tag{7.2}$$

where $\mathcal{L}_\xi$ is the Lie-derivative. Introducing the lapse function $\mathcal{N}$ according to

$$g = -\mathcal{N}^2 \tag{7.3}$$

we get the following transformation law from (7.2)

$$\bar{\delta}\phi^\mu = \dot{\phi}^\mu \xi \quad \text{and} \quad \bar{\delta}\mathcal{N} = (\mathcal{N}\xi)^\cdot \,. \tag{7.4}$$

The action (7.1) leads to the primary constraint $\phi_1 = \pi_g = 0$ which in turn implies the secondary one

$$\phi_2 \equiv \gamma = \frac{1}{2}\left(\pi^\mu \pi_\mu + m^2\right). \tag{7.5}$$

These are FCC. The partial gauge fixing $F_1 = g^{00} + 1 = 0$ and ϕ_1 form a conjugate second class pair and can be eliminated. Applying the standard procedure one finds then the following first order action

$$S = \int \left[\pi_\mu \dot{\phi}^\mu - \mathcal{N}\gamma\right] dt, \tag{7.6}$$

where the π_μ are the momenta conjugate to the ϕ^μ,

$$\{\phi^\mu, \pi_\nu\} = \delta^\mu_\nu. \tag{7.7}$$

The Lagrangean multiplier $\mathcal{N}$ accompanying the constraint γ (which is the super-Hamiltonians) reintroduces the lapse function.

The action (7.6) is invariant with respect to the infinitesimal off mass-shell *gauge* transformations

$$\delta_\lambda \phi^\mu = \{\phi^\mu, \lambda\gamma\} = \pi^\mu \lambda, \quad \delta_\lambda \pi_\mu = \{\pi_\mu, \lambda\gamma\} = 0, \quad \delta_\mathcal{N} = \dot{\lambda}. \tag{7.8}$$

With the identification $\lambda = \mathcal{N}\xi$ these transformations coincide with the diffeomorphism-induced transformations (7.4), but *only on mass shell:*

$$\dot{\phi}^\mu = \mathcal{N}\pi^\mu, \qquad \dot{\pi}^\mu = 0. \tag{7.9}$$

This is a general problem with diffeomorphism invariant theories. The transformations generated by the FCC alone can only be identified with the Lagrangean symmetries (where one used the relation (3.8b)) if we impose the equations of motion (go on shell). On the technical side the difficulty of identifying gauge and diffeomorphism transformations can be traced back to the nonlinear dependence of the constraint on the momentum. This is the important difference between internal and spacetime symmetries. In the following section we shall see how the canonical transformations generated by the FCC must be modified to yield all Lagrangean symmetries.

7.2 Hamiltonian vs. Lagrangean Symmetries

In this chapter I shall consider a general FC system, the first order action of which is given by (5.41). This action describes both systems with a finite or infinite number of degrees of freedom if the following condensed notation (DeWitt 1965) is assumed: For systems with a finite number of degrees of freedom a and i are discrete and for field theories they denote both internal indices and space-coordinates. To distinguish internal from composite indices we shall use tildes for the latter ones. For field theories $\tilde{i} = \{i, x\}$ and $\tilde{a} = \{a, x\}$, where i and a are some discrete (internal) indices. For a scalar field $q^{\tilde{i}}(t) = \varphi^x(t) = \varphi(x,t)$ and for a vector field $q^{\tilde{i}}(t) = A^{i,x}(t) = A^i(x,t)$. We adopt the Einstein convention and assume summation over discrete repeated indices and integration over continuous ones. We shall not distinguish $q^{i,x}$ and q^i_x and use the position of the continuous index just to indicate when we should integrate. Thus, e.g.,

$$\xi^x p_{i,x} \dot{q}^{i,x} = \sum_i \int dx \xi(x) p_i(x) \dot{q}^i(x)$$

but

$$\xi^x p^x_i q^{ix} = \sum_i \xi(x) p_i(x) q^i(x), \qquad \text{no integration.}$$

Sometimes it will be convenient to resolve the composite index $\tilde{i}$ (or $\tilde{a}$) as i, x (or a, x). If the system contains fermions then some of the variables $p, q, \mathcal{N}$ will be of Grassmannian type.

In particular the first order action reads

$$S_e = \int \left(p_{\tilde{i}} \dot{q}^{\tilde{i}} - \mathcal{N}^{\tilde{a}} \gamma_{\tilde{a}} - H \right) dt. \tag{7.10}$$

For FC systems the constraints and Hamiltonian form a closed algebra (possibly extended to fermionic variables, in which case the algebra is graded (Casalbuoni 1976)):

$$\{\gamma_{\tilde{a}}, \gamma_{\tilde{b}}\} = t^{\tilde{c}}_{\tilde{a}\tilde{b}} \gamma_{\tilde{c}} \qquad \text{and} \qquad \{H, \gamma_{\tilde{a}}\} = t^{\tilde{b}}_{\tilde{a}} \gamma_{\tilde{b}}, \tag{7.11}$$

with structure coefficients $t^{\tilde{c}}_{\tilde{a}\tilde{b}}$ and $t^{\tilde{b}}_{\tilde{a}}$. These coefficients may depend on the canonical variables p, q.

The equations of motion resulting from the variation of the action (7.10) with respect to q, p and the Lagrangean multipliers $\mathcal{N}$,

$$\delta S_e = \int \left(\delta p_{\tilde{i}} EM(q^{\tilde{i}}) - \delta q^{\tilde{i}} EM(p_{\tilde{i}}) - \delta \mathcal{N}^{\tilde{a}} \gamma_{\tilde{a}} \right) dt + \text{bound. terms}, \tag{7.12}$$

are

$$\begin{aligned} EM(q^{\tilde{i}}) &\equiv \dot{q}^{\tilde{i}} - \{q^{\tilde{i}}, \mathcal{N}^{\tilde{b}} \gamma_{\tilde{b}} + H\} = 0, \\ EM(p_{\tilde{i}}) &\equiv \dot{p}_{\tilde{i}} - \{p_{\tilde{i}}, \mathcal{N}^{\tilde{b}} \gamma_{\tilde{b}} + H\} = 0, \\ \gamma_{\tilde{a}} &= 0. \end{aligned} \tag{7.13}$$

We use the abbreviations $EM(q)$ and $EM(p)$ for the equations of motion. Of course, on mass shell we have $EM = 0$, but off mass shell either $EM(q)$ or $EM(p)$ (or both) does not vanish.

The first order action is invariant with respect to the infinitesimal transformations (Mukhanov and Wipf 1994)

$$\hat{I}_{\xi,\lambda}\, F(q,p,\mathcal{N}) = F\big(\hat{I}_{\xi,\lambda}\, q, \hat{I}_{\xi,\lambda}\, p, \hat{I}_{\xi,\lambda}\, \mathcal{N}\big), \qquad \hat{I}_{\xi,\lambda} = \hat{1} + \delta_{\xi,\lambda} + \cdots, \tag{7.14a}$$

where

$$\begin{aligned} \delta_{\xi,\lambda} q^{ix} &= EM(q^{ix})\xi^x + \{q^{ix}, \lambda^{\bar{b}}\gamma_{\bar{b}}\}, \\ \delta_{\xi,\lambda} p_i^x &= EM(p_i^x)\xi^x + \{p_i^x, \lambda^{\bar{b}}\gamma_{\bar{b}}\}, \\ \delta_{\xi,\lambda} \mathcal{N}^{\bar{a}} &= \dot{\lambda}^{\bar{a}} - \lambda^{\bar{b}}\mathcal{N}^{\bar{c}} t^{\bar{a}}_{\bar{c}\bar{b}} - \lambda^{\bar{b}} t^{\bar{a}}_{\bar{b}}. \end{aligned} \tag{7.14b}$$

The functions $\xi^x = \xi(x,t)$ and $\lambda^{\bar{a}} = \lambda^a(\mathcal{N},x,t)$ are the 'parameters' of the infinitesimal transformations. The first order Lagrangean is invariant up to a total time-derivative and correspondingly the action up to boundary terms

$$\delta_{\xi,\lambda} S_e = \left(p_{\bar{i}}\delta_{\xi,\lambda}\, q^{\bar{i}} - \lambda^{\bar{a}}\gamma_{\bar{a}}\right)\Big|_{t_i}^{t_f}. \tag{7.15}$$

If the parameters $\xi, \lambda^{\bar{a}}$ vanish at the initial and final times then $\delta_{\xi,\lambda} S = 0$. If we would like the action to be invariant even under transformations with non-zero ξ, λ at the boundaries then we need to add to the action the total derivative of some function $Q(q,p)$ which satisfies the equation

$$\frac{\delta Q}{\delta q^{\bar{i}}}\delta_{\xi,\lambda} q^{\bar{i}} + \frac{\delta Q}{\delta p_{\bar{i}}}\delta_{\xi,\lambda} p_{\bar{i}} = \dot{\lambda}^{\bar{a}}\gamma_{\bar{a}} - p_{\bar{i}}\delta_{\xi,\lambda} q^{\bar{i}}, \tag{7.16}$$

where $\delta/\delta q$ etc. means functional derivative with respect to q.

To construct the finite transformations we need to integrate the infinitesimal transformations. To be successful in this 'exponentiation' it is clear that the following necessary condition should be fulfilled: The algebra of infinitesimal transformations should be closed. To check that let us calculate the commutator of two subsequent infinitesimal transformations (7.14) with parameters ξ_1, λ_1 and ξ_2, λ_2 respectively. If we are making two subsequent infinitesimal transformations, then $\lambda^{\bar{b}}$ of the second transformation will depend on q,p if the structure constants depend on the canonical variables (since $\lambda^{\bar{b}}$ depends on $\mathcal{N}$) and we should keep $\lambda^{\bar{b}}$ inside the Poisson brackets even for purely bosonic theories.

For an arbitrary algebraic function $F(q,p)$ of the canonical variables (for example $F=q$ or $F=p$) a rather lengthy but straightforward calculation yields the commutator

$$\begin{aligned} \big[\hat{I}_{\xi_2\lambda_2}, \hat{I}_{\xi_1\lambda_1}\big]\, F^x(q,p) &= \Big(\frac{\delta F^x}{\delta q_i^z} EM(q_i^z) + (q \to p)\Big)(\dot{\xi}_1^z\xi_2^z - \xi_1^z\dot{\xi}_2^z) \\ &+ \Big((\xi_2^x - \xi_2^y)\lambda_1^{\bar{c}} - \mathcal{N}^{\bar{c}}\xi_2^x\xi_1^y - (1 \leftrightarrow 2)\Big)\Big(\{F^x, \frac{\delta\gamma_{\bar{c}}}{\delta q_j^y}\}EM(q_j^y) + (q \to p)\Big) \\ &- (\xi_2^x\xi_1^y - \xi_1^x\xi_2^y)\Big(\{F^x, \frac{\delta H}{\delta q_j^y}\}EM(q_j^y) + (q \to p)\Big) + \{F^x, \bar{\lambda}^{\bar{c}}\gamma_{\bar{c}}\} \end{aligned} \tag{7.17a}$$

and correspondingly for the Lagrangean multipliers one has

$$\begin{aligned}[\hat{I}_{\xi_2\lambda_2}, \hat{I}_{\xi_1\lambda_1}]\mathcal{N}^{\bar{a}} &= (\hat{I}_{\bar{\lambda}} - 1)\mathcal{N}^{\bar{a}} + \lambda_2^{\bar{d}}\lambda_1^{\bar{c}}\Big(\dot{t}^{\bar{a}}_{\bar{c}\bar{d}} - \{t^{\bar{a}}_{\bar{c}\bar{d}}, \mathcal{N}^{\bar{e}}\gamma_{\bar{e}} + H\}\Big)\\ &- (\lambda_2^{\bar{c}}\xi_1^{\bar{x}} - \lambda_1^{\bar{c}}\xi_2^{\bar{x}})\Big(\frac{\delta}{\delta q_i^{\bar{x}}}(\mathcal{N}^{\bar{b}}t^{\bar{a}}_{\bar{b}\bar{c}} + t^{\bar{a}}_{\bar{c}})EM(q_i^{\bar{x}}) + (q \to p)\Big),\end{aligned} \tag{7.17b}$$

where we have introduced

$$\bar{\lambda}^{\bar{a}} = \lambda_1^{\bar{e}}\lambda_2^{\bar{b}}t^{\bar{a}}_{\bar{b}\bar{e}} + \frac{\delta\lambda_2^{\bar{a}}}{\delta\mathcal{N}^{\bar{b}}}\delta_{\lambda_1}\mathcal{N}^{\bar{b}} - \frac{\delta\lambda_1^{\bar{a}}}{\delta\mathcal{N}^{\bar{b}}}\delta_{\lambda_2}\mathcal{N}^{\bar{b}}. \tag{7.17c}$$

We would like to stress that when we are performing the second transformation in (2.17a,b) which follows the first one, then we must use the transformed variables. In particular, instead of $\lambda_2(\mathcal{N}, x, t)$ we must take $\lambda_2(\hat{I}_{\lambda_1}\mathcal{N}, x, t)$. This explains the appearance of the last terms in (7.17c)

In the particular case where the structure coefficients $t^{\bar{a}}_{\bar{b}\bar{c}}$ do not depend on the canonical variables q, p the parameter $\bar{\lambda}$ also does not depend on them as can be seen from (7.17c). Also, $\dot{t}^{\bar{a}}_{\bar{b}\bar{c}} = 0$ in this case and thus the commutator of two transformations generated by the constraints only ($\xi = 0$) yields again a transformation generated by the constraints. Hence, *if the structure coefficients do not depend on the canonical variables the transformations generated by the constraints form a closed algebra off mass-shell.* On the other hand, if the structure coefficients do depend on the canonical variables that does not automatically imply that the algebra of transformations will not close. Indeed, the q, p-dependence in the formula (7.17c) for $\bar{\lambda}$ can, in principle, be cancelled against an appropriate choice of the $\mathcal{N}$-dependence of λ. Actually this takes place for gravity, where some of the structure coefficients depend on q (Mukhanov and Wipf 1994). Also the last terms in (7.17b) vanish in this case in the subspace

$$\mathcal{M} : EM(q) = \dot{q}^{\bar{i}} - \{q^{\bar{i}}, \mathcal{N}^{\bar{b}}\gamma_{\bar{b}} + H\} = 0 \tag{7.18}$$

and the algebra of transformations generated only by the constraints is closed in this subspace where the Lagrangean system lives. Although it is still closed off mass shell, it is not for all trajectories in phase space.

The algebra of transformations (7.17) can also be closed in all relevant cases even when $\xi \neq 0$ if the $\lambda^{\bar{b}}$ and ξ are related in a certain way. The corresponding transformations are actually the symmetry transformations corresponding to the symmetries of the Lagrangean system when some of the constraints are nonlinear in the momenta.

Now we shall study the question when the symmetry transformations of a Hamiltonian system are also symmetry transformations of the corresponding Lagrangean system. For that the transformations (7.14b) should at least leave the subspace $\mathcal{M}$ (see (7.18)) on which the Lagrangean system lives, invariant. That is, they should leave any trajectory which belongs to the subspace $\mathcal{M}$ in this subspace. The necessary conditions for this can be gotten by varying (7.18) as follows

$$\begin{aligned}\left(\delta_{\xi,\lambda}q^{\bar{i}}\right)^{\cdot} &= \frac{\delta^2(H + \mathcal{N}^{\bar{e}}\gamma_{\bar{e}})}{\delta p_{\bar{i}}\delta q^{\bar{j}}}\,\delta_{\xi,\lambda}q^{\bar{j}}\\ &+ \frac{\delta^2(H + \mathcal{N}^{\bar{e}}\gamma_{\bar{e}})}{\delta p_{\bar{i}}\delta p_{\bar{j}}}\,\delta_{\xi,\lambda}p_{\bar{j}} + \{q^{\bar{i}}, \delta_{\xi,\lambda}\mathcal{N}^{\bar{e}}\gamma_{\bar{e}}\}.\end{aligned} \tag{7.19}$$

Thus the transformations δq, δp and $\delta\mathcal{N}$ should satisfy the equation (7.19) in the subspace $\mathcal{M}$. Only in this case can the transformations in phase space be symmetry transformations of the corresponding Lagrangean system. Substituting (7.14b) into (7.19) this condition simplifies to

$$\frac{\delta^2(H+\mathcal{N}^{\bar{e}}\gamma_{\bar{e}})}{\delta p_i^x \delta p_j^y} EM(p_j^y)\xi^y = \frac{\delta^2\gamma_{\bar{e}}}{\delta p_i^x \delta p_j^y} EM(p_j^y)\lambda^{\bar{e}} \tag{7.20}$$

and imposes a certain functional dependence between ξ and λ^a. If this condition is fulfilled the phase space transformations (7.14) can be interpreted as Lagrangean symmetries. At the same time the number of free functions becomes equal to the number of constraints as it should be.

Gauge Invariance. If the constraints are linear and H is at least quadratic in the momenta then only for $\xi^z = 0$ can equation (7.20) be satisfied [8]. So, in this case the transformations which are generated by the constraints alone will also be symmetry transformations for the corresponding Lagrangean system. We shall call them *gauge transformations.* For example, in Yang-Mills theories all constraints are linear in the momenta and the finite gauge transformations can be recovered as transformations generated only by the constraints ($\xi=0$) in the Hamiltonian formalism [9] (see (6.16) and (6.17)).

Reparametrization invariance. Usually the reparametrization invariance of a Lagrangean system, if it exists, is identified with the gauge invariance in the Hamiltonian formalism. As we shall see they are actually very different and this identification can only be made on mass shell.

If some of the constraints are nonlinear in momenta then it is obvious that the transformations generated by the constraints only ($\xi = 0$) do not satisfy the condition (7.20) and hence can not be symmetry transformations of the corresponding Lagrangean system. However, in all known theories with nonlinear constraints there are canonical coordinates such that $H = 0$ and the condition (7.20) can be satisfied if we impose some functional dependence between λ and ξ in (7.14) so that $\xi \neq 0$ for such theories. Thus the nonlinear constraints generate the symmetry transformations of the corresponding Lagrangean system only in very special combination with 'trivial' ξ-transformations. The reason for that is the following: a transformation generated by a nonlinear constraint takes off mass-shell trajectories away from the subspace $\mathcal{M}$ and the extra compensating transformation returns the trajectories back to $\mathcal{M}$. More explicitly taking $\lambda^{\bar{e}}$ to be $\lambda^{ez} = \mathcal{N}^{ez}\xi^z$ in (7.20) we reduce this equation to

$$\mathcal{N}^{ez}(\xi^y - \xi^z)\frac{\delta^2\gamma_{ez}}{\delta p^{ix}\delta p^{jy}} EM(p^{jy}) = 0. \tag{7.21}$$

One sees at once that if

$$\frac{\delta^2\gamma_{ez}}{\delta p^{ix}\delta p^{jy}} \sim \delta(z-y)$$

---

[8] If H and all constraints are at most linear in momenta, as is the case for the Chern-Simons theories, then the Hamiltonian system is strongly degenerate.

[9] Another interesting class of theories where all constraints are linear in momenta are the constrained Wess-Zumino-Novikov-Witten theories (Feher et.al. 1992).

then even for constraints nonlinear in the momenta the equation (7.21) is satisfied off mass shell ($EM(p) \neq 0$). From that it follows immediately that the transformations (7.14) with $\lambda^{ez} = \mathcal{N}^{ez}\xi^z$ are symmetry transformations for the corresponding Lagrangean system if $H=0$. We shall call the corresponding invariance reparametrization invariance: $\hat{R}_\xi = \hat{I}_{\xi,\lambda^{ez}=\mathcal{N}^{ez}\xi^z}$. The explicit form of the reparametrization transformations read

$$\begin{aligned} \delta_\xi q^{ix} &= \dot{q}^{ix}\xi^x + (\xi^y - \xi^x)\{q^{ix}, \mathcal{N}^{by}\gamma_{by}\} \\ \delta_\xi p_i^x &= \dot{p}_i^x\xi^x + (\xi^y - \xi^x)\{p_i^x, \mathcal{N}^{by}\gamma_{by}\} \\ \delta_\xi \mathcal{N}^{ax} &= (\mathcal{N}^{ax}\xi^x)^\cdot - \xi^y\mathcal{N}^{by}\mathcal{N}^{cz}t^{ax}_{cz,by}. \end{aligned} \tag{7.22}$$

These transformations are the correct ones for theories with non-linear constraints. For example, for the relativistic particle the transformations (7.22) (and not the gauge transformations (7.8) generated by the first class constraints alone) coincide with (7.4) on $\mathcal{M}$.

For the relativistic particle, string and gravity the transformations (7.22) form a *closed algebra off mass shell.* This closure is not ment to be obvious since it depends on the concrete structure of the constraints. We verified it for each concrete system separately (Mukhanov and Wipf 1994).

For the supersymmetric particle there are two constraints which are quadratic in the momenta and (7.22) is replaced by a two-parametric family of symmetry transformations. They are the supersymmetric extensions of the reparametrization transformations and do not coincide (off mass shell) with the gauge transformations generated by the constraints.

An interesting question to which we have no general answer is the following: what are the conditions to exponentiate the infinitesimal transformations to finite ones. For the relativistic particle and string and for gravity the finite transformations for the corresponding Lagrangean systems are just the familiar symmetries. These finite symmetries can then be formulated in the Hamiltonian formulation and this way one can find the finite transformations in the first order formalism. But in general it is not clear whether the closing of the algebra of infinitesimal transformations is sufficient to exponentiate them. We suppose that this cannot be the case since for a free *nonrelativistic particle*, which very probably does not admit any known finite local symmetry the transformations (7.14) with $\lambda = 0$ form a closed algebra. This difficult and very important question (i.e. for the functional integral) what are the conditions such that the infinitesimal transformations can be exponentiated needs further investigation.

Constraints and the equations of motion. There is a very interesting and non-trivial connection between the equations of motion $EM(q) = EM(p) = 0$ and the constraints $\gamma_{\tilde{a}} = 0$. Clearly, since $\dot{\gamma}_{\tilde{a}} = 0$ the classical trajectories will stay on Γ_c. Conversely, in some theories (e.g. gravity) we can get the equations of motion (or some of them as in string theory) if we only demand that the constraints are fulfilled for all t (i.e. everywhere) and that the symmetry transformations do not destroy this property. For example, in gravity and string theory this means that we demand that the constraints are valid on any spacelike hypersurface, because the symmetry transformations (diffeomorphisms) can be interpreted as a change of foliation of space-time. It is very easy to arrive at this conclusion using the developed formalism.

Let us consider how the constraints change under the symmetry transformations (7.14):

$$\begin{aligned}\delta_{\xi,\lambda}\gamma_{\bar{a}} &= \frac{\delta\gamma_{\bar{a}}}{\delta q^{ix}}\delta_{\xi,\lambda}q^{ix} + \frac{\delta\gamma_{\bar{a}}}{\delta p_i^x}\delta_{\xi,\lambda}p_i^x \\ &= \frac{\delta\gamma_{\bar{a}}}{\delta q^{ix}}EM(q^{ix})\xi^x + \frac{\delta\gamma_{\bar{a}}}{\delta p_i^x}EM(p_i^x)\xi^x + \lambda^{\bar{c}}t^{\bar{b}}_{\bar{a}\bar{c}}\gamma_{\bar{b}}.\end{aligned} \tag{7.23}$$

For the known theories the constraints are local in q and p and involve only space derivatives of q up to second and p up to first order. It follows then that the structure of the functional derivatives of the constraints have the form

$$\begin{aligned}\frac{\delta\gamma_{ay}}{\delta q^{ix}} &= A_{ia}\delta(x,y) + B^j_{ia}\frac{\partial}{\partial y^j}\delta(x,y) + D^{jk}_{ia}\frac{\partial^2}{\partial y^j\partial y^k}\delta(x,y) \\ \frac{\delta\gamma_{ay}}{\delta p_i^x} &= E^i_a\delta(x,y) + F^{ij}_a\frac{\partial}{\partial y^j}\delta(x,y),\end{aligned} \tag{7.24}$$

where $A, B, \ldots$ are functions of q^y and p^y. Substituting (7.24) into (7.23) a straightforward calculation yields

$$\begin{aligned}\delta_{\xi,\lambda}\gamma_{ay} &= \left(\dot{\gamma}_{ay} + \mathcal{N}^{\bar{b}}t^{\bar{c}}_{\bar{b},ay}\gamma_{\bar{c}} + t^{\bar{c}}_{ay}\gamma_{\bar{c}}\right)\xi_y \\ &\quad + \lambda^{\bar{c}}t^{\bar{b}}_{ay,\bar{c}}\gamma_{\bar{b}} + \left(B^j_{ia}EM(q^{iy}) + F^{ij}_aEM(p_i^y)\right)\frac{\partial\xi^y}{\partial y^j} \\ &\quad + D^{jk}_{ia}\left(2\frac{\partial EM(q^{iy})}{\partial y^j}\frac{\partial\xi^y}{\partial y^k} + EM(q^{iy})\frac{\partial^2\xi^y}{\partial y^j\partial y^k}\right).\end{aligned} \tag{7.25}$$

It follows that for systems with a *finite number of degrees of freedom* $\delta\gamma \sim \gamma$, so that if we impose the constraints for a phase-space path then the transformed path satisfies also the constraints. For that none of the equations of motion must be satisfied. Thus for a finite number of degrees of freedom the constraints γ_a do not tell us anything about the dynamics of the system.

Also, the gauge transformations ($\xi = 0$) do respect the constraints since $\delta_{\lambda,\xi=0}\gamma \sim \gamma$ as is clear from (7.23). Again we see that the constraints together with the corresponding symmetry transformations tell us nothing about the dynamics.

However, for field theories with non-linear constraints we must take $\xi \neq 0$ in (7.14). In this case if not all of the coefficients B, D and F vanish we can get (some of) the equations of motion by demanding that the symmetry transformations respect the constraints, i.e. that $\delta\gamma = 0$ if $\gamma = 0$ everywhere. Whether some of the coefficients B, D, F are non-zero depends on the concrete structure of the constraints. Only if the constraints contain some space derivatives of q and/or p are some of the coefficients non-zero. For example, in *string theory* only $B \neq 0$ and correspondingly only some equations of motion, namely two of the $EM(q) = 0$ equations can be gotten by demanding that all constraints should be satisfied everywhere for any foliation. In *gravity* the situation is even more interesting, since all of the coefficients are non-zero and thus *all* equations of motion will be automatically satisfied if we demand just that $\gamma_{\bar{a}} = 0$ for any foliation. So the whole dynamics in gravity can be reduced to the requirement that the constraints be invariant under diffeomorphisms.

References

Abers, E.S., Lee, B.W. (1973): Phys. Rep., **9C**, 1

Allcock, T.R. (1975): Phil. Trans. R. Soc. (Lond.), **279**, 33

Anderson, J.L and Bergmann, P.G. (1951): Phys. Rev. **83**, 1018

Arnold, V. (1978): "Mathematical Methods of Classical Mechanics", Graduate Texts in Math. 60, Springer, Berlin

Batalin, I.A. and Vilkovisky, G.A. (1977): Phys. Lett. **69B** 309

Becchi, C., Rouet C. and Stora, R., (1974): Phys. Lett. **B52**, 344; (1975): Commun. Math. Phys., **42**, 127; (1976): Ann. Phys., **98**, 287

Benguria, R., Cordero, P. and Teitelboim, C. (1977): Nucl. Phys., **B122**, 61

Berezin, F.A. (1966): "The Method of Second Quantization", New York, Academic Press

Berezin, F.A. and Marinov, M.S. (1977): Ann. Phys., **104**, 336

Bergmann, P.G., Goldberg, I., Janis, A. and Newman, E. (1956): Phys. Rev. **103** 807

Bergmann, P.G. and Komar, A. (1960): Phys. Rev. Lett., **4**, 432

Bergmann, P.G. (1961): Rev. Mod. Phys., **33**, 510

Casalbuoni, R. (1976): Nuovo Cimento, **33A**, 115 (1976): Nuovo Cimento, **33A**, 389

Cawley, R. (1979): Phys. Rev. Lett., **42**, 413

Chernoff, P. and Marsden, J. (1974): Lecture Notes Math., **425**

DeWitt, B. (1965): "Dynamical Theory of Groups and Fields", Gordon and Breach, New York

Dirac, P.A.M. (1950): Can. J. Math., **2**, 129; (1951): Can. J. Math., **3**, 1; (1958): Proc. R. Soc. (Lond.), **A246** 326; (1964): "Lectures on Quantum Mechanics", Belfer Graduate School of Science, Yeshiva University, New York

Fadeev L.D. (1969): Theor. Math. Phys., **1** 1

Fadeev, L.D. and Popov, V.N. (1967): Phys. Lett., **25B**, 29

Feher, L., O'Raifeartaigh, L., Ruelle, P., Tsutsui, I. and Wipf, A. (1992): Phys. Rep., **222**, 1

Fröhlich, J. and Studer, U. (1993): Rev. Mod. Phys., **65**, 733

Gervais, J.L. and Sakita, B. (1971): Nucl. Phys., **B64**, 632

Gitman, D.M. and Tyutin, I.V. (1990): "Quantization of Fields with Constraints", Berlin, Springer

Govaerts, J. (1991): "Hamiltonian Quantisation and Constrained Dynamics", Leuven University Press

Gribov, V.N. (1978): Nucl. Phys., **B139**, 1

Hagen, C.R. (1984): Ann. Phys., **157**, 342

Hanson, A.J., Regge, T. and Teitelboim, C. (1976): "Constrained Hamiltonian Systems", Accademia Nazionale dei Lincei

Henneaux, M. and Teitelboim, C. (1992): "Quantization of Gauge Systems", Princeton Univ. Press

Iengo, R. and Lechner, K. (1992): Phys. Rep., **213**, 179

Jackiw, R. (1988): Seminair de Mathematiques Superieurs, Montreal, Quebec

Kiefer, C. and Wipf, A. (1993): "Functional Schrödinger Equation for Fermions in External Gauge Fields", preprint ETH-TH/93-17

Kimura, T. (1962): Prog. Theor. Phys., **27**, 747

Mukhanov, V. and Wipf, A. (1994): "On the Symmetries of Hamiltonian Systems", preprint ETH-TH/94-04

Polyakov, A.M. (1988): Mod. Phys. Lett., **A3**, 325

Polychronakos, A.P. (1990): Ann. Phys., **203**, 231

Sachs, I. and Wipf, A. (1992): Helv. Phys. Acta, **65**, 652

Shanmugadhasan, S. (1963): Proc. Camb. Phil Soc., **59, 743**

Singer, I.M. (1978): Comm. Math. Phys., **60**, 7
Sudarshan, E.C.G. and Mukunda, N. (1974): "Classical Dynamics, A Modern Perspective", Wiley, New York
Sundermeyer, K. (1982): "Constrained Dynamics", Springer Lecture Notes **169**
Symanzik, K. (1983): Nucl. Phys., **190** [FS3], 1
Trautman, A. (1964): "Lectures on General Relativity", Brandeis Summer Institute in Theor. Phys., Prentice-Hall, N.J
Tyutin, I.V. (1975): Lebedev preprint FIAN, 39
Utiyama, R. (1959): Progr. Theor. Phys. Suppl., **9**, 19
Witten, E. (1989): Commun. Math. Phys., **121**, 351
Wu, T.T. and Yang, C.N. (1975): Phys. Rev., **D12**, 3845
For additional references see, e.g., Sundermeyer (1982) and Henneaux and Teitelboim (1992)

The Classical Theory of Canonical General Relativity

R. Beig

Introduction

In the realm of classical mechanics where the canonical formalism made its first appearance it largely served the purpose of a tool for tackling dynamical problems. Up until today Hamiltonian methods are successfully used to prove profound results, as is exemplified, say, by the celebrated KAM theory or existence proofs of periodic orbits for general classes of mechanical systems. However the conceptual framework provided by the canonical formalism has also played the rather different role of offering a convenient starting point for quantization. It is largely for this latter reason that canonical versions of field theories have been developed, and General Relativity (G.R.) is no exception. The reformulation of G.R. in Hamiltonian terms, pioneered by Dirac (1964) and further developed by many others, also marks the starting point of the study, in the Hamiltonian framework, of constrained systems of which G.R. is a special case. Treating G.R. as a constrained Hamiltonian system, in turn, sheds new light on the structure even of purely classical G.R. While there is thus no doubt as to the conceptual interest in the canonical version of G.R., Hamiltonian methods, in contrast to their role in classical mechanics, have so far only occasionally been used as a technical tool to tackle specific problems, e.g. in cosmology (Ryan 1972), black hole physics (Sudarsky and Wald 1992) or relativistic celestial mechanics (Schäfer 1990).

The present review of canonical G.R. consists of two sections. In the first section we go through the procedure of first writing the Einstein–Hilbert action in $3+1$–form and then performing the Legendre transformation to phase space variables. The observation that the Legendre map is singular, i.e. not onto and non–invertible on its image, is the starting point of the 'Dirac–Bergmann algorithm' for singular Lagrangians. Running through this procedure we arrive at a reformulation of the field equations of G.R. as a Hamiltonian system with first–class constraints, which is also called the ADM formalism (Arnowitt et al. (1962). The material here is standard and has been treated before in several reviews (see Ashtekar 1988, Hanson et al. 1976, Isenberg and Nester 1980, Sundermeyer 1982, Wald 1984). As opposed to most — but not all (see Ashtekar 1988) — of these works we shall use the language of symplectic geometry throughout. Moreover, a symplectic version of the Dirac–Bergmann procedure due to Gotay et al. (1978) will be employed. We shall, for simplicity, confine ourselves to the vacuum case, and here to spacetimes with compact space sections, making only some general remarks on the asymptotically flat case. We will work 'formally', i.e. discard technicalities coming from the infinite–dimensionality of the phase space of G.R. The final picture at least, i.e. the ADM formalism, does in fact have a firm mathematical foundation, in the spatially compact case, thanks to the work of Fischer, Marsden and Moncrief (see e.g. Fischer and Marsden 1979 and references therein). For rigorous results in the asymptotically flat case, see Andersson (1987).

In the more specialized second section we chose to give a broad outline of the 'reduction problem' for G.R., i.e. the program of writing canonical G.R. as a Hamiltonian system for unconstrained variables ('true degrees of freedom'). We made this choice because it is in the reduction process, where one focusses on the peculiar role of time in canonical G.R., which has recently received great attention in the context of quantum gravity (see Isham, this volume, and Kuchar 1992). Section two starts with an explanation of the general scheme underlying the reduction process. We then illustrate the concept of intrinsic and extrinsic time function involved in such a reduction in terms of the relativistic particle. (The treatment of the relativistic particle using an extrinsic time function we haven't seen before.) Finally we sketch recent progress in the implementation of this program for full G.R. due to the work of Moncrief (1988) and Fischer and Moncrief (1994).

1 From Einstein–Hilbert to ADM

Let M be an orientable and time–oriented spacetime with Lorentz metric g_{ab} of signature $(-+++)$. We assume (M, g) to be globally hyperbolic with Cauchy surfaces diffeomorphic to the compact orientable manifold Σ. Thus $M \cong \Sigma \times I$, I being an open interval. Let us have a particular embedding $\varphi : \Sigma \to M$ with $\varphi(\Sigma)$ a Cauchy surface. The pull–back of g by φ turns Σ into a Riemannian manifold with metric $q = \varphi^* g$. Using local coordinates y^α in Σ ($\alpha = 1, 2, 3$) and x^a in M ($a = 0, 1, 2, 3$), we have that, on $\varphi(\Sigma)$, $x^a = \varphi^a(y^\alpha)$ and

$$q_{\alpha\beta}(y) = \varphi^a{}_{,\alpha}(y)\varphi^b{}_{,\beta}(y)g_{ab}(\varphi(y)). \tag{1}$$

Define a unit normal n_a to $\varphi(\Sigma)$ by the conditions

$$\varphi^a{}_{,\alpha}n_a = 0 \qquad \text{and} \qquad g^{ab}n_a n_b = -1, \qquad n^a \text{ future pointing} \tag{2}$$

and the extrinsic curvature $K_{\alpha\beta}$ by

$$K_{\alpha\beta} = \varphi^a{}_{,\alpha}\varphi^b{}_{,\beta}\nabla_a \bar{n}_b \tag{3}$$

where ∇_a is the covariant derivative with respect to g_{ab} and $\bar{n}_a$ is a covector in a small neighbourhood of $\varphi(\Sigma)$ in M, smoothly extending n_a, but otherwise arbitrary. We now omit the bar on $\bar{n}_a$. We shall make use of some identities connecting g_{ab}, $q_{\alpha\beta}$ and $K_{\alpha\beta}$. The first one is

$$\varphi^a{}_{,\alpha}\varphi^b{}_{,\beta}\varphi^c{}_{,\gamma}\varphi^d{}_{,\delta}R_{abcd} = \mathcal{R}_{\alpha\beta\gamma\delta} + 2K_{\gamma[\alpha}K_{\beta]\delta} \tag{4}$$

where R_{abcd} and $\mathcal{R}_{\alpha\beta\gamma\delta}$ are, respectively, the curvature tensors of g_{ab} and $q_{\alpha\beta}$ ("Gauss equation"). Suitable contractions of Equ. (4) with q imply

$$2G_{ab}n^a n^b = \mathcal{R} + K^2 - K_{\alpha\beta}K^{\alpha\beta}, \tag{5}$$

G_{ab} being the Einstein tensor of g and $\mathcal{R}$ the scalar curvature of q. There is also the "Codazzi equation" given by

$$\varphi^a{}_{,\alpha}\varphi^b{}_{,\beta}\varphi^c{}_{,\gamma}R_{abcd}n^d = 2D_{[\alpha}K_{\beta]\gamma} \tag{6}$$

which upon contraction leads to

$$\varphi^b{}_{,\beta}R_{bd}n^d = \varphi^b{}_{,\beta}G_{bd}n^d = D^\alpha K_{\alpha\beta} - D_\beta K, \tag{7}$$

with indices α, β raised by $q^{\alpha\beta}$, D_α the covariant derivative with respect to $q_{\alpha\beta}$ and $K = q^{\alpha\beta}K_{\alpha\beta}$. Setting the left–hand side in Equ.-s (5) and (7) to zero amounts to the constraint equations of G.R. in vacuo. We also need the easily verified identity

$$R_{ab}n^a n^b = -n^a(\nabla_a\nabla_b - \nabla_b\nabla_a)n^b = (\nabla_a n^a)^2 - (\nabla_b n^a)(\nabla_a n^b) + \text{ divergence.} \tag{8}$$

Using (5) and (8), there follows

$$R = \mathcal{R} + K_{\alpha\beta}K^{\alpha\beta} - K^2 + \text{ divergence.} \tag{9}$$

Let us now have a whole family of embeddings $\varphi^a = \varphi^a(y^a, \lambda)$ with λ in some small interval $I = [0, \varepsilon)$. In addition we assume $\varphi_\lambda = \varphi(y, \lambda)$ to be a diffeomorphism of $\Sigma \times I$ into a piece of M bounded in the past by $\varphi_0(\Sigma)$, with each $\varphi_\lambda(\Sigma)$ a Cauchy surface. (It is worth pointing out here that the final evolution equations, by virtue of the known existence and uniqueness theorems for the Einstein equations, can be solved also when the latter assumption is dropped. For example $\varphi(y, \lambda)$ could be "boost–like", i.e. $\dot{\varphi}(y, \lambda)$ could change from being a future–pointing timelike vector to past–pointing along a hypersurface S of Σ. Then the lapse function N, to be defined presently, changes sign across S. Or $\dot{\varphi}(y, \lambda)$ could be everywhere tangential to $\varphi_0(\Sigma)$ so that N is identically zero on Σ and φ_λ corresponds to the action of a one–parameter family of diffeomorphisms of Σ on $\varphi_0(\Sigma)$.) Given g_{ab} and φ_λ, we define fields $N > 0$ and N^α on Σ, called lapse and shift, respectively, by

$$\dot{\varphi}^a(y, \lambda) = \frac{\partial}{\partial \lambda}\varphi^a(y, \lambda) = N n^a + \varphi^a{}_{,\alpha} N^\alpha \tag{10}$$

with $n^a = n^a(y, \lambda)$ for all λ satisfying Equ.-s (2). Conversely, given g_{ab}, lapse and shift and the initial condition that $\varphi^a(y, 0) = (0, y)$, one could imagine trying to solve the highly implicit system consisting of equations (10) and (2) in terms of $\varphi^a(y, \lambda)$. Interestingly, this need never arises. The reason is that the spacetime metric in G.R. is not given beforehand. Rather one solves an evolution problem with N, N^α as external variables, which gives one the metric g_{ab} in the coordinates $(x^a) = (x^0, x^\alpha) = (\lambda, y^\alpha)$ (globally: the pull–back of g under φ). Thus $\varphi(y, \lambda)$, solving Equ. (10), would simply be the identity map.

Let us compute

$$\varphi^* g = \dot{\varphi}^a \dot{\varphi}^b g_{ab} d\lambda^2 + 2\dot{\varphi}^a \varphi^b{}_{,\alpha} g_{ab} d\lambda dy^\alpha + \varphi^a{}_{,\alpha}\varphi^b{}_{,\beta} g_{ab} dy^\alpha dy^\beta. \tag{11}$$

Thus

$$\begin{aligned} (\varphi^* g)_{\alpha\beta} &= q_{\alpha\beta} \\ (\varphi^* g)_{00} &= -N^2 + N_\alpha N_\beta q^{\alpha\beta} \\ (\varphi^* g)_{0\alpha} &= q_{\alpha\beta} N^\beta. \end{aligned} \tag{12}$$

We have in particular $N^2 - N_\alpha N^\alpha > 0$. Taking the λ–derivative of $q_{\alpha\beta}$, one obtains the identity

$$\dot{q}_{\alpha\beta} = 2N K_{\alpha\beta} + D_\alpha N_\beta + D_\beta N_\alpha = 2N K_{\alpha\beta} + \mathcal{L}_{\vec{N}} q_{\alpha\beta}. \tag{13}$$

Doing the same with $K_{\alpha\beta}$, using Equ. (4) and imposing $\varphi^a{}_{,\alpha}\varphi^b{}_{,\beta}G_{ab}=0$, we find that

$$\dot{K}_{\alpha\beta} = -N\mathcal{R}_{\alpha\beta} + 2NK_\alpha{}^\gamma K_{\beta\gamma} - NKK_{\alpha\beta} + D_\alpha D_\beta N + \mathcal{L}_{\vec{N}}K_{\alpha\beta}. \quad (14)$$

Equ.-s (13,14) furnish a system of evolution equations for $(q_{\alpha\beta}, K^{\alpha\beta})$, given $(N,\vec{N})$. Equ.-s (14) together with the constraint equations implied by (5) and (7) exhaust the information contained in the Einstein equations. In particular, there are no evolution equations for lapse and shift. For the existence of solutions it is important that the constraint equations, if they have been imposed for $\lambda = 0$, are satisfied automatically for $\lambda > 0$. This is the case and follows from the contracted Bianchi identity $\nabla^a G_{ab} = 0$. In the canonical version of the theory, this phenomenon will reappear as the "first–class–property" of the phase–space form of the constraints.

We now write the Einstein–Hilbert action, namely

$$S[g] = \int R(-\det g)^{1/2} d^4x \quad (15)$$

in the (λ, y)–coordinates. Using $(-\det g)^{1/2} = N(\det q)^{1/2}$ and observing that the term arising from the divergence in Equ. (9) vanishes because of our assumptions on Σ, this becomes

$$S[g] = \int L d\lambda \quad (16)$$

with the Lagrangian L given by

$$L = \int_\Sigma N[\mathcal{R} + K_{\alpha\beta}K^{\alpha\beta} - K^2](\det q)^{1/2} d^3y. \quad (17)$$

Here L should be seen as a functional $L : TC \to \mathbf{R}$ with configuration space C made up of the fields

$$\begin{aligned} C &= \{q_{\alpha\beta}, N, \vec{N}\} \\ TC &= \{q_{\alpha\beta}, N, \vec{N}; \dot{q}_{\alpha\beta}, \dot{N}, \dot{\vec{N}}\} \end{aligned} \quad (18)$$

on Σ. Thus $K_{\alpha\beta}$ in (17) should be interpreted as being eliminated in favour of $\dot{q}_{\alpha\beta}$, by means of (13). Note that L is independent of $(\dot{N}, \dot{\vec{N}})$. The corresponding phase space $\Gamma = T^*C$ is

$$\Gamma = \{q_{\alpha\beta}, N, \vec{N}; p^{\alpha\beta}, \pi, \vec{\pi}\} \quad (19)$$

where the momentum variables $p^{\alpha\beta}$, π, $\vec{\pi}$, being covectors on C, are densities. The canonical symplectic potential ω on Γ is given by

$$\omega = \int_\Sigma (p^{\alpha\beta}\delta q_{\alpha\beta} + \pi\delta N + \vec{\pi}\delta\vec{N})d^3y. \tag{20}$$

Note that tangent vectors on Γ consist of functions $(\delta q_{\alpha\beta}, \delta N, \delta\vec{N}; \delta p^{\alpha\beta}, \delta\pi, \delta\vec{\pi})$, so ω is a one–form on Γ with components linear in momenta.

The Legendre map $\mathcal{L} : TC \to \Gamma$ associated with the Lagrangian L is defined, using an obvious shorthand notation, $\mathcal{L} : (Q, \dot{Q}) \mapsto (Q, P)$, by

$$P = \frac{\delta L}{\delta\dot{Q}}. \tag{21}$$

In the present case we find

$$\begin{aligned} \pi &= \vec{\pi} = 0 \\ p^{\alpha\beta} &= (\det q)^{1/2}(K^{\alpha\beta} - Kq^{\alpha\beta}). \end{aligned} \tag{22}$$

Thus $\mathcal{L}$ is not onto, since the canonical momenta of lapse and shift have to be zero. We define the "energy function" $E : TC \to \mathbf{R}$ by

$$E = P\dot{Q} - L = \int_\Sigma p^{\alpha\beta}\dot{q}_{\alpha\beta}d^3y + \int_\Sigma \pi\dot{N}d^3y + \int_\Sigma \vec{\pi}\dot{\vec{N}}d^3y - L, \tag{23}$$

where Equ.'s (13,22) are used to understand E as a functional on TC. Clearly

$$\mathcal{L}(TC) = \{q_{\alpha\beta}, N, \vec{N}; p^{\alpha\beta}, 0, \vec{0}\} =: \Gamma_1 \tag{24}$$

$\Gamma_1 \subset \Gamma$ is called the primary constraint set.

Now try to define a function H on Γ_1 by

$$H \circ \mathcal{L} = E. \tag{25}$$

It is known in general (see Wipf, this volume), and easily seen here, that equation (25) does in fact define a single–valued function H. After integration by parts, H is given by

$$\begin{aligned} H &= \int_\Sigma N\left[(\det q)^{-1/2}p^{\alpha\beta}p_{\alpha\beta} - \frac{1}{2}(\det q)^{-1/2}p^2 - (\det q)^{1/2}\mathcal{R}\right]d^3y \\ &\quad - 2\int_\Sigma N^\alpha(D_\beta p_\alpha{}^\beta)d^3y \qquad (26) \\ &= \int_\Sigma N\mathcal{H}d^3y + \int_\Sigma N^\alpha\mathcal{H}_\alpha d^3y. \qquad (27) \end{aligned}$$

Note that H is linear in $(N, \vec{N})$. The canonical 2-form Ω, pulled back to Γ_1, i.e. $\Omega_1 = \Omega|_{\Gamma_1} = -d\omega|_{\Gamma_1}$ is equal to

$$\Omega_1[(\delta' q, \delta' p, \delta' N, \delta' \vec{N}), (\delta q, \delta p, \delta N, \delta \vec{N})] = \int_\Sigma (\delta' q_{\alpha\beta} \delta p^{\alpha\beta} - \delta' p^{\alpha\beta} \delta q_{\alpha\beta}) d^3 y. \tag{28}$$

Note that the order of arguments as in Equ. (19) has been changed. Now view Ω_1 as a map from $X \in T\Gamma_1$ to $X \rfloor \Omega_1 \in T^*\Gamma_1$, where $X \rfloor \Omega_1$ is taken to mean contraction on the first index. Since Ω_1 does not contain δN, $\delta \vec{N}$-terms, Ω_1 has a null space $T\Gamma_1^\perp$, consisting of vectors Y of the form $Y = (0, 0, N, \vec{N})$. Thus (Γ_1, Ω_1) is only pre-symplectic. We now do the best we can to solve the Hamilton equations

$$X \rfloor \Omega_1 = dH \qquad \text{on } \Gamma_1. \tag{29}$$

Solving (29) for X is not possible on all of Γ_1, but only at points at which dH does not contain δN, $\delta \vec{N}$-terms, or equivalently at points where $Y(H) = 0$ for all $Y \in T\Gamma_1^\perp$. This, in turn, requires that

$$\mathcal{H} = 0, \qquad\qquad \mathcal{H}_\alpha = 0. \tag{30}$$

Equ. (30) are of course nothing but the initial-value constraints of G.R. The solutions X of (29) exist at points of $\Gamma_2 \subset \Gamma_1$, where the secondary constraint set Γ_2 is defined by $\mathcal{H} = \mathcal{H}_\alpha = 0$. Our task of finding consistent equations of motion would be complete if these X's would be tangent to Γ_2. (We are assuming here that Γ_2 is a smooth submanifold of Γ_1. This is known to be the case at points of Γ_2, the evolved spacetimes of which have no Killing symmetries (Moncrief 1975).) Because then points on Γ_2 will stay on Γ_2 under the flow of X. Luckily this is the case, as we now show.

By explicit computations we find that X, solving (29), is of the form

$$X = (\dot{q}_{\alpha\beta}, \dot{p}^{\alpha\beta}, \dot{N}, \dot{\vec{N}})$$

at $(q_{\alpha\beta}, p^{\alpha\beta}, N, \vec{N})$ with

$$\dot{q}_{\alpha\beta} = 2N(\det q)^{-1/2} q_{\alpha\gamma} q_{\beta\delta} \left(p^{\gamma\delta} - \frac{1}{2} q^{\gamma\delta} p \right) + \mathcal{L}_{\vec{N}} q_{\alpha\beta} \tag{31}$$

$$\begin{aligned} \dot{p}^{\alpha\beta} \; &= \; -2N(\det q)^{-1/2} \left(p^{\alpha\gamma} p^\beta{}_\gamma - \frac{1}{2} p p^{\alpha\beta} \right) + N(\det q)^{1/2} G^{\alpha\beta} \\ &\quad + (\det q)^{1/2} (D^\alpha D^\beta N - q^{\alpha\beta} \Delta N) + \mathcal{L}_{\vec{N}} p^{\alpha\beta}, \end{aligned} \tag{32}$$

whereas $\dot{N}$, $\dot{\vec{N}}$ are arbitrary on Γ_1. Equ.'s (31,32) are equivalent to the Einstein evolution equations (13,14) on Γ_2. After lengthy calculations we find that

$$\begin{aligned} X(\mathcal{H}_\alpha) &= \mathcal{L}_{\vec{N}}\mathcal{H}_\alpha + \mathcal{H}D_\alpha N & (33)\\ X(\mathcal{H}) &= ND^\alpha\mathcal{H}_\alpha + 2(D^\alpha N)\mathcal{H}_\alpha + \mathcal{L}_{\vec{N}}\mathcal{H}. & (34)\end{aligned}$$

In particular $X(\mathcal{H}_\alpha)$ and $X(\mathcal{H})$ vanish at points of Γ_1 having $\mathcal{H}$ and $\mathcal{H}_\alpha$ zero on all of Σ. Thus, as promised, the Dirac–Bergmann algorithm terminates here. Note that the Hamiltonian H vanishes identically on Γ_2. It is now customary to change the viewpoint somewhat. Since $(N,\vec{N})$ has to satisfy neither constraints nor evolution equations, it is convenient to altogether abandon them as dynamical variables. More specifically, we take the set $C' = \{\text{metrics } q_{\alpha\beta} \text{ on } \Sigma\}$ as configuration space and $T^*C' = \{q_{\alpha\beta}, p^{\alpha\beta}\}$ as phase space with $\omega' = \int_\Sigma p_{\alpha\beta}\delta q^{\alpha\beta} d^3y$ as the basic 1–form. Then $\Omega' = -d\omega'$ is non–degenerate. We now omit the primes on C', ω' and Ω'. The constraint set $\bar{\Gamma} \subset \Gamma = T^*C$ is again defined by (30). In $H = H(N,\vec{N})$ we now consider $(N,\vec{N})$ as external variables, in the sense that $(N,\vec{N})$ are functions on Γ which take values in the set of scalar– respectively vector–fields on Σ. Furthermore, we do not require $N^2 - N_\alpha N^\alpha > 0$ any longer. The equations of motion (29) take the form

$$X\rfloor\Omega = dH(N,\vec{N}). \tag{35}$$

$X_{(N,\vec{N})} = (\dot{q}_{\alpha\beta}, \dot{p}^{\alpha\beta})$ is now unique and is, on $\bar{\Gamma}$, again given by (31,32). This is so despite the $(q_{\alpha\beta}, p^{\alpha\beta})$–dependence of $(N,\vec{N})$, since

$$H(N,\vec{N}) = \int_\Sigma (N\mathcal{H} + N^\alpha\mathcal{H}_\alpha)d^3y, \tag{36}$$

and the terms in dH involving functional derivatives of $(N,\vec{N})$ drop out on $\bar{\Gamma}$. Consistency is again guaranteed by Equ.'s (33,34). We now rewrite these equations by multiplying (33) with $\vec{N}'$ and (34) with N' and integrating over Σ. Integrating by parts in the resulting expressions and defining the Poisson bracket by

$$\{F,G\} = -\mathcal{L}_X G \qquad \text{with} \qquad X\rfloor\Omega = dF \tag{37}$$

we find

$$\begin{aligned} \{H(\vec{N}'), H(\vec{N})\} &= H(\mathcal{L}_{\vec{N}'}\vec{N}) & (38)\\ \{H(\vec{N}'), H(N)\} &= H(\mathcal{L}_{\vec{N}'}N) & (39)\\ \{H(N'), H(N)\} &= H(N'q^{\alpha\beta}D_\beta N - Nq^{\alpha\beta}D_\beta N') & (40)\end{aligned}$$

where $H(N) := H(N,\vec{0})$ and $H(\vec{N}) := H(0,\vec{N})$. Equ.'s (38,39,40) are called the Dirac commutation relations. Because of the peculiar nature of the Hamiltonian (being a linear combination of the functions defining the constraint $\bar{\Gamma} \subset \Gamma$), they express at the same time the consistency of the Hamiltonian equation and the fact, that the constraint set $\bar{\Gamma}$ is **first–class**, i.e. the Poisson brackets of the constraint functions defining $\bar{\Gamma}$ are zero, when restricted to $\bar{\Gamma}$ (see Wipf, this volume). Notice from (38,39) that the Poisson action of $H(\vec{N})$ has an obvious interpretation in terms of infinitesimal diffeomorphisms of Σ. To find it recall that diffeomorphisms of Σ have a natural action on C which, in turn, lifts to a natural action on Γ. Infinitesimally this becomes the action on $(q_{\alpha\beta}, p^{\alpha\beta})$ by Lie derivative with respect to vector fields $\vec{N}$ on Σ. The corresponding infinitesimal canonical transformations of Γ are precisely generated by

$$H(\vec{N}) = -2\int_{\Sigma} N^{\beta}(D_{\alpha}p_{\beta}{}^{\alpha})d^3y = \int_{\Sigma}(\mathcal{L}_{\vec{N}}q_{\alpha\beta})p^{\alpha\beta}d^3y. \tag{41}$$

In particular $H(\vec{N})$ defines a "moment map" for this action (see Abraham and Marsden 1978), sending points of Γ into elements of the dual of the Lie algebra of vector fields on Σ. (Again, allowing as we do, a dependence of $\vec{N}$ on (q,p) does not influence the statements when, finally, everything is restricted to points of $\bar{\Gamma}$.) The other constraint functions, namely $H(N)$, which are the ones generating 'true time evolutions', do not have such an obvious geometric interpretation in the present framework. This can be seen from the fact that, due to Equ. (5), they are quadratic in momenta rather than linear. Furthermore the differential operator acting on N and N' entering the right hand side of (40) has non–constant coefficients on Γ, because of the presence of the inverse metric $q^{\alpha\beta}$. This fact, which distinguishes G.R. from gauge theories, is usually expressed by saying that the constraints satisfy an "open algebra".

Suppose we pull back Equ. (35) to $\bar{\Gamma}$. This gives

$$X \rfloor \bar{\Omega} = 0 \tag{42}$$

where $\bar{\Omega} = \Omega|_{\bar{\Gamma}}$. Conversely, in finite dimensions at least, every X satisfying (42), is of the form $X = X_{(N,\vec{N})}$ implied by Equ. (31,32). (Proving this in the present, infinite–dimensional situation requires much further work, see Fischer and Marsden 1979.) Thus we can, if we wish, forget about the Hamiltonian $H(N,\vec{N})$ and characterize the X's simply as the elements of the null distribution $\mathcal{D}$ of the pre–symplectic form $\bar{\Omega}$ on $\bar{\Gamma}$. (It is this picture which one also obtains using the so–called covariant–phase–space method, Ashtekar et al. 1991.)

Let us sum up. We have cast G.R. into the form of a constrained system $\bar{\Gamma} \subset \Gamma$ which is first–class. This means that a vector X at a point of $\bar{\Gamma}$ such that $(X \rfloor \Omega)|_{\bar{\Gamma}}$ is zero, is in fact tangent to $\bar{\Gamma}$. Thus the vector X belongs to the integrable distribution $\mathcal{D}$ spanned by the elements of $T\bar{\Gamma}$ which annihilate $\bar{\Omega} = \Omega|_{\bar{\Gamma}}$. The leaves of $\mathcal{D}$ play the role of "gauge subspaces" of $\bar{\Gamma}$. For G.R. in the spatially compact case the phase space Γ consists of the pairs $\{q_{\alpha\beta}, p^{\alpha\beta}\}$, i.e. the cotangent bundle of the set of all Riemannian metrics q on a fixed, compact three–manifold Σ. Solving the last equation in (22) for $K_{\alpha\beta}$ in terms of $(q_{\alpha\beta}, p^{\alpha\beta})$, we can view $q_{\alpha\beta}$ and $K_{\alpha\beta}$ as induced metric respectively extrinsic curvature of some spacelike embedding φ of Σ into a Lorentzian manifold $(M = \Sigma \times I, g)$. With this interpretation of Γ the subset $\bar{\Gamma} \subset \Gamma$ corresponds to those pairs $(q_{\alpha\beta}, K_{\alpha\beta})$ which arise from embeddings φ of Cauchy slices in (M, g) with g obeying the Einstein vacuum equations. The vector fields $X_{(N,\vec{N})}$ which generate $\mathcal{D}$ give rise, via Equ. (10), to one–parameter families of such embeddings. Thus the gauge subspaces defined by a given $(q_{\alpha\beta}, p^{\alpha\beta})$, i.e. the orbit of $(q_{\alpha\beta}, p^{\alpha\beta})$ under $X_{(N,\vec{N})}$ for all $(N, \vec{N})$ corresponds to Cauchy slices in one and the same vacuum spacetime (M, g). The gauge distribution $\mathcal{D}$ contains a subdistribution $\mathcal{D}'$ consisting of all vector fields of the form $X_{(0,\vec{N})}$: These yield the natural action of diffeomorphisms of Σ on $(q_{\alpha\beta}, p^{\alpha\beta})$. On the other hand, vector fields of the form $X_{(N,\vec{0})}$ (which according to Equ. (40) do not form a subdistribution) correspond to "true time evolutions" of $\varphi(\Sigma)$.

One can now divide $\bar{\Gamma}$ by the gauge distribution $\mathcal{D}$, thus obtaining a space $\tilde{\Gamma}$ with a non–degenerate 2–form $\tilde{\Omega}$. The quotient space $(\tilde{\Gamma}, \tilde{\Omega})$ is called the reduced phase space. In the case of G.R. this space is simply the space of non–isometric vacuum spacetimes. In particular, in factoring out $\mathcal{D}$, we have eliminated dynamics from the theory.

We now come to the issue of 'observables'. By a definition which is, at least in the case of gauge theories, generally accepted, an observable is a function O on $\bar{\Gamma}$ which has a well–defined projection down to $\tilde{\Gamma}$. In other words, O is constant along $\mathcal{D}$. Thus, in the present circumstances, O would have to be constant along all points (q, p) corresponding to metric and extrinsic curvature of Cauchy slices in a given spacetime, i.e. a 'constant of motion'. If we restrict ourselves to observables which are "local", i.e. integrals over Σ of densities which only depend on finitely many derivatives of $q_{\alpha\beta}$ and $p^{\alpha\beta}$, it has recently been shown (Torre 1993) that none exist — except the trivial ones which are constant on $\bar{\Gamma}$, such as $H(N)$ and $H(\vec{N})$. Thus observables have to be objects which are constructed from the present variables in a non-local manner. Given the spacetime interpretation of the theory, it would be easy to name such variables. For example, when (M, g) is the spacetime evolving from $(q, p) \in \bar{\Gamma}$, one could take

$$O := \sup L(\gamma)$$

where $L(\gamma)$ is the length of an inextendible timelike curve in (M, g) and the supremum is taken over all such curves. Clearly, O thus defined is an observable, but, as a function of (q, p) not only non-local but highly implicit, involving the full solution to the Cauchy problem. Perhaps the issue of observables in the above sense can be better tackled using a different representation of phase space, such as the one provided by the Ashtekar–Rovelli–Smolin loop variables (Ashtekar 1991).

The situation is different if we consider the phase–space of asymptotically flat Cauchy data. Here we take, for example, Σ to be diffeomorphic to $\mathbf{R}^3$ and assume the $q_{\alpha\beta}$'s to decay to a flat metric near infinity at some rate and the $p^{\alpha\beta}$'s to go to zero, the idea being that these correctly described data on Cauchy slices which approach hyperplanes near the spatial infinity of an asymptotically flat spacetime. As long as $(N, \vec{N})$ are required to go to zero at infinity, the previous analysis goes through unchanged, except for technicalities involved with the function spaces. Again one finds that $\bar{\Gamma} \subset \Gamma$ is first–class, with the null distribution $\mathcal{D}$ spanned by $X_{(N,\vec{N})}$. The boundary conditions on $(N, \vec{N})$ mean that the corresponding families φ_λ of embeddings of Σ approach the identity near infinity. So, clearly, one wants to relax these conditions on $(N, \vec{N})$ in order to include asymptotic space and time translations, rotations and boosts, in short: the full Poincaré group in φ_λ. Then the expressions $H(N)$ and $H(\vec{N})$ found previously will either diverge at infinity or not be differentiable functions of (q, p): the partial integrations necessary to compute their functional derivatives will give rise to non–vanishing surface terms. The idea, originally due to de Witt (1967) and systematically carried out by Regge and Teitelboim (1974) with corrections and a reformulation in symplectic terms in Beig and Ó Murchadha (1987), is to add certain surface integrals to $H(N)$ and $H(\vec{N})$ so that they are finite even with the relaxed boundary conditions and their derivatives exist. The modified functions $H'(N)$ and $H'(\vec{N})$ are not any longer constant on $\bar{\Gamma}$ and thus the associated Hamiltonian vector fields do not lie in $\mathcal{D}$. For example, $H'(N)$ for N going to 1 at infinity is the ADM energy, which is certainly not constant on $\bar{\Gamma}$. (In fact, critical points of $H'(N)|_{\bar{\Gamma}}$ would presumably have to be initial data for stationary solutions of the vacuum equations (see Sudarsky and Wald 1992) and thus, by the Lichnerowicz theorem, data for the Minkowski space time.) Furthermore, the modified functions $H'(N)$ and $H'(\vec{N})$, restricted to $\bar{\Gamma}$, are not all different: two pairs $(N, \vec{N})$ and $(N_1, \vec{N}_1)$ give the same value for H' if they "differ by gauge", i.e. by contributions going to zero at infinity. Thus, modulo gauge, we obtain 10 different Hamiltonians on $\bar{\Gamma}$, corresponding of course to total linear and angular momentum. One now checks that these Hamiltonians Poisson–commute with elements of $\mathcal{D}$ on $\bar{\Gamma}$, i.e. are observables. In addition their Poisson brackets relative to Ω, evaluated on $\bar{\Gamma}$, satisfy the commutation relations of the Lie algebra of the Poincaré group. Hence the same follows for their projections to the reduced phase space $(\tilde{\Gamma}, \tilde{\Omega})$. For a rigorous analysis of some of these issues see Andersson (1987).

2 The Reduction Program

We have seen in the first lecture that the process of complete gauge reduction, when applied to G.R. in the spatially compact case, is a bit like throwing the baby out with the bath: it deprives us of the possibility to talk about time evolution in Hamiltonian terms. However, there is an alternative, incomplete gauge reduction, sometimes called deparametrization, which **does** lead to a description of time evolution in terms of an unconstrained Hamiltonian system. To describe it, let us first assume that the constrained system $(\bar{\Gamma}, \bar{\Omega} = -d\bar{\omega})$ is such that the null distribution $\mathcal{D}$ of $\bar{\Omega}$ is 1–dimensional. The typical example is the relativistic particle which we will discuss presently. In General Relativity the distribution $\mathcal{D}$, being parametrized by $(N, \vec{N})$, is of course infinite–dimensional. So applying the present scheme will have to involve gauge fixing $(N, \vec{N})$ up to one parameter.

Now suppose that we are given a **time function** T. This is a function $T : \bar{\Gamma} \to \mathbf{R}$ with non–zero gradient, the level surfaces of which are transversal to the foliation defined by $\mathcal{D}$. Thus we now have two slicings of $\bar{\Gamma}$: one defined by $\mathcal{D}$ with one–dimensional leaves and one defined by T, the leaves of which are surfaces (locally) diffeomorphic to $\tilde{\Gamma}$. Given T we can view $\bar{\Gamma}$ as a product of an interval $I \subset \mathbf{R}$, the range of T, with a space $\hat{\Gamma}$ which is diffeomorphic to $\tilde{\Gamma}$: We choose a map ψ with

$$\psi : I \times \hat{\Gamma} \to \bar{\Gamma}, \tag{43}$$

satisfying

$$T(\psi(\tau, \cdot)) = \tau, \qquad \tau \in I. \tag{44}$$

Clearly, there are many such maps. We now assume, that we have managed to choose ψ in a very particular way. Namely, pulling back $\bar{\omega}$ to $I \times \hat{\Gamma}$

by ψ we have the decomposition

$$\psi^* \bar{\omega} = \vartheta_\tau - h d\tau, \tag{45}$$

where ϑ_τ means $\psi^*\bar{\omega}$ restricted to $\tau = \text{const}$ and h is given by $-h = \frac{\partial}{\partial \tau} \rfloor (\psi^* \bar{\omega})$. Note that, by fiat, $d\vartheta_\tau$ is non–degenerate on $\widehat{\Gamma}$. In general, both ϑ_τ and h will depend on τ. But imagine, we have, by some luck, found ψ such that ϑ_τ resulting from the potential $\bar{\omega}$, possibly after a redefinition $\bar{\omega} \mapsto \bar{\omega} + dF$, is in fact independent of τ.

The Hamiltonian h will in general still depend on τ. It then follows that the triple $(I \times \widehat{\Gamma}, \vartheta, h)$ is a time–dependent Hamiltonian system, the motions of which are equivalent to a gauge–fixed description of $\mathcal{D} \subset \bar{\Gamma}$. To prove this, first note that the equation $X \rfloor d\bar{\omega} = 0$ defining $\mathcal{D}$ results from varying the homogeneous action

$$\bar{S}[\bar{\gamma}] = \int \bar{\gamma}^* \bar{\omega}, \qquad \bar{\gamma} : \mathbf{R} \to \bar{\Gamma} \tag{46}$$

i.e.

$$\delta \bar{S}[\bar{\gamma}] = 0 \Longleftrightarrow \dot{\bar{\gamma}} \rfloor d\bar{\omega} = 0. \tag{47}$$

On the other hand, varying the pulled–back action

$$S[\gamma] = \int \gamma^* (\vartheta - h d\tau), \qquad \gamma : \tau \in I \mapsto \tau \times \widehat{\Gamma} \tag{48}$$

yields for $\dot{\gamma}$ the equation

$$\dot{\gamma} \rfloor (d\vartheta - dh \wedge d\tau) = 0 \tag{49}$$

together with, of course,

$$\dot{\gamma} \rfloor d\tau = 1. \tag{50}$$

Thus, by construction, the time–dependent Hamiltonian system (49,50) is just a gauge–fixed version of (47). Let us call $(I \times \widehat{\Gamma}, \vartheta, h)$ a deparametrization of $(\bar{\Gamma}, \bar{\Omega})$ for the time function T.

We now ask whether, given a time function T, a deparametrization can always be found. The answer is, locally at least, "yes". To see this, take for $\widehat{\Gamma}$ the leaf of the T-foliation given by, say, $T = 0$ and take for $\psi(\tau, z)$ with $\tau \in I$ and $z \in \widehat{\Gamma}$ the unique point of $\bar{\Gamma}$ lying in the leaf $T = \tau$ and in the same leaf of the $\mathcal{D}$–foliation as z. Thus, by construction, $\dot{\psi} = \frac{\partial}{\partial \tau} \psi$ satisfies $\dot{\psi} \rfloor d\bar{\omega} = 0$. Consider the function F on $\bar{\Gamma}$, defined in the chart given by ψ^{-1} as

$$-F(\tau, z) = \int_0^\tau \dot{\psi}(\tau', z) \rfloor \bar{\omega}(\psi(\tau', z)) d\tau', \tag{51}$$

and pick a new potential $\bar{\omega}'$ on $\bar{\Gamma}$ by

$$\bar{\omega}' = \bar{\omega} + dF. \tag{52}$$

Then we find that $\vartheta'_\tau = \psi^*(\bar{\omega} + dF)|_\tau$ satisfies

$$\frac{\partial}{\partial \tau}\vartheta'_\tau = \psi^*(\dot{\psi}\rfloor d\bar{\omega})\Big|_\tau = 0. \tag{53}$$

Thus a map ψ of the required kind has been found. Having found one deparametrization, we can find many others, related to each other by "time–dependent canonical transformations" (see Abraham–Marsden 1978). These will lead to different $\vartheta's$ and different Hamiltonians. In particular, it is always possible to find a ψ so that the Hamiltonian h is zero ("transformation of $(I \times \widehat{\Gamma}, \vartheta, h)$ to equilibrium"). In fact, it is easy to see that the ψ we have just considered has exactly this property that h vanishes and so "there is no motion". Now this seems disappointing at first: Given any time function, there are, not one, but many deparametrizations, and among them one can, in principle find one for which the dynamics is completely trivial. The catch lies in the word "in principle": in constructing the above ψ giving the trivial deparametrization of $(\bar{\Gamma}, \bar{\Omega})$ we have, firstly, scaled X in $\mathcal{D}$ by requiring $X\rfloor d\tau = 1$ and, secondly, defined ψ by the flow of X. Thus this particular deparametrization requires that we have solved the problem of motion beforehand. However, there is a large class of constrained systems $(\bar{\Gamma}, \bar{\Omega})$ and, on them, natural classes of time functions for which, in turn, one can construct natural deparametrizations without prior solution of the motion. Often, in these systems, $(\bar{\Gamma}, \bar{\Omega})$ arises as a constraint in some symplectic manifold (Γ, Ω) with Γ a cotangent bundle, $\Gamma = T^*C$ where these constraints are defined by constraint functions which are at most quadratic in momenta. The most elementary example is of course the relativistic particle.

Example 1, The Relativistic Particle in Minkowski space time: Here C is $\mathbf{R}^4$ and ω on $\Gamma = T^*C$ given by $\omega = p_a dq^a$. $\bar{\Gamma} \subset \Gamma$ is defined by

$$\mathcal{H} = g^{ab}p_a p_b + m^2 = 0, \qquad m > 0, \qquad p_0 < 0, \tag{54}$$

with g_{ab} a flat Lorentz metric with signature $(- + ++)$. Choose as time function T the coordinate q^0 on C of some standard inertial system $(q^0, q^\alpha; p_0, p_\alpha)$. Such a choice of time function is called 'intrinsic'. Next compute the Poisson bracket of T (note that T is defined on all of Γ, not just $\bar{\Gamma}$) with $\mathcal{H}$. We find

$$\{T, \mathcal{H}\} = -2p_0 > 0. \tag{55}$$

So T increases along the Hamiltonian flow X of $\mathcal{H}$, and thus the level surfaces of T are transversal to $\mathcal{D}$, which is spanned by X. Since, also,

$$\{T, \mathcal{H}\} = \frac{\partial}{\partial p_0}\mathcal{H}, \tag{56}$$

p_0 can be eliminated from $\mathcal{H} = 0$. Taking for ψ the natural embedding $(\tau, x^\alpha, p_\beta) \to \bar{\Gamma}$, we find for $\psi^*\bar{\omega}$ that

$$\psi^*\bar{\omega} = p_\alpha dq^\alpha - (m^2 + \vec{p}^2)^{1/2} d\tau. \tag{57}$$

Thus $\vartheta|_\tau = p_\alpha dq^\alpha$ and $h = (m^2 + \vec{p}^2)^{1/2}$. In this particular case h is time-independent. This is of course due to the fact that the time function is associated with a (timelike) Killing vector of the metric g_{ab} entering the constraint. This ends our first example. For "full" General Relativity, i.e. discarding minisuperspace-models, a deparametrization based on an intrinsic time function has, to my knowledge, never been carried very far. More successful so far has been an "extrinsic time function". To explain this concept, we take as a simple model case the

Example 2: Relativistic Particle in the Milne-Universe: Here we take $\mathcal{C}$ to be the wedge $t < -|x| < 0$ of $(1+1)$-dimensional Minkowski spacetime with metric $g = -dt^2 + dx^2$. Define coordinates $(\rho > 0, \xi)$

$$t = -\rho \cosh \xi \qquad\qquad x = \rho \sinh \xi, \tag{58}$$

so that, now, g becomes

$$g = -d\rho^2 + \rho^2 d\xi^2. \tag{59}$$

Note that ξ is a spacelike boost-parameter. The basic one-form ω on Γ is, of course, given by

$$\omega = p_\xi d\xi + p_\rho d\rho \tag{60}$$

with p_ξ, p_ρ the momenta conjugate to ξ and ρ. The mass-shell constraint defining $\bar{\Gamma}$ now takes the form

$$\mathcal{H} = -p_\rho^2 + \rho^{-2} p_\xi^2 + m^2 = 0. \tag{61}$$

We restrict ourselves to p_ρ's with $p_\rho > m$. $p_\rho = m$ would correspond to radial particles moving towards the origin $t = x = 0$. We also restrict ourselves to $p_\xi > 0$, i.e. particles moving to the right (p_ξ is conserved along the Hamiltonian flow of $\mathcal{H}$). We now propose to use as time function T the momentum p_ρ

$$T = p_\rho. \tag{62}$$

We find that

$$\{T, \mathcal{H}\} = 2\frac{p_\rho^2 - m^2}{\rho} > 0 \qquad \text{on } \bar{\Gamma} \tag{63}$$

so T is a good time function. Since T depends on momenta, it is called an "extrinsic" time. Physically the following goes on. As the particles move from left to right they measure the relative energy of radial observers. This energy is nothing but $p_a x^a/\rho = p_\rho$. As they approach the right half of the past light cone of $t = x = 0$, i.e. $x = -t > 0$, these radial observers become lightlike and $T = p_\rho$ increases indefinitely.

Now, for ψ we simply take the natural embedding of $(\tau = p_\rho; \xi, p_\xi)$ into $\bar{\Gamma}$. We find, changing ω into $\bar{\omega} = \omega - d(\rho p_\rho)$ and eliminating ρ using (61), that

$$\psi^* \bar{\omega} = p_\xi d\xi - \frac{p_\xi}{(\tau^2 - m^2)^{1/2}} d\tau, \qquad m < \tau < \infty. \tag{64}$$

So ψ is a good deparametrization since $p_\xi d\xi$ is independent of τ and

$$h = \frac{p_\xi}{(\tau^2 - m^2)^{1/2}} \tag{65}$$

is the time–dependent Hamiltonian. This ends Example 2.

There is a sense in which Example 2 is more akin to G.R. than Example 1, namely in that the configuration space $\mathcal{C}$ is, not a linear space, but a cone in a linear space. Similarly, the configuration space $\mathcal{C}$ of G.R. is the cone of all positive–definite metrics on Σ:

$$\mathcal{C} = \{\text{metrics } q_{\alpha\beta} \text{ on } \Sigma\}$$

$$T^*\mathcal{C} = \{q_{\alpha\beta}, p^{\alpha\beta}\} \qquad \text{with} \qquad \omega = \int_\Sigma q_{\alpha\beta} \delta p^{\alpha\beta} d^3 y,$$

as at the end of the first lecture, and $\bar{\Gamma} \subset \Gamma$ is defined by

$$\mathcal{H} = -(\det q)^{1/2} \mathcal{R} + (\det q)^{-1/2} p_{\alpha\beta} p^{\alpha\beta} - \frac{1}{2} (\det q)^{-1/2} p^2 = 0 \tag{66}$$

$$\mathcal{H}_\alpha = -2 D_\beta p_\alpha{}^\beta = 0. \tag{67}$$

Following York (1972), consider the subset $\Gamma_\tau \subset \bar{\Gamma}$, defined by the additional condition that the mean (= trace of the extrinsic) curvature is constant ("CMC–condition"), namely

$$p = q_{\alpha\beta} p^{\alpha\beta} = -2(\det q)^{1/2} \tau, \qquad \tau = \text{constant}. \tag{68}$$

Condition (68) is also called York's extrinsic time gauge. That (68) deserves the name "time gauge", rests on the following fact: There is an open set $\mathcal{U}$ of points in $\bar{\Gamma}$ and an interval $I \subset \mathbf{R}$, such that for each $(q_{\alpha\beta}, p^{\alpha\beta}) \in \mathcal{U}$ and each $\tau \in I$ there exists an element $(\bar{q}_{\alpha\beta}, \bar{p}^{\alpha\beta}) \in \mathcal{U}$ which lies in the same gauge subspace as $(q_{\alpha\beta}, p^{\alpha\beta})$ and satisfies (68). In spacetime language, there is a nonempty, open set of spacetimes having a CMC surface (see Choquet-Bruhat 1976 and also Bartnik 1988). At present it is not known how large the set $\mathcal{U}$ is (I thank R. Bartnik for information on this issue). Having found one CMC surface in a spacetime, we can find a whole foliation by CMC surfaces, along which τ varies monotonically. To see this, one writes the condition $((\det q)^{-1/2}p)^{\cdot} = C$ = constant, using (31,32), for arbitrary lapse N and zero shift $\vec{N}$. One obtains an elliptic equation for N which, generically, has a unique solution for $C \neq 0$ and only the trivial solution $N = 0$ for $C = 0$. Thus, indeed, in $\Gamma' = \bigcup_{\tau \in I} \Gamma_\tau$, we have "gauge–fixed the lapse–freedom" up to the parameter τ. Since the CMC–condition is invariant under spatial diffeomorphism, $\Gamma_\tau \subset \bar{\Gamma}$ is still first–class, with the null space $\mathcal{D}'$ of $\bar{\Omega}|_{\Gamma_\tau}$ spanned by the shift–transformations. We can now take the quotient $\widehat{\Gamma}_\tau$ of Γ_τ by $\mathcal{D}'$ to obtain, for each τ, a copy of the reduced phase space $\widetilde{\Gamma}$. The space $\widehat{\Gamma}' = \bigcup_{\tau \in I} \widehat{\Gamma}_\tau$ will be the space playing the role of $\bar{\Gamma}$ in the model situation depicted at the beginning of this section. What we have just said amounts to the statement that $\bar{\Omega}$, projected down to $\widehat{\Gamma}'$ has a one dimensional null distribution and that the map T, sending each point in $\widehat{\Gamma}'$ to it's τ–value is a good time function. One now has to find a deparametrization, i.e. the analogue of the map ψ in the model–situation. To construct it, consider the space $\Gamma_{TT} \subset \Gamma$, defined by

$$p = 0, \qquad D_\alpha p_\beta{}^\alpha = 0. \tag{69}$$

The constraint $\Gamma_{TT} \subset \Gamma$ is first class, since, setting $E(\mu) = \int_\Sigma \mu p d^3y$, we have

$$-2\left\{\int_\Sigma \mu^\alpha D_\beta p_\alpha{}^\beta d^3y, \int_\Sigma \mu p d^3y\right\} = \{H(\vec{\mu}), E(\mu)\} = E(\mathcal{L}_{\vec{\mu}}\mu) \tag{70}$$

for any scalar field μ and vector field μ^α on Σ. Now define the York–map $\mathcal{Y}_\tau : \Gamma_{TT} \to \Gamma_\tau$ by the following construction: For $(q,p) \in \Gamma_{TT}$, define

$$q'_{\alpha\beta} = \phi^4 q_{\alpha\beta} \tag{71}$$

$$p'^{\alpha\beta} = \phi^{-4} p^{\alpha\beta} - \frac{2\tau}{3}\phi^2 (\det q)^{1/2} q^{\alpha\beta}, \tag{72}$$

where ϕ is the unique solution of the Lichnerowicz equation

$$L_q\phi := \left(-\Delta_g + \frac{\mathcal{R}}{8}\right)\phi = \frac{1}{8}(\det q)^{-1} p_{\alpha\beta}p^{\alpha\beta}\phi^{-7} - \frac{\tau^2}{12}\phi^5. \tag{73}$$

It is straightforward to check that, by virtue of (69) and (71,72,73), the pair $(q'_{\alpha\beta}, p'^{\alpha\beta})$ is actually in Γ_τ. Furthermore one can show (Isenberg, Marsden 1984), that $\mathcal{Y}_\tau$ is surjective. However $\mathcal{Y}_\tau$ cannot be injective. The reason for this is that the conditions (69) are invariant under conformal rescaling $(q_{\alpha\beta}, p^{\alpha\beta}) \mapsto (\sigma^4 q_{\alpha\beta}, \sigma^{-4}p^{\alpha\beta})$, and all such conformally related pairs are mapped into the same point of Γ_τ by $\mathcal{Y}_\tau$. Thus one has to gauge–fix the conformal freedom allowed by (69). This one does by solving the Yamabe problem for $q_{\alpha\beta}$, i.e. by finding σ, such that $\mathcal{R}[\sigma^4 q] =$ constant (see Lee and Parker 1987), where either the value of this constant or the volume Vol $[\sigma^4 q]$ is fixed. (The sign of the constant $\mathcal{R}[\sigma^4 q]$ is conformally invariant. We ignore problems associated with this procedure in the positive–scalar–curvature case where $\sigma^4 q_{\alpha\beta}$ can be non–unique.) Let us call the resulting, smaller space $\Gamma_c \subset \Gamma_{TT}$. Now, as shown in Isenberg and Marsden (1984), $\mathcal{Y}_\tau$ restricted to Γ_c does become one–to–one. Furthermore this map is actually (pre–) symplectic in that it carries the pre–symplectic two–form on Γ_c into that on Γ_τ. We can now factor out the remaining gauge freedom on both the domain and the target space of $\mathcal{Y}_\tau$ by dividing by diffeomorphisms of Σ. In this way we find a map

$$[\mathcal{Y}_\tau] : \Gamma_c/\text{Diff}\,\Sigma \to \Gamma_\tau/\text{Diff}\,\Sigma = \widehat{\Gamma}_\tau.$$

Next observe that $\Gamma_c/\text{Diff}\,\Sigma$ carries a natural symplectic potential. In fact, $\Gamma_c/\text{Diff}\,\Sigma$ is essentially the cotangent bundle of conformal superspace, i.e. the space of Riemannian metrics on Σ, with conformal rescalings and diffeomorphisms factored out (see Fischer and Marsden 1977). The image of the corresponding symplectic potential under $[\mathcal{Y}_\tau]$ is thus a symplectic potential $\bar{\omega}_\tau$ for $\bar{\Omega}|_{\Gamma_\tau}$ projected down to $\widehat{\Gamma}_\tau$. Everything so far has been done for constant τ. We now take for ψ the map $[\mathcal{Y}_\tau]$, viewed as a map

$$\psi : \{\tau \in I\} \times \Gamma_c/\text{Diff}\,\Sigma \to \widehat{\Gamma}' = \bigcup_{\tau\in I} \widehat{\Gamma}_\tau.$$

The collection of symplectic potentials $\bar{\omega}_\tau$ gives, in the obvious way, rise to a (pre–) symplectic potential ω' on $\widehat{\Gamma}'$. We are thus exactly in the model situation: Pulling back ω' on $\widehat{\Gamma}'$ by ψ to $\Gamma_c/\text{Diff}\,\Sigma$ and restricting to $\tau =$ constant we obtain, by construction, the natural symplectic potential which is clearly τ–independent. Thus ψ is a good deparametrization. It remains to determine the $d\tau$–component of $\psi^*\omega'$, i.e. (minus) the

Hamiltonian h. It turns out that h is nothing but the pull–back under ψ of Vol $(q'_{\alpha\beta})$, viewed as a function on $\widehat{\Gamma}'$. Notice that h, despite it's simple geometrical interpretation, as a function on $\Gamma_c/\mathrm{Diff}\ \Sigma$ is actually quite complicated, involving the conformal factor ϕ, which solves (73) for g in the $\mathcal{R} =$ constant–gauge and so, in particular, depends on τ. These constructions have been carried out in detail by Moncrief (1989) in the case of 2+1–gravity and have for 3+1–gravity been announced in Fischer and Moncrief (1993). To sum up, vacuum General Relativity for spacetimes with compact space sections can be viewed as a time–dependent Hamiltonian system on the cotangent bundle of conformal superspace with the volume function playing the role of the Hamiltonian.

If we accept as fundamental the above deparametrization of G.R., based on a rigid spacetime slicing, namely the one by CMC–surfaces, we are led to a less restrictive notion of observable: Namely, every function on the cotangent bundle of conformal superspace gives rise to an observable, and this observable will in general be time dependent. Some of these observables, as the Hamiltonian from above, will in fact be integrals of local functions on $\bar{\Gamma}$, thus evading the conclusion of Torre's theorem. It thus appears that the question as to what should be the correct notion of observables in canonical G.R., which is clearly important for any quantum theory of gravity, is not fully understood even on the classical level.

Acknowledgements

This work has been supported by Fonds zur Förderung der wissenschaftlichen Forschung in Österreich, Projekt Nr. P9376-PHY.

References

[1] Abraham, R, Marsden, J E (1978), Foundation of Mechanics, Benjamin, Reading.

[2] Andersson, L (1987), Momenta and Reduction for General Relativity, J. Geom. and Physics **4**, 289–314.

[3] Arnowitt, R, Deser, S, and Misner, C W (1962), The dynamics of general relativity, in Gravitation: an introduction to current research, ed. by L Witten, Wiley, New York.

[4] Ashtekar, A (1988), New Perspectives in Canonical Gravity (with invited contributions), Bibliopolis, Napoli.

[5] Ashtekar, A, Bombelli, L, and Reula, O (1991), The Covariant Phase Space of Asymptotically Flat Gravitational Fields, in Mechanics, Analysis and Geometry: 200 Years after Lagrange, ed. by M Francaviglia, North–Holland, Amsterdam.

[6] Ashtekar, A (1991), Lectures on Non–Perturbative Canonical Gravity, World Scientific, Singapore.

[7] Bartnik, R (1988), Remarks on constant mean curvature surfaces in cosmological spacetimes, Commun. Math. Phys. **117**, 615–624.

[8] Beig, R, and Ó Murchadha, N (1987), The Poincaré Group as the Symmetry Group of Canonical General Relativity, Ann. Phys. **174**, 463–498.

[9] Choquet–Bruhat, Y (1976), Maximal submanifolds and manifolds with constant mean, extrinsic curvature of a Lorentzian manifold, Ann. Scuola Norm. Pisa, Ser IV, **3**, 361–376.

[10] deWitt, B (1967), Quantum Theory of Gravity. I. The Canonical Theory, Phys. Rev. **160**, 1113–1148.

[11] Dirac, P A M (1964), Lectures in Quantum Mechanics, Academic Press, New York.

[12] Fischer, A, and Marsden, J (1977), The manifold of conformally equivalent metrics, Canadian Journal of Mathematics **1**, 193–209.

[13] Fischer, A, and Marsden, J (1979), Topics in the dynamics of general relativity, in Proceedings of the International School of Physics "Enrico Fermi", Course LXVII, Isolated Gravitating Systems in General Relativity, ed. J. Ehlers, North–Holland, Amsterdam.

[14] Fischer, A, and Moncrief, V (1994), Reducing Einstein's Equations to an Unconstrained Hamiltonian System on the Cotangent Bundle of Teichmüller Space, in Physics on Manifolds, Proceedings of the International Colloquium in honour of Yvonne Choquet–Bruhat, eds. M Flato, R Kerner and A Lichnerowicz, Kluwer, Dordrecht.

[15] Gotay, M, Nester, J M, and Hinds, G (1978), Presymplectic manifolds and the Dirac–Bergmann theory of constraints, Journ. Math. Phys. **19**, 2388–2399.

[16] Hanson, A, Regge, T, and Teitelboim, C (1976), Constrained Hamiltonian Systems, Accademia Nazionale dei Lincei, Roma.

[17] Isenberg, J, and Nester, J (1980), Canonical Gravity, in General Relativity and Gravitation, Vol. 1, ed. A Held, Plenum, New York.

[18] Isenberg, J, and Marsden, J E (1984), The York map is a canonical transformation, J. Geom. and Physics **1**, 85–105.

[19] Kuchař, K (1992), Time and Interpretations of Quantum Gravity, in Proceedings of the 4th Canadian Conference on General Relativity and Relativistic Astrophysics, eds. G Kunstatter, D Vincent and J Williams, World Scientific, Singapore.

[20] Lee, J, and Parker, T (1987), The Yamabe Problem, Bull. Am. Math. Soc. **17**, 37–91.

[21] Moncrief, V (1975), Spacetime symmetries and linearization stability of the Einstein equations I, Journ. Math. Phys. **16**, 493–498.

[22] Moncrief, V (1989), Reduction of the Einstein equations in 2 + 1 dimensions to a Hamiltonian system over Teichmüller space, Journ. Math. Phys. **30**, 2907–2914.

[23] Ó Murchadha, N, and York, J W Jr. (1973), Existence and uniqueness of the Hamiltonian constraint of general relativity on compact manifolds, Journ. Math. Phys. **14**, 1551–1557.

[24] Regge, T, and Teitelboim, C (1974), Role of Surface Integrals in the Hamiltonian Formulation of General Relativity, Ann. Phys. **88**, 286–318.

[25] Ryan, M (1972), Hamiltonian Cosmology, Lect. Notes in Phys. **13**, Springer, Berlin.

[26] Schäfer, G (1990), Reduced Hamiltonian formalism for general–relativistic adiabatic fluids and applications, Astron. Nachr. **311**, 213–217.

[27] Sudarsky, D, and Wald, R M (1992), Extrema of Mass, Stationarity and Staticity, and Solutions to the Einstein–Yang–Mills Equations, Phys. Rev. **D46**, 1453–1474.

[28] Sundermeyer, K (1982), Constrained Dynamics, Lect. Notes in Phys. **169**, Springer, Berlin.

[29] Torre, C G (1993), Gravitational observables and local symmetries, Phys. Rev. **D48**, R2373–R2376.

[30] Wald, R M (1984), General Relativity, Univ. of Chicago Press, Chicago.

[31] York, J W (1972), Role of Conformal Three–Geometry in the Dynamics of Gravitation, Phys. Rev. Lett. **28**, 1082–1085.

Ashtekar Variables in Classical General Relativity

Domenico Giulini

Introduction

The task I was given for this lecture is to present Ashtekar's connection variables for classical General Relativity in a pedagogical manner. My intention is to describe a possible route towards a reformulation of General Relativity in terms of Ashtekar's connection variables. I try to give self-contained and hopefully painless derivations of the crucial steps, without entering too much into those details not directly relevant for this purpose. There exists a comprehensive monograph on this subject (Ashtekar 1991) which contains many applications, as well as a periodically updated bibliography (Brügmann 1993), so that the interested reader should have no difficulties to find his/her way into the subject and the current developments. I make no attempt to give a full account of the current status.

In these lectures I proceed as follows: In the first chapter, I very briefly review some elementary concepts from differential geometry, mainly to fix notation and conventions. Chapter two introduces the variational principle which we use to derive the field-equations of General Relativity. Chapter three considers complex General Relativity and shows how its field equations can be obtained from a variational principle involving only the self-dual part of the connection. In chapter four the (3+1)-decomposition is presented in as much detail as seemed necessary for an audience that does not consist entirely of canonical relativists. It is then applied to complex General Relativity in chapter five, where for the first time Ashtekar's connection variables are introduced. The Hamiltonian of complex General Relativity is presented in terms of connection variables. In chapter six the constraints that follow from the variational principle are analyzed and their Poisson brackets are presented. In chapter seven we discuss the reality conditions that have to be imposed by hand to select real solutions, and briefly sketch the geometric interpretation of the new variables. In chapter 8 we indicate how the Hamiltonian has to be amended by surface integrals in the case of open initial data hypersurfaces with asymptotically flat data. It ends with a demonstration of the positivity of the mass at spatial infinity for maximal hypersurfaces. Throughout we will make no use of spinors.

Chapter 1. Some Basic Differential Geometry

Let M be a connected orientable and time orientable Lorentz 4-manifold with topology $\Sigma \times \mathbf{R}$, where Σ can be any connected orientable 3-manifold. Our signature convention is "mostly plus", $(-,+,+,+)$. Greek indices refer to coordinate bases, latin indices to frame bases. If they are taken from the beginning of the alphabet, i.e., $(\alpha, \beta, ..; a, b, ..)$, their range is $\{0,1,2,3\}$, whereas for the middle of the alphabet, i.e., $(\mu, \nu, ..; i, j, ..)$, their range is only $\{1,2,3\}$. Square brackets including a string of n indices denote full antisymmetrization including the factor $1/n!$. In the same way, round brackets denote symmetrization. The Lorentz metric is denoted by g (components g_{ab}, $g_{\alpha\beta}$). η_{ab} denotes the matrix $diag(-1,1,1,1)$. The components of the curvature tensor are written $R^a{}_{bcd}$, $R_{ab} = R^c{}_{acb}$ are the components of the Ricci-tensor and $R = g^{ab}R_{ab}$ denotes the Ricci scalar.

The structure group for the real frame bundle on M is $GL(4,R)$, and $SO(1,3)$ if one restricts to oriented orthonormal frames. We shall adopt this restriction throughout which is equivalent to imposing an orientation and a metric structure. We denote the Lie algebra of $SO(1,3)$ by $so(1,3)$. Due to the assumption that M is topologically $\Sigma \times \mathbf{R}$ and orientable, the frame bundle is necessarily trivial so that we can always assume the existence of a globally defined tetrad. This is also true if one considers the complexified tangent bundle, which we shall need later on. We stress that this is particular to four dimensions and will generally not hold in dimensions three or higher than four. Let

$$\{e_a\} \quad \text{orthonormal tetrad} \tag{1.1}$$

$$\{e^a\} \quad \text{orthonormal co-tetrad dual to (1.1)}, \tag{1.2}$$

$$\text{so that} \quad g(e_a, e_b) = \eta_{ab} \tag{1.3}$$

$$\text{and} \quad e^a(e_b) = \delta^a_b \,. \tag{1.4}$$

The volume form, ϵ, on M induced by g is given by ($\wedge$ denotes the antisymmetric tensor product)

$$\epsilon = e^0 \wedge e^1 \wedge e^2 \wedge e^3 = \frac{1}{4!}\, \epsilon_{abcd}\, e^a \wedge e^b \wedge e^c \wedge e^d \,, \tag{1.5}$$

$$\text{where} \quad \epsilon_{abcd} = 4!\, \delta^0_{[a}\delta^1_b\delta^2_c\delta^3_{d]} \,. \tag{1.6}$$

Once we have restricted the structure group to $SO(1,3)$ we also restrict the connections to be metric-preserving. The connection 1-form can be represented by a globally defined $so(1,3)$-valued 1-form, ω^a_b, and the curvature by an $so(1,3)$-valued 2-form Ω^a_b. If ∇ denotes the covariant derivative, one has

$$\nabla_{e_a} e_b = \omega^c_b(e_a)\, e_c \tag{1.7}$$
$$\eta_{ac}\omega^c_b =: \omega_{ab} = -\omega_{ba}\,, \tag{1.8}$$

and

$$(\nabla_{e_a}\nabla_{e_b} - \nabla_{e_b}\nabla_{e_a} - \nabla_{[e_a,e_b]})\, e_c = \Omega^d_c(e_a, e_b)\, e_d = R^d{}_{cab}\, e_d \tag{1.9}$$
$$\eta_{ac}\Omega^c_b =: \Omega_{ab} = -\Omega_{ba}\,. \tag{1.10}$$

Throughout we identify the elements of the Lie algebra with tangent space tensors σ^b_a such that $\eta_{ac}\sigma^c_b = \sigma_{ab} = -\sigma_{ba}$. The torsion is represented by an R^4-valued 2-form T^a which is defined through

$$\nabla_{e_a} e_b - \nabla_{e_b} e_a - [e_a, e_b] = T^c(e_a, e_b)\, e_c\,. \tag{1.11}$$

Let V denote a vector space that carries a representation, ρ, of $SO(1,3)$, and λ a V-valued n-form on M which under change of tetrads transforms via ρ. Then the exterior covariant derivative, D, on vector-valued forms is defined via

$$D\lambda := d\lambda + \rho(\omega) \wedge \lambda\,, \tag{1.12}$$
$$\text{so that} \quad D^2\lambda = \rho(\Omega) \wedge \lambda\,. \tag{1.13}$$

Here the symbol $\rho(\omega) \wedge \lambda$ is to be understood in the following way: The representation ρ of $SO(1,3)$ on V induces a representation of its Lie algebra $so(1,3)$ on V, which we also denote by ρ. Via this representation ω ($so(1,3)$-valued) acts on λ (V-valued), while as forms the exterior product is taken. The exterior covariant derivatives with respect to two different connections ω and ω' (denoted by D^ω and $D^{\omega'}$) are related via

$$D^{\omega'}\lambda = D^\omega\lambda + \rho(\omega' - \omega) \wedge \lambda\,. \tag{1.14}$$

Applying (1.11-13) to the R^4-valued 1-form $\{e^a\}$ (carrying the defining representation) one obtains the first Cartan structure equation and the first Bianchi identity:

$$De^a = de^a + \omega^a_b \wedge e^b = T^a\,, \tag{1.15}$$
$$D^2 e^a = \Omega^a_b \wedge e^b = DT^a\,. \tag{1.16}$$

By direct calculation, using the definition of the curvature, and regarding Ω^a_b as an $so(1,3)$-valued 2-form, we obtain the second Cartan structure equation and second Bianchi identity:

$$d\omega^a_b + \omega^a_c \wedge \omega^c_b = d\omega^a_b + \tfrac{1}{2}[\omega,\omega]^a_b = \Omega^a_b\,, \tag{1.17}$$

$$D\Omega^a_b = d\Omega^a_b + [\omega,\Omega]^a_b = 0\,. \tag{1.18}$$

Here [,] denotes the commutator if the product of the form degrees is even, and the anticommutator if it is odd. Note also that it would neither be algebraically correct nor meaningful to write (1.17) as $D\omega^a_b$ (in the adjoint representation): firstly, this expression would not have the factor 1/2 that appears in (1.17) and, secondly, ω^a_b does not transform with any linear transformation under change of tetrads, as would be necessary for the covariant derivative to be meaningful. Rather, from (1.7) one easily proves that under a change of frames: $e_a \mapsto e'_a = R^b_a e_b$, with $SO(1,3)$-valued matrix function R, one has (in matrix notation):

$$\omega \overset{R}{\longmapsto} \omega' = R^{-1}\omega R + R^{-1}dR\,. \tag{1.19}$$

Finally we mention the Hodge-duality map, $*$, which provides a linear isomorphism between n forms and $(4-n)$-forms at each point on M. It can be defined via

$$*(e^{a_1} \wedge \ldots \wedge e^{a_n}) := \frac{1}{(4-n)!}\epsilon^{a_1\ldots a_n}{}_{a_{n+1}\ldots a_4} e^{a_{n+1}} \wedge \ldots \wedge e^{a_4}\,, \tag{1.20}$$

and linear extension, where the indices on ϵ are raised using η^{ab} (the inverse matrix to η_{ab}, which has the same entries). Since $\epsilon_{a_1a_2a_3a_4}$ is invariant under $SO(1,3)$, the exterior covariant derivative and the $*$-operation are compatible in the following sense:

$$D*(e^{a_1} \wedge \ldots \wedge e^{a_n}) = \frac{1}{(4-n)!}\epsilon^{a_1\ldots a_n}{}_{a_{n+1}\ldots a_4} D(e^{a_{n+1}} \wedge \ldots \wedge e^{a_4})\,. \tag{1.21}$$

Applying $*$ twice results in plus or minus the identity. In fact, on any n-form λ one has

$$*(*\lambda) = -(-1)^{n(4-n)}\lambda\,. \tag{1.22}$$

Analogous formulae hold in any spacetime dimension. Here, the first minus sign on the right hand side of (1.21) is the sign of the determinant of η_{ab}. In four dimensions and Lorentz signature $*$ squares to -1 on two forms, so that eigenforms of $*$ only exist on the complexified tangent bundle with eigenvalues $+i$ (self-dual 2-forms) and $-i$ (anti-self-dual 2-forms). Given two n-forms λ and σ at some point in M:

$$\lambda = \frac{1}{n!}\lambda_{a_1\ldots a_n} e^{a_1} \wedge \ldots \wedge e^{a_n} , \tag{1.23}$$

$$\sigma = \frac{1}{n!}\sigma_{a_1\ldots a_n} e^{a_1} \wedge \ldots \wedge e^{a_n} , \tag{1.24}$$

we define their inner product, induced by g, via

$$\langle \lambda, \sigma \rangle := \frac{1}{n!}\lambda_{a_1\ldots a_n} g^{a_1 b_1} \ldots g^{a_n b_n} \sigma_{b_1\ldots b_n} = \lambda_{a_1\ldots a_n} \sigma^{a_1\ldots a_n} , \tag{1.25}$$

and have

$$\lambda \wedge *\sigma = \langle \lambda, \sigma \rangle \, \epsilon . \tag{1.26}$$

Chapter 2. The Variational Principle

We first show how the well known Einstein-Hilbert action* is written in terms of the curvature 2-form Ω_{ab} and the co-tetrads e^a.

$$\begin{aligned}
\text{Action } = S :=& \int \Omega_{ab} \wedge *(e^a \wedge e^b) \\
=& \tfrac{1}{2} \int R_{abcd}(e^c \wedge e^d) \wedge *(e^a \wedge e^b) \\
=& \tfrac{1}{2} \int R_{abcd} \langle e^c \wedge e^d , e^a \wedge e^b \rangle \, \epsilon \\
=& \tfrac{1}{2} \int R_{abcd}(\eta^{ac}\eta^{bd} - \eta^{ad}\eta^{bc}) \, \epsilon \\
=& \int R \epsilon \; = \; \text{Einstein-Hilbert Action.}
\end{aligned} \tag{2.1}$$

We now take the co-tetrads e^a and connection 1-forms ω^a_b as independent variables with respect to which we vary the action S. We stress that since we restricted the connection 1-form to be $so(1,3)$-valued we have

* Throughout this article we write all actions in their simplest form neglecting prefactors. One way to normalize the action within pure gravity is to evaluate the energy of the Schwarzschild solution and require it to be equal to the standard expression. The Einstein-Hilbert action, as written down in (2.1), gives the right energy expression in units where $16\pi G/c^4 = 1$. Note also that in the Hamiltonian formulation the canonical momenta scale with the prefactor.

put in metricity of the connection by hand. Varying the curvature with respect to the connection 1-form yields:

$$\delta\Omega_{ab} = d(\delta\omega_{ab}) + [\omega, \delta\omega]_{ab} = D(\delta\omega_{ab})\,, \tag{2.2}$$

so that the variation of S with respect to ω, denoted by δ^ω, is given by

$$\delta^\omega S = \tfrac{1}{2}\int \delta\omega_{ab} \wedge \epsilon^{ab}{}_{cd}\, D(e^c \wedge e^d)\,, \tag{2.3}$$

where we have performed an integration by parts and used the prescription that the variations vanish at the boundary of the integration domain. The requirement of stationarity is thus equivalent to:

$$\delta^\omega S = 0 \Leftrightarrow D(e^a \wedge e^b) = 0 = T^a \wedge e^b - e^a \wedge T^b\,. \tag{2.4}$$

It is not difficult to show directly from the rightmost expression that this implies vanishing torsion. But, keeping an eye towards later generalizations, we shall proceed differently. We write $\omega = \Gamma + \Lambda$ where Γ is the Levi-Civita connection, i.e., the unique metric, torsion free connection. If we denote the exterior covariant derivative with respect to Γ (ω) by D^Γ (D^ω), we have, using (1.13) and the vanishing torsion of Γ,

$$\begin{aligned} D^\omega(e^a \wedge e^b) =& D^\Gamma(e^a \wedge e^b) + \Lambda^a_c \wedge e^c \wedge e^b + \Lambda^b_c \wedge e^a \wedge e^c \\ =& (\Lambda^a_{[cd]}\delta^b_e + \Lambda^b_{[ce]}\delta^a_d)\; e^c \wedge e^d \wedge e^e\,, \end{aligned} \tag{2.5}$$

where we used the expansion $\Lambda^a_b = \Lambda^a_{cb} e^c$. This last expression vanishes precisely if the cyclic sum in $c - d - e$ of the coefficient does:

$$\Lambda^a_{[cd]}\delta^b_e + \Lambda^b_{[ce]}\delta^a_d + \Lambda^a_{[de]}\delta^b_c + \Lambda^b_{[dc]}\delta^a_e + \Lambda^a_{[ec]}\delta^b_d + \Lambda^b_{[ed]}\delta^a_c \; = 0\,. \tag{2.6}$$

Contraction on b and e yields:

$$\Lambda^a_{[cd]} + \Lambda^b_{[cb]}\delta^a_d - \Lambda^b_{[db]}\delta^a_c = 0\,. \tag{2.7}$$

Another contraction on a and d implies $\Lambda^a_{[ca]} = 0$, which, after reinsertion into (2.7), implies $\Lambda^a_{[cd]} = 0$, i.e. symmetry of Λ^a_{bc} in the lower index pair. In its purely covariant form: $\eta_{bd}\Lambda^d_{ac} =: \Lambda_{abc}$, we obtain a tensor which is symmetric in the first and third, and antisymmetric in the second and third slot, where the antisymmetry follows from the metricity of ω and Γ. But such a tensor is necessarily zero as three consecutive applications of interchangements of the first with third followed by second with third index show. Note that $\Lambda^a_{[bc]} = 0$ is just the condition for ω and Γ having the same torsion, so that the last arguments may be seen as a uniqueness proof for metric, torsion free connections. As result of the variation δ^ω we thus obtain

$$\delta^\omega S = 0 \Longleftrightarrow \omega = \Gamma = \text{Levi-Civita connection.} \tag{2.8}$$

For the Levi-Civita connection Γ we can invert the first Cartan structure equation (1.15) and express it purely in terms of the co-tetrad e^a:

$$\Gamma^{ab} = -* \left[(*de^a) \wedge e^b - (*de^b) \wedge e^a + \tfrac{1}{2}(e^a \wedge e^b) \wedge *(de^c \wedge e^d)\, \eta_{cd} \right] . \tag{2.9}$$

Next, we take the variation with respect to the co-tetrad e^a, which we denote by δ^e. The variation of the integrand yields:

$$\begin{aligned} \delta^e \left(\Omega_{ab} \wedge *(e^a \wedge e^b) \right) &= \left(\epsilon^{ab}{}_{cd} \Omega_{ab} \wedge e^c \right) \wedge \delta e^d \\ &= \tfrac{1}{2} \left(\epsilon^{ab}{}_{cd}\, R_{abgh}\, e^g \wedge e^h \wedge e^c \right) \wedge \delta e^d \\ &= \tfrac{1}{2} \left(\epsilon_{abcd}\, R^{ab}{}_{gh}\, \epsilon^{ghce}\, \eta_{ef} * e^f \right) \wedge \delta e^d \\ &= -\tfrac{1}{2} \left(3!\, \delta^g_{[a} \delta^h_b \delta^e_{d]}\, R^{ab}{}_{gh}\, \eta_{ef} * e^f \right) \wedge \delta e^d \\ &= 2 \left(R_{fd} - \tfrac{1}{2} \eta_{fd} R \right) * e^f \wedge \delta e^d , \end{aligned} \tag{2.10}$$

so that

$$\delta^e S = 0 \Longleftrightarrow \frac{\delta S}{\delta e^b} = 2 \left(R_{ab} - \tfrac{1}{2} \eta_{ab} R \right) * e^a = 0 \tag{2.11}$$

$$\Longleftrightarrow G_{ab} = R_{ab} - \tfrac{1}{2} \eta_{ab} R = 0 . \tag{2.12}$$

If we finally insert into this variation formula the result of the first variation, $\omega = \Gamma$, where Γ is to be understood as a function of the co-tetrad according to (2.9), then we obtain equations for the co-tetrads which are equivalent to Einstein's vacuum equations.

Chapter 3. Complex GR and Self-Dual Variational Principle

We now extend the differential-geometric framework and consider the *complexified* tensor bundle

$$T_{\mathbf{C}} = \bigoplus_{r,s} T^r_s(M) \otimes \mathbf{C}\,, \tag{3.1}$$

over the *real* manifold M. Here $T^r_s(M)$ denotes the (real) tensor bundle of r-fold contravariant and s-fold covariant tensors over M. The structure group is now $SO(1,3)_{\mathbf{C}} \cong SO(4)_{\mathbf{C}}$. The fields e^a, ω^a_b, Ω^a_b, T^a and η_{ab} extend to sections in the complexified bundle. Everything we have said so far is still valid in the complex case. In particular, the variational principle and the resulting equations of motion transcribe verbatim with all quantities now being complex. The theory so obtained may be called complex General Relativity. Since the complex tensor bundle was obtained by complexification of a real tensor bundle, there is a natural reality structure. In other words, we know how to complex conjugate, namely by complex conjugation of the components with respect to a real basis which is naturally given to us since we work over a real manifold. We can thus speak of real sections in the complex tensor bundle.

On the space of complex 2-forms we can now consider eigenforms to the $*$-operator (remember, on 2-forms: $** = -$ identity)

$$* \lambda = i\lambda \quad \text{self-dual}\,, \tag{3.2a}$$

$$* \lambda = -i\lambda \quad \text{anti-self-dual}\,. \tag{3.2b}$$

Similarly, in the complexified Lie algebra $so(1,3)_{\mathbf{C}}$, we can define a dualization operation (denoted by $\star$):

$$\star\sigma_{ab} := \tfrac{1}{2}\epsilon_{ab}{}^{cd}\sigma_{cd} \tag{3.3}$$

that also squares to minus the identity and thus decomposes the complexified Lie algebra into self- and anti-self-dual part:

$$so(1,3)^{(+)}_{\mathbf{C}} = \{\sigma \in so(1,3)_{\mathbf{C}} \;/\; \star\sigma = i\sigma\} \quad \text{self-dual}\,, \tag{3.4a}$$

$$so(1,3)^{(-)}_{\mathbf{C}} = \{\sigma \in so(1,3)_{\mathbf{C}} \;/\; \star\sigma = -i\sigma\} \quad \text{anti-self-dual}\,. \tag{3.4b}$$

Using the standard formula for multiplying two ϵ-tensors one easily shows the following relations for commutators:

$$\star[\sigma, \star\tau] = -[\sigma,\tau] = \star[\star\sigma,\tau]\,, \tag{3.5}$$

$$\text{hence} \quad \star[\sigma,\tau] = [\star\sigma,\tau] = [\sigma,\star\tau] \tag{3.6}$$

$$\text{and} \quad [\star\sigma,\star\tau] = -[\sigma,\tau]\,. \tag{3.7}$$

If we denote the projectors onto the self-dual and anti-self-dual parts of $so(1,3)_{\mathbb{C}}$ by $P^{(+)}$ and $P^{(-)}$:

$$P^{(\pm)} = \tfrac{1}{2}(1 \mp i\star)\,, \tag{3.8}$$

then (3.5-7) imply

$$P^{(\pm)}[\sigma,\tau] = [P^{(\pm)}\sigma,\tau] = [\sigma,P^{(\pm)}\tau] = [P^{(\pm)}\sigma,P^{(\pm)}\tau]\,. \tag{3.9}$$

This tells us that $so(1,3)^{(\pm)}_{\mathbb{C}}$ form ideals in $so(1,3)_{\mathbb{C}}$ with projection homomorphisms $P^{(\pm)}$. We thus have a decomposition of the Lie algebra into two ideals:

$$so(1,3)_{\mathbb{C}} = so(1,3)^{(+)}_{\mathbb{C}} \oplus so(1,3)^{(-)}_{\mathbb{C}}\,. \tag{3.10}$$

$so(1,3)$ and $so(1,3)^{(\pm)}_{\mathbb{C}}$ are simple Lie algebras whereas $so(1,3)_{\mathbb{C}}$ is only semi-simple.

We can now decompose ω and Ω according to (3.9):

$$\omega = {}^{(+)}\omega + {}^{(-)}\omega = P^{(+)}\omega + P^{(-)}\omega\,, \tag{3.11}$$

$$\Omega = {}^{(+)}\Omega + {}^{(-)}\Omega = P^{(+)}\Omega + P^{(-)}\Omega = \Omega({}^{(+)}\omega) + \Omega({}^{(-)}\omega)\,, \tag{3.12}$$

where $\Omega({}^{(\pm)}\omega)$ is the curvature built from ${}^{(\pm)}\omega$ according to (1.17). It should be noted that ${}^{(\pm)}\omega$ are not connections on the original bundle with structure group $SO(1,3)_{\mathbb{C}}$ (e.g. a general gauge transformation in $SO(1,3)_{\mathbb{C}}$ will not leave it $so(1,3)^{(\pm)}_{\mathbb{C}}$-valued), but on associated bundles whose structure groups are given by the subgroups in $SO(1,3)_{\mathbb{C}}$ that correspond to the Lie sub-algebras $so(1,3)^{(\pm)}_{\mathbb{C}}$. Noting that $\Omega_{ab}\wedge *(e^a\wedge e^b) = \star\Omega_{ab}\wedge(e^a\wedge e^b)$, we can decompose the action of complex General Relativity:

$$\begin{aligned} S_{\mathbb{C}} = S^{(+)}_{\mathbb{C}} + S^{(-)}_{\mathbb{C}} &= \int {}^{(+)}\Omega_{ab}\wedge *(e^a\wedge e^b) + \int {}^{(-)}\Omega_{ab}\wedge *(e^a\wedge e^b) \\ &= i\int {}^{(+)}\Omega_{ab}\wedge(e^a\wedge e^b) - i\int {}^{(-)}\Omega_{ab}\wedge(e^a\wedge e^b)\,. \end{aligned} \tag{3.13}$$

The variation of the actions $S_{\mathbb{C}}$ and $S^{(\pm)}_{\mathbb{C}}$ with respect to ω, respectively ${}^{(\pm)}\omega$, will in each case result in a determination of these connections in terms of the co-tetrad fields. The variation with respect to the co-tetrad fields, evaluated at the particular value of the connections just determined, then result in equations for the co-tetrad fields only. Their solutions will then determine connections ω (if $S_{\mathbb{C}}$ is varied) or ${}^{(\pm)}\omega$ (if $S^{(\pm)}_{\mathbb{C}}$ is varied) in the corresponding bundles. The crucial property of these different variational principles is that they result in the *same* equations for the co-tetrads. Let us summarize this in the following

Theorem. *The stationary points of $S_{\mathbf{C}}$ and $S_{\mathbf{C}}^{(\pm)}$ lie over the same co-tetrad fields. All three actions assume the value zero at their stationary points.*

Proof. The variation of $S_{\mathbf{C}}$ proceeds identically to the variation of the real action S. With $S_{\mathbf{C}}^{(\pm)}$ we also proceed as in the real case. We shall only deal with $S_{\mathbf{C}}^{(+)}$ since the anti-self-dual case is entirely analogous. First varying ${}^{(+)}\omega$ we obtain (where we denote the exterior covariant derivative with respect to ${}^{(\pm)}\omega$ by ${}^{(\pm)}D$)

$$\begin{aligned}\delta^{^{(+)}\omega} S_{\mathbf{C}}^{(+)} &= \int {}^{(+)}D(\delta{}^{(+)}\omega_{ab}) \wedge *(e^a \wedge e^b)\\ &= i \int \delta{}^{(+)}\omega_{ab} \wedge {}^{(+)}D(e^a \wedge e^b)\\ &= 0 \Leftrightarrow {}^{(+)}D(e^a \wedge e^b)^{(+)} = 0\end{aligned} \tag{3.14}$$

where $(e^a \wedge e^b)^{(+)}$ denotes the self-dual part of $(e^a \wedge e^b)$ considered as a Lie algebra valued 2-form. We set ${}^{(+)}\omega = {}^{(+)}\Gamma + {}^{(+)}\Lambda = \Gamma - {}^{(-)}\Gamma + {}^{(+)}\Lambda$ and have, using that Γ has vanishing torsion and that the anti-self-dual ${}^{(-)}\Gamma^{ab}$ commutes with the self-dual $(e^a \wedge e^b)^{(+)}$, analogous to formula (2.5),

$$\begin{aligned}\delta^{^{(+)}\omega} S^{(+)} = 0 \Leftrightarrow &{}^{(+)}\Lambda^a_{\ c} \wedge (e^c \wedge e^b)^{(+)} + {}^{(+)}\Lambda^b_{\ c} \wedge (e^a \wedge e^c)^{(+)}\\ &= {}^{(+)}\Lambda^a_{\ c} \wedge (e^c \wedge e^b) + {}^{(+)}\Lambda^b_{\ c} \wedge (e^a \wedge e^c)\\ &= ({}^{(+)}\Lambda^a_{\ [cd]}\delta^b_e + {}^{(+)}\Lambda^b_{\ [ce]}\delta^a_d)\, e^c \wedge e^d \wedge e^e\,,\end{aligned} \tag{3.15}$$

where we used the expansion ${}^{(+)}\Lambda^a_{\ b} = {}^{(+)}\Lambda^a_{\ cb} e^c$. We can now conclude exactly as in (2.5) and (2.6), with ${}^{(+)}\Lambda$ replacing Λ, and obtain the analogous formula to (2.7). This leads to

$$\delta^{^{(+)}\omega} S^{(+)} = 0 \Longleftrightarrow {}^{(+)}\omega = {}^{(+)}\Gamma = \text{self-dual part of the Levi-Civita connection.} \tag{3.16}$$

As in the real case, ${}^{(+)}\omega$ is now determined as function of the co-tetrad by taking the self dual part of (2.9).

Next we perform the variation with respect to the co-tetrad. The variational derivative is easily obtained:

$$\frac{\delta S_{\mathbf{C}}^{(+)}}{\delta e^b} = 2i\,{}^{(+)}\Omega_{ab} \wedge e^a\,. \tag{3.17}$$

Evaluated at ${}^{(+)}\omega = {}^{(+)}\Gamma$, this is equal to

$$\left.\frac{\delta S_{\mathrm{C}}^{(+)}}{\delta e^b}\right|_{{}^{(+)}\omega={}^{(+)}\Gamma} = i\left(\Omega_{ab}\wedge e^a - i\tfrac{1}{2}\epsilon^{cd}{}_{ab}\Omega_{cd}\wedge e^a\right) \tag{3.18}$$

$$= \tfrac{1}{2}\epsilon^{cd}{}_{ab}\Omega_{cd}\wedge e^a \tag{3.19}$$

$$= \left(R_{ab} - \tfrac{1}{2}\eta_{ab}R\right) * e^a \tag{3.20}$$

$$= \frac{1}{2}\left.\frac{\delta S_{\mathrm{C}}}{\delta e^b}\right|_{\omega=\Gamma}, \tag{3.21}$$

where from (3.18) to (3.19) we have used the first Bianchi identity (1.16) and the step from (3.19) to (3.20) is entirely analogous to the calculation leading to (2.10). Equation (3.20) then follows from comparison with (2.11), simply transcribed to the complex case. Finally, we observe that at the stationary points the actions $S_{\mathrm{C}}^{(\pm)}$ and S_{C} assume the value zero. Note also the factor $\frac{1}{2}$ in (3.21), which is just the $\frac{1}{2}$ from the projection operation (formula (3.8)) onto the self-dual part. It implies that the properly normalized self-dual action should have twice the prefactor of the action S_{C}. This concludes our proof of the theorem •.

Chapter 4. The 3+1-Decomposition

Let $M \cong \mathbf{R}\times\Sigma$ be foliated by the images of a one-parameter family of embeddings, $\mathcal{E}_t : \Sigma \to M$, and let Σ_t denote the image of Σ under $\mathcal{E}_t$ in M. We have the following vector field over $\mathcal{E}_t$:

$$\frac{\partial}{\partial t} := \frac{d}{dt}\mathcal{E}_t \qquad p \mapsto \frac{d}{dt}\mathcal{E}_t(p) \in T_{\mathcal{E}_t(p)}(M), \tag{4.1}$$

whose spacetime interpretation is to generate the motion of Σ through M. We decompose it with respect to an orthonormal tetrad $\{e_\perp, e_k\}$ which is adapted to the foliation. The dual co-tetrad is $\{e^\perp, e^k\}$. Note that the e_k's are chosen tangential to Σ, which does not imply the e^k's to be tangent as well (i.e. a section in the cotangent bundle of Σ). Rather, the dual basis to $\{e_k\}$, intrinsic to Σ, is given by the projections of the e^k's parallel to Σ and will instead be called $\{\theta^k\}$. We will not use it until the last section. We now have:

$$\frac{\partial}{\partial t} = Ne_\perp + N^k e_k, \tag{4.2}$$

where N and $N^k e_k$ are the so-called lapse function and shift vector field respectively. Let further $\{x^\mu\}$ be a coordinate system on Σ so that $\{t, x^\mu\}$ is an adapted coordinate system on M and where $\partial/\partial t$ is given by (4.2).

We shall from now on write a $\perp$ for the zeroth frame index and t for the zeroth coordinate to distinguish the two and to remind of the special choice made.

The relation between the spatial coordinate and frame bases is given by

$$\frac{\partial}{\partial x^\mu} = e^k_\mu e_k\,, \quad e_k = e^\mu_k \frac{\partial}{\partial x^\mu}\,. \tag{4.3}$$

where $\{e^k_\mu\}$ and $\{e^\mu_k\}$ are relative inverse matrices. We have (N^μ denote the components of the shift vector with respect to the coordinate basis)

$$\begin{pmatrix} \partial_t \\ \partial_\mu \end{pmatrix} = \begin{pmatrix} N & N^k \\ 0 & e^k_\mu \end{pmatrix} \begin{pmatrix} e_\perp \\ e_k \end{pmatrix} \tag{4.4}$$

$$\begin{pmatrix} e_\perp \\ e_k \end{pmatrix} = \begin{pmatrix} 1/N & -N^\mu/N \\ 0 & e^\mu_k \end{pmatrix} \begin{pmatrix} \partial_t \\ \partial_\mu \end{pmatrix} \tag{4.5}$$

and

$$(dt\,,\, dx^\mu) = \left(e^\perp\,,\, e^k\right) \begin{pmatrix} 1/N & -N^\mu/N \\ 0 & e^\mu_k \end{pmatrix} \tag{4.6}$$

$$\left(e^\perp\,,\, e^k\right) = (dt\,,\, dx^\mu) \begin{pmatrix} N & N^k \\ 0 & e^k_\mu \end{pmatrix}. \tag{4.7}$$

The four-dimensional Lorentz metric then decomposes as follows:

$$\begin{aligned} ds^2 &= \eta_{ab}\, e^a \otimes e^b = g_{\alpha\beta}\, dx^\alpha \otimes dx^\beta \\ &= -e^\perp \otimes e^\perp + \sum_{k=1}^{3} e^k \otimes e^k \\ &= -N^2 dt \otimes dt + \sum_{k=1}^{3} (N^k dt + e^k_\mu dx^\mu) \otimes (N^k dt + e^k_\nu dx^\nu)\,, \end{aligned} \tag{4.8}$$

where the coordinate components of the spatial part are given by

$$h_{\mu\nu} = \sum_{k=1}^{3} e^k \otimes e^k(\partial_\mu, \partial_\nu) = \sum_{k=1}^{3} e^k_\mu e^k_\nu \tag{4.9}$$

$$\text{and inverse} \quad h^{\mu\nu} = \sum_{k=1}^{3} e^\mu_k e^\nu_k\,, \tag{4.10}$$

$$\text{so that} \quad e^\mu_k = h^{\mu\nu} \delta_{kl} e^l_\nu \quad \text{and} \quad e^k_\mu = h_{\mu\nu} \delta^{kl} e^\nu_l\,. \tag{4.11}$$

In words: e^k_μ is obtained from e^ν_l by raising the (latin) frame index with the Kronecker delta δ^{kl} ($= \delta_{kl}$) and lowering the (greek) coordinate index with $h_{\mu\nu}$. Using (4.10-11) and the definition $N_\mu := h_{\mu\nu} N^\nu$, we can write the Lorentz metric and its "inverse" in the following form:

$$\begin{aligned} g_{\alpha\beta}\, dx^\alpha \otimes dx^\beta &= -e_\perp \otimes e_\perp + \sum_{k=1}^{3} e_k \otimes e_k \\ &= -(N^2 - N^\mu N_\mu)\, dt \otimes dt + N_\mu\, (dt \otimes dx^\mu + dx^\mu \otimes dt) + h_{\mu\nu}\, dx^\mu \otimes dx^\nu \end{aligned} \tag{4.12}$$

$$\begin{aligned} g^{\alpha\beta}\, \partial_\alpha \otimes \partial_\beta &= -e^\perp \otimes e^\perp + \sum_{k=1}^{3} e^k \otimes e^k \\ &= -\frac{1}{N^2} \partial_t \otimes \partial_t + \frac{N^\mu}{N^2} (\partial_t \otimes \partial_\mu + \partial_\mu \otimes \partial_t) + \left(h^{\mu\nu} - \frac{N^\mu N^\nu}{N^2} \right) \partial_\mu \otimes \partial_\nu \end{aligned} \tag{4.13}$$

or simply:

$$\{g_{\alpha\beta}\} = \begin{pmatrix} -N^2 + N^\mu N_\mu & N_\mu \\ N_\mu & h_{\mu\nu} \end{pmatrix} \tag{4.14}$$

$$\{g^{\alpha\beta}\} = \begin{pmatrix} -1/N^2 & N^\mu/N^2 \\ N^\mu/N^2 & h^{\mu\nu} - N^\mu N^\nu/N^2 \end{pmatrix} . \tag{4.15}$$

The square-root of the modulus of the determinant of $\{g_{\alpha\beta}\}$, denoted by $\sqrt{g}$, is most easily obtained from (4.7,9) and the definition of the volume 4-form (1.5) (the square-root of the determinant of $\{h_{\mu\nu}\}$ is called $\sqrt{h}$):

$$\begin{aligned} \epsilon &= e^\perp \wedge e^1 \wedge e^2 \wedge e^3 = N \det\{e^k_\mu\}\ dt \wedge dx^1 \wedge dx^2 \wedge dx^3 \\ &= N\sqrt{h}\ dt \wedge dx^1 \wedge dx^2 \wedge dx^3 \\ &= \sqrt{g}\ dt \wedge dx^1 \wedge dx^2 \wedge dx^3 , \quad \text{hence} \quad \sqrt{g} = N\sqrt{h} \end{aligned} \tag{4.16}$$

Note that

$$\sqrt{h} = \det\{e^k_\mu\} = 1/\det\{e^\mu_k\} . \tag{4.17}$$

If we were to follow the standard route and put the action principle (2.1) into Hamiltonian form, we would decompose the co-tetrad variables according to (4.7), i.e., we would use the variables

$$\left(e^k_\mu , N^\mu , N \right) . \tag{4.18}$$

A first step in arriving at an algebraically simpler form of the Hamiltonian action integral is to take the frame rather than co-frame components as variables and redefine these variables by partially densitizing them (i.e. multiplying by powers of $\sqrt{h}$). The new variables are:

$$E_k^\mu := \sqrt{h}\, e_k^\mu = \frac{1}{\det\{e_k^\mu\}} e_k^\mu \tag{4.19}$$

$$N'^\mu := N^\mu\,, \tag{4.20}$$

$$N' := \frac{1}{\sqrt{h}} N = \det\{e_k^\mu\}\, N\,. \tag{4.21}$$

We have

$$\{dx^\alpha(e_a)\} = \begin{pmatrix} 1/N & 0 \\ -N^\mu/N & e_k^\mu \end{pmatrix} = \frac{1}{\sqrt{h}} \begin{pmatrix} 1/N' & 0 \\ -N'^\mu/N' & E_k^\mu \end{pmatrix} =: \frac{1}{\sqrt{h}} E_a^\alpha \tag{4.22}$$

and

$$\sqrt{g} = N\sqrt{h} = N' h \tag{4.23}$$

$$h = [\det\{e_k^\mu\}]^{-2} = \det\{E_k^\mu\}\,. \tag{4.24}$$

Now we have all the necessary formulae to write the action

$$S = \int \Omega_{ab} \wedge *(e^a \wedge e^b) \tag{4.25}$$

in 3+1-decomposed form using the densitized variables just introduced. Since from now on the old variables will never appear, we remove the dashes from these variables. But remember: from now on E_k^μ are the components of a triad of density weight-one, N^μ is a spatial vector field and N is a spatial scalar of density weight minus one. The action integrand then decomposes as follows:

$$\begin{aligned}
&\Omega_{ab} \wedge *(e^a \wedge e^b) = \tfrac{1}{2} R_{ab\alpha\beta}\, dx^\alpha \wedge dx^\beta \wedge *(e^a \wedge e^b) \\
&= \tfrac{1}{2} R_{ab\alpha\beta} \langle dx^\alpha \wedge dx^\beta\,,\, e^a \wedge e^b \rangle \epsilon \\
&= \tfrac{1}{2} R_{ab\alpha\beta}\, E_c^\alpha E_d^\beta \langle e^c \wedge e^d\,,\, e^a \wedge e^b \rangle \frac{1}{h} \epsilon \\
&= R^{cd}{}_{\alpha\beta} E_c^\alpha E_d^\beta\, N\, dt d^3x \\
&= \left(N\, R^{kl}{}_{\mu\nu}\, E_k^\mu E_l^\nu - 2N^\mu\, R^{\perp k}{}_{\mu\nu}\, E_k^\nu + 2R^{\perp k}{}_{t\nu}\, E_k^\nu\right) dt d^3x\,,
\end{aligned} \tag{4.26}$$

where we simply write dtd^3x for $dt \wedge dx^1 \wedge dx^2 \wedge dx^3$. The first thing to observe is the purely polynomial dependence on all field variables N, N^μ, E_k^μ and ω_α^{ab}. The reason for this lies in the usage of the densitized variables (4.19-21). In a next step we would like to calculate the Hamiltonian. For this we have to calculate the momenta by differentiating with respect to the time derivatives of the field variables. Obviously, N, N^μ and E_k^μ appear undifferentiated. As far as the connection variables are concerned, the first and second terms in (4.26) only contain ω_μ^{kl} and $\omega_\mu^{\perp l}$ and spatial derivatives thereof. The third term, however, contains ω_μ^{kl}, ω_t^{kl}, $\omega_t^{\perp k}$ and spatial derivatives thereof, but also $\omega_\nu^{\perp k}$ with its time derivative. No other quantities appear with their time derivatives.

In calculating the Hamiltonian we would then have to go through a constraint analysis along the standard algorithm as outlined in (Dirac 1967) and spelled out in more detail in (Hanson, Regge and Teitelboim 1976) or (Henneaux and Teitelboim 1992). We also recommend (Gotay, Nester, Hinds 1978) and (Marmo, Mukunda and Samuel 1983) for a more geometric approach. A good summary may be found in the lecture by Andreas Wipf in this volume. Here we cannot go through the full constraint analysis as it would present itself when started at this point. We merely want to point out one of its important results that may bee seen as being responsible for the strategy of complexification. More details may be found in (Romano 1993).

In our case, the constraint analysis would start with the primary constraints of vanishing momenta for all field variables except $\omega_\nu^{\perp k}$, and then continue with secondary constraints to ensure preservation of these during Hamiltonian evolution. This system of constraints then turns out not to be first class. Solving the purely second-class constraints reintroduces a complicated dependence on the field variables, which essentially brings us back to the standard ADM treatment of canonical General Relativity. This undesirable feature can be located as being due to the nondynamical nature of the field components ω_μ^{kl}, i.e. their vanishing momenta. The way complex General Relativity circumvents this problem is via the self dual representation (the anti-self-dual representation would just be as good), where

$$\tfrac{1}{2}\epsilon^{\perp k}{}_{lm}\,{}^{(+)}\omega_\alpha^{lm} = i\,{}^{(+)}\omega_\alpha^{\perp k}\,, \tag{4.27}$$

so that the time derivatives of ${}^{(+)}\omega_\mu{}^{lm}$ now appear through those of ${}^{(+)}\omega_\mu{}^{\perp k}$. To see how much is achieved this way we have to explicitly 3+1-decompose the self dual action of complex General Relativity. This is done in the next section.

Chapter 5. 3+1-Decomposition of Self-Dual Action in Ashtekar's Variables

From now on we only consider complex General Relativity with self dual action. We omit the (+) on ${}^{(+)}\omega$ and simply write ω for the self dual connection. Unless stated otherwise, all the connections and curvatures are self-dual from now on. We have

$$\omega_\alpha^{\perp k} = -i\tfrac{1}{2}\epsilon^{\perp k}{}_{lm}\,\omega_\alpha^{lm} \tag{5.1}$$
$$\omega_\alpha^{lm} = -i\,\epsilon^{lm}{}_{\perp p}\,\omega_\alpha^{\perp p} \tag{5.2}$$
$$R^{\perp k}{}_{\alpha\beta} = -i\tfrac{1}{2}\,\epsilon^{\perp k}{}_{lm}\,R^{lm}{}_{\alpha\beta}\,, \tag{5.3}$$

which we use to convert all $R^{\perp k}{}_{\alpha\beta}$ into $R^{nm}{}_{\alpha\beta}$ and $\omega_\alpha^{\perp k}$ into ω_α^{nm} in the expression for the 3+1-decomposed self-dual action. Moreover, we write antisymmetric complex self-dual matrices as 3-dimensional complex vectors, i.e., if λ^{kl} is such a matrix, we set:

$$\begin{aligned}\lambda^m &= -i\tfrac{1}{2}\epsilon^{\perp m}{}_{kl}\,\lambda^{kl}\,,\\ \lambda^{kl} &= -i\epsilon^{kl}{}_{\perp m}\,\lambda^m =: -i\epsilon^{kl}{}_m\lambda^m\,,\end{aligned} \tag{5.4}$$

and abbreviate the 3-tuple $(\lambda^1, \lambda^2, \lambda^3)$ by $\vec{\lambda}$. We also use "scalar-" and "vector-product" notations:

$$\sum_{k=1}^{3}\lambda^k\sigma^k =: \vec{\lambda}\cdot\vec{\sigma} \tag{5.5}$$
$$i\epsilon^k{}_{nm}\,\lambda^n\sigma^m := (\vec{\lambda}\times\vec{\sigma})^k\,. \tag{5.6}$$

This allows us to use convenient identities like

$$\begin{aligned}\vec{\lambda}\cdot(\vec{\sigma}\times\vec{\rho}) &= (\vec{\lambda}\times\vec{\sigma})\cdot\vec{\rho}\,,\\ \vec{\lambda}\times(\vec{\sigma}\times\vec{\rho}) &= \vec{\rho}(\vec{\lambda}\cdot\vec{\sigma}) - \vec{\sigma}\,(\vec{\lambda}\cdot\vec{\rho})\,.\end{aligned} \tag{5.7}$$

Note that due to the factor of i (which we included for later convenience) in (5.6), the right hand side of the second formula (5.7) carries the opposite sign to the familiar rule in vector calculus. Using this notation, we have e.g. ($\lambda^k_l = \delta_{ln}\lambda^{kn}$, that is, indices are raised and lowered with the Kronecker delta)

$$\lambda^k_l\sigma^l = (\vec{\lambda}\times\vec{\sigma})^k \tag{5.8}$$
$$[\lambda,\sigma]^{kl} := (\lambda^k_r\sigma^{rl} - \sigma^k_r\lambda^{rl}) = \lambda^k\sigma^l - \lambda^l\sigma^k\,, \tag{5.9}$$
$$\text{hence}\quad -i\tfrac{1}{2}\epsilon^{\perp m}{}_{kl}\,[\lambda,\sigma]^{kl} = (\vec{\lambda}\times\vec{\sigma})^m\,. \tag{5.10}$$

Using these formulae we can simply convert the curvature components into a convenient 3-dimensional notation. The connection becomes a 3-vector valued 1-form $\vec{\omega}$, and by $\vec{\omega} \times \vec{\omega}$ we shall denote the vector valued 2-form whose components are $i\epsilon^k{}_{lm}\omega^l \wedge \omega^m$.

$$R^{\perp k}{}_{\mu\nu} = (d\vec{\omega} + \vec{\omega} \times \vec{\omega})^k_{\mu\nu} \tag{5.11}$$

$$R^{kl}{}_{\mu\nu} = -i\epsilon^{kl}{}_p \, (d\vec{\omega} + \vec{\omega} \times \vec{\omega})^p_{\mu\nu} \tag{5.12}$$

$$R^{\perp k}{}_{t\nu} = \partial_t \omega^k_\nu - \partial_\nu \omega^k_t - 2(\vec{\omega}_\nu \times \vec{\omega}_t)^k \tag{5.13}$$

The next step is to bring these expressions into a 3-dimensional *covariant* form, that is, into a form involving only geometric quantities on the 3-dimensional manifold Σ. Note, for example, that the expression on the right hand side of (5.11) differs by a relative factor of 2 from what would be the curvature of a 3-dimensional connection $\vec{\omega}$; it cannot be, since $\vec{\omega}_\mu$ does not even define a 3-dimensional connection. But $2\vec{\omega}$ does! (connections form an affine but not a linear space and cannot generally be added and multiplied with numbers). An elementary way to see this is the following (during this argument we reemploy the old notation using ${}^{(+)}\omega$): the spatial part of the self-dual projection for the curvature 1-form is given by, ${}^{(+)}\omega_\mu{}^{kl} = \frac{1}{2}(\omega_\mu{}^{kl} - i\epsilon^{kl}{}_{\perp n}\omega_\mu{}^{\perp n})$. $\omega_\mu{}^{kl}$ is easily seen to form the components of a spatial connection and $\omega_\mu{}^{\perp n} = K^n_\mu$, form the components of the 3-dimensional tensor of extrinsic curvature (as we shall independently show below in (7.11-12)). Thus their sum (but not half of it) form a 3-dimensional connection. A geometrically more satisfying argument, which we suppress at this point, can be given using general bundle language. It is however important to realize that the spacetime structure group $SO(1,3)_{\mathbb{C}}$ is reduced to the spatial structure group $SO(3)_{\mathbb{C}}$ which is that subgroup that stabilizes the normal $e_\perp$. It is only with respect to this reduction that the components $\omega_\mu{}^{kl}$ form a 3-connection and $\omega_\mu{}^{\perp n}$ a tensor.

We thus define

$$\vec{A}_\mu := 2\vec{\omega}_\mu \tag{5.14}$$

$$\vec{\Lambda} := -2\vec{\omega}_t \tag{5.15}$$

(the minus sign in (5.15) being chosen for later convenience) and have

$$R^{\perp k}{}_{\mu\nu} = \tfrac{1}{2} F^k_{\mu\nu} \tag{5.16}$$

$$R^{kl}{}_{\mu\nu} = -i\tfrac{1}{2}\epsilon^{kl}{}_p F^p_{\mu\nu}\,, \tag{5.17}$$

$$\text{where} \quad \vec{F}_{\mu\nu} = (d\vec{A} + \tfrac{1}{2}\vec{A} \times \vec{A})_{\mu\nu} \tag{5.18}$$

$$\text{and} \quad R^{\perp k}{}_{t\nu} = \tfrac{1}{2}(\partial_t A^k_\nu + D_\nu \Lambda^k)\,, \tag{5.19}$$

$$\text{where} \quad D_\nu \vec{\Lambda} := \partial_\nu \vec{\Lambda} + \vec{A}_\nu \times \vec{\Lambda}\,. \tag{5.20}$$

We finally also abbreviate $(E_1^\mu, E_2^\mu, E_3^\mu) = \vec{E}^\mu$ and regard the action as functional in the variables $\vec{A}_\mu$ $\vec{E}^\mu$, N and N^μ. These are *Ashtekar's variables*, or *connection variables*, since $\vec{A}_\mu$ is a connection 1-form for a $SO(3)_{\mathbb{C}}$-connection over the 3-dimensional manifold Σ. Given the relations above and the decomposition (4.26) of the action (4.25) (now considered for the complex case), we can write the complex action in the following form:

$$S_{\mathbb{C}} = \int \Big\{ (\partial_t \vec{A}_\mu) \cdot \vec{E}^\mu \\ - \Big[\tfrac{1}{2} N \, \vec{F}_{\mu\nu} \cdot (\vec{E}^\mu \times \vec{E}^\nu) + N^\mu \, \vec{F}_{\mu\nu} \cdot \vec{E}^\nu + \vec{\Lambda} \cdot D_\mu \vec{E}^\mu \Big] \Big\} \, dt d^3x \, , \quad (5.21)$$

where the last term has already been integrated by parts without keeping the surface term. The action is now seen to be already in Hamiltonian form, i.e., of the form

$$\int (\dot{q}p - \mathcal{H}(q,p)) \, dt \, , \quad (5.22)$$

thus identifying the pair of canonically conjugate variables as

$$\left(\vec{A}_\mu \, , \, \vec{E}^\mu \right) \quad (5.23)$$

and the Hamiltonian

$$\mathcal{H} = \int_\Sigma \Big\{ \tfrac{1}{2} N \, \vec{F}_{\mu\nu} \cdot (\vec{E}^\mu \times \vec{E}^\nu) + N^\mu \, \vec{F}_{\mu\nu} \cdot \vec{E}^\nu + \vec{\Lambda} \cdot (D_\nu \vec{E}^\nu) \Big\} \, d^3x \, . \quad (5.24)$$

In making this direct identification of the canonical variables from the Hamiltonian variational principle we have jumped over a large part of constraint analysis. Note that originally both, the connection and the tetrad, were configuration-space variables and now turn out to be canonical conjugates. This would have been seen in the constraint analysis as an outcome of the explicit reduction of certain second-class constraints (e.g. Romano 1993). On the other hand, we know of the equivalence of the original action with the reformulated action (5.21), which can be employed to circumvent a tedious piece of explicit reduction of second-class constraints by starting directly from (5.21). This leaves us with a purely first-class system, as will be shown in the following chapter.

Chapter 6. Constraint Algebra

If we vary the action with respect to the fields N, N^μ and $\vec{\Lambda}$ we obtain equations in the canonical variables without time derivatives, that is, we obtain the constraints:

$$H := \tfrac{1}{2}\vec{F}_{\mu\nu} \cdot (\vec{E^\mu} \times \vec{E^\nu}) = 0\,, \tag{6.1}$$

$$H_\mu := \vec{F}_{\mu\nu} \cdot \vec{E^\nu} = 0\,, \tag{6.2}$$

$$\vec{H} := D_\mu \vec{E^\mu} = 0\,. \tag{6.3}$$

The Hamiltonian is just the sum of smeared constraints with the smearing functions N, N^μ and $\vec{\Lambda}$:

$$\mathcal{H} = \int_\Sigma \left\{ NH + N^\mu H_\mu + \vec{\Lambda} \cdot \vec{H} \right\} d^3x =: \mathcal{S}[N] + \mathcal{D}[N^\mu] + \mathcal{G}[\vec{\Lambda}]\,, \tag{6.4}$$

where the expressions on the right hand side are just abbreviations for the integrals in their appearing order. The constraint equations on phase space are properly interpreted as saying that for specified test-function spaces for N, N^μ and $\vec{\Lambda}$ the expressions $\mathcal{S}(N)$, $\mathcal{D}(N^\mu)$ and $\mathcal{G}(\vec{\Lambda})$ have to vanish *for all* test fields N, N^μ and $\vec{\Lambda}$. We call them the scalar-, diffeomorphism-, and gauge-constraints respectively. Note that there are infinitely many of each of them.

The actions of the scalar-, gauge-, and diffeomorphism-constraint on the phase space are easily determined. Let $\{,\}$ denote the Poisson bracket with respect to the symplectic structure given to us with the action (5.21) (in the standard way, namely by telling us that $\vec{A}_\mu$ and $\vec{E^\mu}$ are Darboux coordinates on phase space for the symplectic 2-form):

$$\{\vec{A}_\mu, \mathcal{S}[N]\} = N\,\vec{E^\nu} \times \vec{F}_{\mu\nu} \tag{6.5}$$

$$\{\vec{E^\mu}, \mathcal{S}[N]\} = D_\nu(N\,\vec{E^\nu} \times \vec{E^\mu})\,, \tag{6.6}$$

$$\text{and} \quad \{\vec{A}_\mu, \mathcal{G}[\vec{\Lambda}]\} = -D_\mu \vec{\Lambda} \tag{6.7}$$

$$\{\vec{E^\mu}, \mathcal{G}[\vec{\Lambda}]\} = \vec{\Lambda} \times \vec{E^\mu}\,, \tag{6.8}$$

$$\text{and} \quad \{\vec{A}_\mu, \mathcal{D}[N^\mu]\} = L_N \vec{A}_\mu + \{\vec{A}_\mu, \mathcal{G}[\vec{A}(N)]\} \tag{6.9}$$

$$\{\vec{E^\mu}, \mathcal{D}[N^\mu]\} = L_N \vec{E^\mu} + \{\vec{E^\mu}, \mathcal{G}[\vec{A}(N)]\}\,, \tag{6.10}$$

where L_N denotes the Lie derivative with respect to the vector field $N^\mu \partial_\mu$. Note that Lie derivatives of densities differ from those of the undensitized quantities by divergence terms in N^μ. The Lie derivative of the connection $\vec{A}_\mu$ is defined componentwise by treating the connection as three one-forms. Note that this definition makes only sense with respect to a global trivialization of the underlying bundle (which we do have here), but not generally. We also set $\vec{A}(N) = \vec{A}_\mu N^\mu$.

The first observation we make is, that due to the simple polynomial form of the Hamiltonian we never had to invert the matrix $\{e^\mu_k\}$ in order to calculate the Hamiltonian flow. This formally gives rise to the possibility of generalized, degenerate solutions, which have no counterpart in the metric formulation.

Formulae (6.7-10) show that $\mathcal{G}$ generates $SO(3)_\mathbb{C}$ frame rotations and that $\mathcal{D}$ generates diffeomorphisms. It might appear that $\mathcal{D}$ also generates frame rotations, if one defines a "pure" diffeomorphism by having only a Lie derivative on the right hand side of (6.9-10). But such a definition does not make unambiguous sense since it is gauge-dependent. In fact, the right hand side of (6.9-10) *is* gauge independent. Nevertheless, with respect to a chosen trivialization, such a definition of a pure diffeomorphism is possible and might even be convenient for applications. In the present case, this amounts to a redefinition of the diffeomorphism generator:

$$H_\mu \mapsto H'_\mu := H_\mu - \vec{A}_\mu \cdot \vec{H}\,, \quad \text{or} \quad \mathcal{D}'[N^\mu] = \mathcal{D}[N^\mu] - \mathcal{G}[\vec{A}(N)] \tag{6.11}$$

$$\text{so that} \quad \{\vec{A}_\mu, \mathcal{D}'[N^\mu]\} = L_N \vec{A}_\mu \tag{6.9'}$$

$$\{\vec{E}^\mu, \mathcal{D}'[N^\mu]\} = L_N \vec{E}^\mu\,. \tag{6.10'}$$

With a little bit of further work one can calculate the Poisson brackets for all constraint functions:

$$\{\mathcal{G}[\vec{\Lambda}], \mathcal{G}[\vec{M}]\} = \mathcal{G}[\vec{\Lambda} \times \vec{M}] \tag{6.12}$$

$$\{\mathcal{G}[\vec{\Lambda}], \mathcal{D}[N^\mu]\} = 0 \tag{6.13}$$

$$\text{or} \quad \{\mathcal{G}[\vec{\Lambda}], \mathcal{D}'[N^\mu]\} = -\mathcal{G}[L_N \vec{\Lambda}] \tag{6.13'}$$

$$\{\mathcal{G}[\Lambda], \mathcal{S}(N)\} = 0 \tag{6.14}$$

$$\{\mathcal{D}[N^\mu], \mathcal{D}[M^\mu]\} = \mathcal{D}[L_N M^\mu] - \mathcal{G}[\vec{F}(N, M)] \tag{6.15}$$

$$\text{or} \quad \{\mathcal{D}'[N^\mu], \mathcal{D}'[M^\mu]\} = \mathcal{D}'[L_N M^\mu] \tag{6.15'}$$

$$\{\mathcal{D}[N^\mu], \mathcal{S}[M]\} = \mathcal{S}[L_N M] + \mathcal{G}[M N^\mu \vec{F}_{\mu\nu} \times \vec{E}^\nu] \tag{6.16}$$

$$\text{or} \quad \{\mathcal{D}'[N^\mu], \mathcal{S}[M]\} = \mathcal{S}[L_N M] \tag{6.16'}$$

$$\{\mathcal{S}[N], \mathcal{S}[M]\} = \mathcal{D}[K^\mu] = \mathcal{D}'[K^\mu] + \mathcal{G}[\vec{A}(K)] \tag{6.17}$$

$$\text{where} \quad K^\mu := \vec{E}^\mu \cdot \vec{E}^\nu (M \partial_\nu N - N \partial_\nu M)\,. \tag{6.18}$$

First of all, we observe that the right sides are all proportional to constraint functions with coefficients that generally depend on phase space. Phase space dependencies appear in (6.15)(6.16) and (6.17). In the first two cases these can be removed by using $\mathcal{D}'$ instead of $\mathcal{D}$ (equations (6.15') and (6.16')). So the only phase space dependence appears in the Poisson bracket of two scalar constraints in the bilinear combination $\vec{E}^\mu \cdot \vec{E}^\nu$ on the right hand side of (6.17). This prevents $\mathcal{S}, \mathcal{D}, \mathcal{G}$ from generating an (infinite dimensional) Lie algebra. For example, we cannot calculate higher Poisson brackets using only the relations (6.12-17), since we would need to know their Poisson-bracket with the term (6.18), which cannot be inferred from the relations (6.12-17). For this we would need to know the explicit functional forms of the generators.

Relation (6.17) also tells us that the scalar constraints do not close among themselves, as is the case with the $\mathcal{G}$ and the $\mathcal{D}'$ generators individually. Moreover, the Poisson brackets for $\mathcal{G}$ with any other generator is proportional to $\mathcal{G}$. We say that $\mathcal{G}$ generates an ideal whereas the $\mathcal{D}'$ only generate a subalgebra (note the abuse of language here). For them to form an ideal too we should have the right hand side of (6.16') to be proportional to a diffeomorphism-, not a scalar-constraint. The geometric meaning of $\mathcal{G}$ generating an ideal is the following: the phase space flow generated by $\mathcal{D}$ (or $\mathcal{D}'$) and $\mathcal{S}$ restricted to the constraint hypersurface $\mathcal{G} = 0$ is tangential to this surface and therefore leaves it invariant. This allows us to reduce the constraints in steps, that is, to *first* go to the $\mathcal{G} = 0$ hypersurface and *then* reduce the action generated by all other constraints. That such a separate phase space reduction for the gauge generators exists is clear from the existence of a formulation of the theory in terms of gauge invariant variables, like the metric formulation. However, a similar procedure with the diffeomorphism constraint is impossible, since its flow does not leave the $\mathcal{S} = 0$ hypersurface invariant. On the other hand, for example, if one were to determine functions on phase space (or subsets thereof) which are invariant under the action of the constraints, it would be sufficient to look for functions annihilated by a generating set of constraints. But due to the intertwinement of diffeomorphism- with scalar-constraints, generating sets much smaller than the defining set may actually be found. For example, functions on the constraint hypersurface $\mathcal{S}[N] = 0 = \mathcal{D}[N^\mu]$, with zero derivative along the Hamiltonian vector fields of $\mathcal{S}[N]$ for all N, have also zero derivative along the Hamiltonian vector fields of $\mathcal{D}[K^\mu]$, due to (6.17-18). But it is easy to see that every vector field on the spatial manifold can locally be written as sum of vector fields of the form (6.18) so that $\mathcal{S}[N]$-invariant functions are automatically $\mathcal{D}[N^\mu]$-invariant. Similar observations of this type where first made in (Moncrief 1972) and (Moncrief and Teitelboim 1972).

Chapter 7. Reality Conditions and Geometric Interpretation

So far we have reformulated complex General Relativity. But, of course, ultimately we are interested in the ordinary real case. A central question therefore is how to recover the real theory within the formalism developed, that is, how to derive a real 3-metric from the complex equations of motion for $\vec{A}_\mu$ and $\vec{E}^\mu$ with real lapse and shift functions N and N^μ. In the real case it is also easy to give some geometric interpretation to the variables $\vec{A}_\mu$ and $\vec{E}^\mu$. We shall briefly discuss these points in this chapter.

As outlined in chapter 3, we are working on complex tensor bundles with reality structure. Real sections have real component functions with respect to real bases and real bases are provided since we work on a real manifold. Note that we only require the 3-metric and not the (densitized) tetrad to be real. The former is derived from the latter according to:

$$h^{\mu\nu} = \frac{1}{h}\vec{E}^\mu \cdot \vec{E}^\nu , \tag{7.1}$$

$$\text{so that} \quad h^{\mu\nu} = \left[\det\{\vec{E}^\kappa \cdot \vec{E}^\delta\}\right]^{-\frac{1}{2}} \vec{E}^\mu \cdot \vec{E}^\nu . \tag{7.2}$$

Suppose now that $\vec{E}^\mu \cdot \vec{E}^\nu$ is real. The Hamiltonian flow should not disturb this reality, that is

$$\{\vec{E}^\mu \cdot \vec{E}^\nu , \mathcal{S}[N] + \mathcal{D}'[N^\mu] + \mathcal{G}[\vec{\Lambda}]\} = \text{real}, \tag{7.3}$$

for all $\vec{\Lambda}$ and all *real* N and N^μ. The gauge generator Poisson-commutes with $\vec{E}^\mu \cdot \vec{E}^\nu$ (compare (6.8)) and therefore drops out of discussion. The diffeomorphisms generator simply yields (compare (6.10'))

$$\{\vec{E}^\mu \cdot \vec{E}^\nu , \mathcal{D}'[N^\mu]\} = L_N \left(\vec{E}^\mu \cdot \vec{E}^\nu\right) \tag{7.4}$$

which is real for real N^μ (the components with respect to the real basis ∂_μ). So the diffeomorphism generator as well does not disturb reality. We can thus focus on the scalar constraint. We have

$$\begin{aligned}\{\vec{E}^\mu \cdot \vec{E}^\nu , \mathcal{S}[N]\} = &\left[\vec{E}^\mu \cdot (\vec{E}^\nu \times \vec{E}^\sigma) + \vec{E}^\nu \cdot (\vec{E}^\mu \times \vec{E}^\sigma)\right] \partial_\sigma N \\ &+ N\,(D_\sigma \vec{E}^\sigma) \cdot (\vec{E}^\mu \times \vec{E}^\nu + \vec{E}^\nu \times \vec{E}^\mu) \\ &+ N \left[\vec{E}^\sigma \cdot (\vec{E}^\mu \times D_\sigma \vec{E}^\nu + \vec{E}^\nu \times D_\sigma \vec{E}^\mu)\right] .\end{aligned} \tag{7.5}$$

The first two expressions vanish identically, so that the resulting reality condition is:

$$\left(\vec{E}^\sigma \times D_\sigma \vec{E}^{(\mu}\right) \cdot \vec{E}^{\nu)} = \text{real}. \tag{7.6}$$

One may check that there are no further constraints, i.e., the Poisson bracket of this expression with the Hamiltonian is real due to the reality conditions already posed. So we have the result that the Hamiltonian flow preserves reality provided the initial data are further constrained by the reality of $h^{\mu\nu}$ and (7.6).

Now we turn to the geometric interpretation of the canonical variables. We begin with the momenta $\vec{E}^{\mu}$. We write (7.1) in the form

$$\det\{h_{\kappa\delta}\}h^{\mu\nu} = \vec{E}^{\mu}\cdot\vec{E}^{\nu} \tag{7.7}$$

and consider e.g. the 3-3 component of this equation. The left hand side is just the subdeterminant $h_{11}h_{22} - h_{12}^2$ which for real $h_{\mu\nu}$ measures the area content of the parallelogram in tangent-space spanned by the vectors ∂_1 and ∂_2. The area content of an embedded 2-surface (coordinatized by σ and τ):

$$(\sigma,\tau) \mapsto x^{\mu}(\sigma,\tau)\,, \tag{7.8}$$

is thus given by (no summation convention here):

$$\int \sum_{\lambda,\mu,\nu} \sqrt{\vec{E}^{\lambda}\cdot\vec{E}^{\lambda}}\;\varepsilon_{\lambda\mu\nu}\frac{\partial x^{\mu}}{\partial\tau}\frac{\partial x^{\nu}}{\partial\sigma}\,d\tau d\sigma\,, \tag{7.9}$$

where the integral is taken over the surface. $\varepsilon_{\lambda\mu\nu}$ denotes the 3-form density on Σ of weight -1, whose components in *all* coordinate systems are 0 if two indices coincide and ± 1 if they form an even, respectively odd permutation of 123. In that sense the variables $\vec{E}^{\mu}$ are geometrically connected to areas rather than length.

Next we consider the variables $\vec{A}_{\mu}$. For this, we restore our old notation using ω for the connection 1-form and ${}^{(+)}\omega$ for the self-dual part, and have

$$A^k_{\mu} = 2\,{}^{(+)}\omega^k_{\mu} = -i\tfrac{1}{2}\epsilon^{\perp k}{}_{lm}(\omega^{lm}_{\mu} - i\epsilon^{lm}{}_{\perp p}\omega^{\perp p}_{\mu})\,. \tag{7.10}$$

If the equations of motion are satisfied we have $\omega = \Gamma$ (the Levi-Civita connection). Let its spacetime covariant derivative be ${}^{(4)}\nabla$ and ${}^{(3)}\nabla$ its induced space covariant derivative on Σ (compatible with $h_{\mu\nu}$). We recall the definition of extrinsic curvature:

$${}^{(4)}\nabla_{e_k}e_l =: {}^{(3)}\nabla_{e_k}e_l + K_{kl}\,e_{\perp} \tag{7.11}$$

$$\text{and have}\quad \omega^{\perp k}_{\mu} = K^k_{\mu} = \delta^{kl}e^i_{\mu}K_{il}\,. \tag{7.12}$$

Inserting this into (7.10) yields

$$A^k_{\mu} = \Gamma^k_{\mu} + K^k_{\mu}\,, \tag{7.13}$$

$$\text{where}\quad \Gamma^k_{\mu} := -i\tfrac{1}{2}\epsilon^{\perp k}{}_{ln}\Gamma^{ln}_{\mu}\,. \tag{7.14}$$

The spatial components, Γ_μ^{ln}, are of course just the components for the Levi-Civita connection on Σ compatible with $h_{\mu\nu}$. Relation (7.13) clearly shows the non-triviality of the transformation to the connection variables: the connection 1-form $\vec{A}$ involves the metric and the extrinsic curvature and therefore both conjugate sets of variables of the metric formulation. We also note that (7.13) gives rise to an alternative way of expressing a reality condition which proves useful in a variety of applications, in particular when compared with the standard metric formulation. The reality of K_μ^k and Γ_μ^{ln} (hence Γ_μ^k is purely imaginary) implies (a bar denotes complex conjugation)

$$A_\mu^k - \bar{A}_\mu^k = 2\Gamma_\mu^k . \tag{7.15}$$

Finally, we point out some further interesting decomposition of the connection 1-form $\vec{A}$. First of all, using the dual basis $\{\theta^k\}$ of $\{e_k\}$ on Σ (as introduced before equation (4.2)), we have $A^k = A_\mu^k dx^\mu = A_l^k\theta^l$. The latin indices are now raised and lowered using δ_{kl}, ie., $A_{kl} = \delta_{kn}A_l^n$. Correspondingly, (7.13) reads: $A_{kl} = \Gamma_{kl} + K_{kl}$. Due to the symmetry of K_{kl} (which follows from (7.11) and the vanishing torsions of ${}^{(4)}\nabla$ and ${}^{(3)}\nabla$) the antisymmetric part of A_{kl} just involves Γ_{kl}. Defining the quantity $\Gamma_l := \frac{i}{2}\epsilon_l{}^{nm}\Gamma_{nm}$, which is real, we have

$$A_{kl} = A_{(kl)} - i\epsilon_{kl}{}^n\Gamma_n = \Gamma_{(kl)} + K_{kl} - i\epsilon_{kl}{}^n\Gamma_n . \tag{7.16}$$

We further set

$$A := \delta^{kl}A_{kl} = \delta^{kl}\Gamma_{(kl)} + \delta^{kl}K_{kl} =: \Gamma + K , \tag{7.17}$$

where K, the trace of the extrinsic curvature, is zero, iff Σ is a maximal hypersurface (i.e. a stationary point of the area functional). Using the first Cartan structure equation for the basis $\{\theta^k\}$ and the torsion-free Levi-Civita connection on Σ, one easily shows the following identities:

$$\Gamma_k\theta^k = \tfrac{1}{2} * \delta_{kl}(\theta^k \wedge *d\theta^l) , \tag{7.18}$$

$$\Gamma = \tfrac{1}{2}i * \delta_{kl}(\theta^k \wedge d\theta^l) , \tag{7.19}$$

where $*$ is now the 3-dimensional Hodge-duality map on Σ (defined analogously to (1.20)). $\Gamma_k\theta^k$ and Γ are a *frame dependent* 1-form and a *frame dependent* function respectively (recall the transformation law for connections (1.19)) and have been employed in (Nester 1989) to fix the gauge freedom in the choice of frames by imposing the gauge conditions:

$$d(\Gamma_k\theta^k) = 0 , \tag{7.20}$$

$$d\Gamma = 0 . \tag{7.21}$$

If we insert the Levi-Civita connection with respect to an arbitrary rotated frame by using formula (1.19) to parameterize it by the rotation matrix R, these conditions can be seen to form a system of non-linear elliptic differential equations for R which under certain circumstances fixes it up to a constant rotation. Since we assume Σ to be connected, (7.20) is clearly equivalent to $\Gamma = const.$. If we make the additional assumption that Σ has trivial first DeRahm Cohomology (equivalently, the space of harmonic 1-forms is zero dimensional. A sufficient condition for this is that Σ has finite fundamental group.), then all closed 1-forms are exact so that $\Gamma_k = e_k(\rho)$ for some globally defined function ρ on Σ. In this integrated form the gauge conditions have been shown to be equivalent to a massive 2-spinor equation (the mass being given by the value Γ), which is *linear* and elliptic (Dimakis and Müller-Hoissen 1989). For asymptotically flat Σ, the fall-off conditions require $\Gamma = 0$, in which case existence and uniqueness of the massless spinor equation has been demonstrated, up to possible isolated singularities, in (Ashtekar and Horowitz 1983). For a neighborhood of flat 3-space, existence and uniqueness can be shown directly (Nester 1989, also Nester 1992).

Another interesting question that plays a significant rôle in the investigation of the initial data problem is, what spacetimes can be evolved from maximal (i.e. $K = 0$) hypersurfaces Σ. It has been shown (Bartnik 1984) that all spacetimes sufficiently close (in a sense specified by Bartnik) to Minkowski space allow such maximal slicings. Conversely, it is also known that there are even topological restrictions on Σ in order to allow for any maximal initial data (Witt 1987). The manifolds ruled out are those which do not allow for metrics with everywhere positive Ricci scalar (as is directly seen from the Hamiltonian constraint in the standard metric representation). They all have infinite fundamental group.

At the end of the next section we will show how the special gauge conditions (7.20-21) can be used to prove a restricted version of the positive mass theorem which applies to maximal Σ (i.e. spacetimes admitting such hypersurfaces).

Chapter 8. Surface Integrals

If the spatial manifold Σ is not closed one has to specify fall-off conditions for the fields $\vec{A}_\mu$ and $\vec{E}^\mu$. These must be chosen wide enough to accommodate physically interesting solutions, with, say, non-vanishing energy, momentum and angular momentum etc. at spatial infinity (we call these quantities the Poincaré charges). In this case the Hamiltonian must be amended by surface integrals in order to make it a continuously differentiable function on phase space. Note that the point is not that otherwise we would get the wrong equations of motion, but rather that we would not get any equations at all. Also, the variational principle needs a careful restatement in order to ensure stationary points in the right class of functions. However, in order to arrive at the right Hamiltonian it is not necessary to restart from a modified variational principle to calculate the modified Hamiltonian by keeping all non-vanishing boundary integrals. A much simpler strategy is to first calculate a formal Hamiltonian "by brute force" without keeping track of any surface integrals, and then amend it "by hand" with the right surface terms that make it a continuously differentiable function. For the metric formulation of General Relativity this is explained and carried through in (Regge and Teitelboim 1979) and in more detail in (Beig and Ó Murchadha 1987), which also corrects a mistake in the first reference concerning the generator of boosts. Using the connection variables, an analysis similar to the one by Beig and Ó Murchadha was performed in (Glößner 1992) and also in (Thiemann 1993). An alternative approach not within the canonical framework is presented in (Ashtekar and Romano 1993). Here we cannot give the full details of the general procedure but only want to outline the underlying ideas.

We already have the (formal) Hamiltonian without the surface terms. Let us restrict to $\vec{\Lambda} = 0 = N^\mu$ so that the (formal) Hamiltonian equals the scalar constraint:

$$\mathcal{S}[N] = \tfrac{1}{2} \int_\Sigma N \vec{F}_{\mu\nu} \cdot (\vec{E}^\mu \times \vec{E}^\nu)\, d^3x\,. \tag{8.1}$$

This expression is clearly differentiable with respect to $\vec{E}^\mu$ (since it enters undifferentiated), but if we try to differentiate it with respect to $\vec{A}_\mu$ we encounter the surface integral

$$-\int_{S_\infty} N\, \vec{A}_\mu \cdot (\vec{E}^\mu \times \vec{E}^\nu)\, d\sigma_\nu\,, \tag{8.2}$$

where S_∞ is the 2-sphere at spatial infinity. The integral over it is understood in the usual way: take the integral over 2-spheres in the finite, such that the result depends on the radius, and *then* take the limit as the radius approaches infinity. Here we assume of course that it be finite. Now, if this surface term is not zero, the functional $S[N]$ is not differentiable with respect to $\vec{A}_\mu$. We restore differentiability by simply subtracting (8.2) from (8.1) and obtain:

$$\mathcal{H}_s[N] := \tfrac{1}{2}\int_\Sigma N\vec{F}_{\mu\nu}\cdot(\vec{E^\mu}\times\vec{E^\nu})\,d^3x + \int_{S_\infty} N\,\vec{A}_\mu\cdot(\vec{E^\mu}\times\vec{E^\nu})\,d\sigma_\nu\,, \quad (8.3)$$

which we call the scalar Hamiltonian. The constraints are still given by (6.1), or, in smeared form, by the requirement that the first term in (8.3) vanishes. On the constraint surface the scalar Hamiltonian is thus non-vanishing and assumes the values of minus the surface integral (8.2). In the non-closed case the Hamiltonian is not the sum of the constraints and does not vanish on the constraint surface. Rather, it defines non-trivial functions which descend to functions on the reduced phase space, i.e., to non-trivial observables. For the case at hand one expects this quantity to be the total energy of the isolated system described by asymptotically flat data, and one may wonder whether the quantity above is actually real. This can be demonstrated quite easily. But to do so, we have to first make slightly more precise what we define as asymptotically flat data. Asymptotically we require $N(r\to\infty)\to const.$ (so we leave out boosts from the discussion given here since for them $N\propto x^\mu$ for large r.)

$$E^\mu_k = \delta^\mu_k + \frac{\alpha^\mu_k(\theta,\varphi)}{r} + O\left(\frac{1}{r}\right)\,, \quad (8.4)$$

$$A^k_\mu = \frac{\beta^k_\mu(\theta,\varphi)}{r^2} + O\left(\frac{1}{r^2}\right)\,, \quad (8.5)$$

where α^μ_k and β^k_μ are functions on the 2-sphere which also obey the conditions to be of even-, respectively odd parity under the antipodal map of the sphere. We refer to (Beig, Ó Murchadha 1987) for a discussion of these conditions in the metric case. $O(\frac{1}{r^n})$ stands for terms with asymptotic radial fall-off faster than $\propto 1/r^n$. Using formulae (7.13-14) we can write the surface contribution to (8.3) "on shell" (i.e., the equations of motion being satisfied) in the form

$$\int_{S_\infty} N\,E^\mu_k E^\nu_l(\Gamma^{kl}_\nu - i\epsilon^{kl}{}_{\perp p}K^p_\nu)\,d\sigma_\mu\,. \quad (8.6)$$

In a real basis the first term is real and the second purely imaginary. But the second vanishes due to the symmetry of K_{kl} and the asymptotic conditions (8.4). The first term can be shown to be just one-half times the ADM energy expression (which means that the properly normalized action is twice ours in units where $16\pi G/c^4 = 1$, as was already remarked in the discussion following equation (3.21)).

Let us also have a look at the diffeomorphism constraint. In the same way as above one immediately sees that differentiability with respect to $\vec{A}_\mu$ is obstructed by the surface integral

$$-2\int_{S_\infty} N^{[\mu}\vec{E}^{\nu]}\cdot\vec{A}_\mu \, d\sigma_\nu \,, \tag{8.7}$$

which we must subtract to obtain the differentiable diffeomorphism-Hamiltonian

$$\mathcal{H}_d[N^\mu] := \int_\Sigma N^\mu \vec{F}_{\mu\nu}\vec{E}^\nu \, d^3x + 2\int_{S_\infty} N^{[\mu}\vec{E}^{\nu]}\cdot\vec{A}_\mu \, d\sigma_\nu \,. \tag{8.8}$$

Note that due to the parity conditions on β^k_μ the surface integral is also finite for asymptotic rotations, where $N^\mu \propto \varepsilon_{\mu\nu\lambda}n^\nu x^\lambda$ with constant n^μ. "On shell", (8.8) becomes

$$\mathcal{H}_d[N^\mu] = 2\int_{S_\infty} (K^k_\nu - i\epsilon^{\perp k}{}_{lm}\Gamma^{lm}_\nu)N^{[\mu}E^{\nu]}_k \, d\sigma_\mu \,. \tag{8.9}$$

Again it can be shown that the imaginary part vanishes. The real part is

$$-\int_{S_\infty} N^\nu (K^\mu_\nu - \delta^\mu_\nu K^\lambda_\lambda) \, d\sigma_\mu \tag{8.10}$$

which is again one-half times the ADM-expression for momentum and angular momentum.

Let us finally mention the gauge constraint. Here the only surface contribution comes from the requirement of differentiability with respect to $\vec{E}^\mu$ and is given by:

$$\int_{S_\infty} \vec{\Lambda}\cdot\vec{E}^\mu \, d\sigma_\mu \,. \tag{8.11}$$

But here one probably does not want to have an extra charge associated with the asymptotic gauge (frame) rotations for which the following fall-off is sufficient (n_μ is the normal to the asymptotic 2-sphere)

$$\Lambda^k \delta^\mu_k n_\mu \propto O\left(\frac{1}{r^2}\right) \tag{8.12}$$

Essentially the same condition follows, for example, from the requirement of invariance of angular momentum under gauge transformations, i.e., frame rotations. Since the Poincaré charges are given by integrals with non manifestly gauge-invariant integrands (due to the explicit apparence of $\vec{A}_\mu$) one generally has to explicitly check compatibility of the chosen fall-off conditions with the requirement of gauge invariance.

We remark that it is indeed possible to write down surface integrals that make the Hamiltonian continuously differentiable and where the lapse and shift function spaces are chosen wide enough to accommodate for all asymptotic Poincaré charges. As expected, it turns out that the boost generator with asymptotic behaviour $N \propto x^\mu$ requires the most subtle discussion to establish finiteness and differentiability. This is as in the metric case – compare (Beig and Ó Murchadha) –, and has in fact a similar solution (Glößner 1992). There is thus an important difference between the closed and open case. In the former, all of dynamics is entirely generated by constraints, whereas in the latter the constraints generate only asymptotically (at spatial infinity) trivial changes, in contrast to the Hamiltonian, which also generates dynamics of quantities at infinity. No effective changes remain in the closed case after having divided out those generated by the constraints. In contrast, in the asymptotically flat case, there are still residual motions which are generated by the Poincaré group.

We want to end this section by demonstrating positivity of the surface integral in (8.3) for the special cases where Σ is maximal (i.e. K=0). In doing this we essentially follow (Nester 1994) with some adaptions in notation. The idea is to rewrite the scalar Hamiltonian (8.3) into a form in which positivity properties can be deduced, possibly by suitably choosing the lapse density N in the interior of Σ. But since on the constraint set the surface integral equals in value the scalar Hamiltonian and is independent of the lapse density in the interior of Σ, this will also show positivity of the surface integral. We start by simply rewriting the scalar Hamiltonian (8.3) in terms of forms by using the dual basis $\{\theta^k\}$, equation (4.19) and $\sqrt{h}d^3x = \theta^1 \wedge \theta^2 \wedge \theta^3$:

$$\mathcal{H}_s[N] = i\int_\Sigma N\sqrt{h}\delta_{kl}F^k \wedge \theta^l - i\int_{S_\infty} N\sqrt{h}\delta_{kl}A^k \wedge \theta^l\,. \tag{8.13}$$

Note that $N\sqrt{h}$ is a scalar function and equal to the lapse before the redefinition (4.21) was made. Converting the surface integral into a volume integral and expanding the forms in terms of θ^ks yields

$$\begin{aligned}\mathcal{H}_s[N] =& \frac{1}{2}\int_\Sigma N\sqrt{h}\left(A^{kl}A_{lk} - A^2\right)\sqrt{h}d^3x \\ &- \int_\Sigma N\sqrt{h}\left(A^{kl}\Gamma_{lk} - A\Gamma\right)\sqrt{h}d^3x \\ &- \int_\Sigma \epsilon^{klm}e_k(N\sqrt{h})\, iA_{lm}\sqrt{h}d^3x\end{aligned} \tag{8.14}$$

using the notation defined at the end of the previous section. We now impose the reality condition (7.15), just transcribed to the components A_{kl} with respect to the real basis $\{\theta^k\}$, and decompose A_{kl} into symmetric and antisymmetric part according to (7.16) and obtain the real expression:

$$\begin{aligned}\mathcal{H}_s[N] =& \frac{1}{2}\int_\Sigma N\sqrt{h}\left(\bar{A}^{(kl)}A_{(kl)} - \bar{A}A\right)\sqrt{h}d^3x \\ &+ \int_\Sigma N\sqrt{h}\Gamma_k\Gamma^k\sqrt{h}d^3x \\ &-2\int_\Sigma e_k(N\sqrt{h})\Gamma^k\sqrt{h}d^3x\,.\end{aligned} \tag{8.15}$$

Finally, we impose the special orthonormal frame gauge (7.20-21) in integrated form: $\Gamma = 0$ and $\Gamma_k = 2e_k(\ln(\lambda))$, where the way of writing the gradient is just chosen for convenience with an everywhere positive and asymptotically constant function λ. We then arrive at

$$\begin{aligned}H_s[N] =& \frac{1}{2}\int_\Sigma N\sqrt{h}\left(\bar{A}^{(kl)}A_{(kl)} - K^2\right)\sqrt{h}d^3x \\ &- 4\int_\Sigma \delta^{kl}e_k\left(\frac{N\sqrt{h}}{\lambda}\right)e_l(\lambda)\sqrt{h}d^3x\,.\end{aligned} \tag{8.16}$$

The second integral can be made to vanish with the choice $N\sqrt{h} = \lambda$, in which case the Hamiltonian simply reads

$$\mathcal{H}_s[\lambda] = \frac{1}{2}\int_\Sigma \lambda\left(\bar{A}^{(kl)}A_{(kl)} - K^2\right)\sqrt{h}d^3x\,. \tag{8.17}$$

For maximal hypersurfaces Σ one has $K = 0$ and the integrand is non-negative and zero only for $A_{(kl)} = 0$, which implies (real and imaginary part vanish separately) $\Gamma_{(kl)} = 0$ and $K_{kl} = 0$. Geometrically, the latter equation just says that Σ is totally geodesic.

References

Ashtekar, A. (1991). Lectures on Non-Perturbative Canonical Gravity. Advanced Series in Astrophysics and Cosmology, Vol. 6, World Scientific, Singapore - New Jersey - London - Hong Kong.

Ashtekar, A., Horowitz, G. (1983). J. Math. Phys., **25**, 1473.

Ashtekar, A., Romano, J.D. (1992). Class. Quant. Grav., **9**, 1069.

Bartnik, R. (1984). The Existence of Maximal Surfaces in Asymptotically Flat Spacetimes. In: Asymptotic Behaviour of Mass and Spacetime Geometry, F. Flaherty (ed.), Lecture Notes in Physics, 202, pp 57-60, Springer Verlag.

Beig, R., Ó Murchadha, N. (1987). Ann. Phys. (N.Y.), **174**, 463.

Brügmann, B. (1993). Bibliography of Publications related to Classical and Quantum Gravity in terms of Ashtekar Variables. Los Alamos bulletin-board preprint gr-qc/9303015.

Dimakis, A., Müller-Hoissen, F. (1989). Phys. Lett. A, **142**, 73.

Dirac, P.A.M. (1967). Lectures on Quantum Mechanics. Yeshiva University, New York: Academic Press.

Glößner, P. (1992). Über die Phasenraumconstraints der Algemeinen Relativitätstheorie in metrischer Formulierung und in der Zusammenhangsdarstellung Ashtekars. Diploma Thesis, Freiburg University, December 1992.

Gotay, M.J., Nester, J.M., Hinds, G. (1978). J. Math. Phys **19**, 2388.

Hanson, A., Regge, T., Teitelboim, C. (1976). Constrained Hamiltonian Systems. Rome: Acad. Naz. dei Lincei.

Henneaux, M., Teitelboim, C. (1992). Quantizations of Gauge Systems. Princeton University Press, Princeton, New Jersey.

Marmo, G., Mukunda, N., Samuel, J. (1983). Revista del Nuovo Cimento **6**, 1.

Moncrief, V. (1972). Phys. Rev. D, **5**, 277.

Moncrief, V., Teitelboim, C. (1972). Phys. Rev. D, **6**, 966.

Nester, J. (1989). J. Math. Phys., **30**, 624.

Nester, J. (1992). J. Math. Phys., **33**, 910.

Nester, J. (1994). Los Alamos bulletin-board preprint gr-qc/9401004.

Regge, T., Teitelboim, C. (1979). Ann. Phys. (N.Y.), **88**, 286.

Romano, J.D. (1993). Gen. Rel. Grav. **25**, 759.

Thiemann, T. (1993). Los Alamos bulletin-board preprint gr-qc/9310039.

Witt, D. (1987). Toplogical Obstructions to Maximal Slices. Santa Barbara Preprint, UCSB-1987.

Quantization of Systems with Constraints

Petr Hájiček

1 Introduction

Models used in contemporary theoretical physics are typically systems with constraints. Methods of quantization of these models are especially interesting for those who try to quantize gravity. Even in this particular field, the literature is enormous (some recent reviews are [1], [2] and [3]). Thus, we have to select. First, we just try to explain two approaches to the Wheeler-DeWitt equation that have gained certain popularity in the recent decade. Chapter 2 will be devoted to the so-called "algebraic method", as developed by A. Ashtekar and his collaborators ([4], [5], [6] and [7]), and Chapter 3 to the so-called "Schrödinger picture method". (The latter method is older than the former; it was not given any name because it was the only one in the field for some time, but now, we have to choose at least a provisional name to distinguish it from the former.) The Schrödinger picture method was exposed very clearly by K. Kuchař [8]; a rigorous formulation of it has been suggested in [7], another one given in [9]. Second, we restrict ourselves to finite dimensional models (that is, "mechanical", as opposed to "field-theoretical" ones).

2 The Algebraic Approach to the Wheeler-DeWitt Equation

2.1 Introduction

The present chapter was originally planned as a pedagogical introduction to the algebraic quantization method. That's why everything is explained with the help of a quite trivial model: the free, relativistic, massive particle in parametrized form. The simplicity of the model justified the hope that the construction of the quantum mechanics can be completed so that a Hilbert space with a full system of observables and a unitary time evolution will result. This insisting on completeness brought, however, unexpected fruits. We will find precise structures underlying the physical space (rigged Hilbert spaces), scalar product (generalized Green's identity), and time evolution (Dirac's forms of relativistic dynamics; choice of Hamiltonian). We will find that different time choices lead, in general, to unitarily inequivalent quantum mechanics. These mechanics are, however, compatible in the sense that common observables are represented in a unitarily equivalent way. This means that the physical questions which can be sensibly asked within each of these inequivalent mechanics will have the same answers. The non-equivalence concerns only the time evolution: in different mechanics, the time evolution is generated by different observables. Let us observe that we will manage, for this simple model, to solve all factor order problems in the ideal way: all (relevant) commutators will exactly reflect the corresponding classical Poisson brackets.Thus, the non-equivalence of the different time evolutions exists already within the classical theory.

We have also tried to suggest how all these ideas could be generalized to less trivial systems. Here, however, much work remains to be done.

2.2 Basical Properties of the Model

Consider the action functional

$$S = \int d\tau (p_\mu \dot{x}^\mu - NC).$$

Here, inertial system coordinates x^μ, $\mu = 0, 1, 2, 3$, span the configuration space $\mathcal{C} \cong \mathbf{R}^4$ of the system, p_μ are the conjugate momenta, the phase space Γ is given by $T^*\mathcal{C}$,

$$C = X(-p_0^2 + \mathbf{p}^2 + m^2), \tag{1}$$

is a super-Hamiltonian constraint,

$$\mathbf{p}^2 := p_1^2 + p_2^2 + p_3^2,$$

$m > 0$ is the mass, and N a Lagrange multiplier. The super-Hamiltonian is determined only up to an arbitrary (differentiable) non-zero factor X,

$$X : \Gamma \to \mathbf{R} \setminus \{0\}.$$

The constraint surface $\tilde{\Gamma} = \{C = 0\} \subset \Gamma$ is seven dimensional and consists of two components, $\tilde{\Gamma}_\pm$, defined by $p_0 = \mp\omega(p)$, respectively, where

$$\omega(\mathbf{p}) = \sqrt{\mathbf{p}^2 + m^2}.$$

The general solution of the equations of motion can be written as follows

$$P_0(x^k - X^k) + P_k(x^0 - X^0) \ = 0, \ \ k = 1,2,3 \tag{2}$$

$$p_\mu = P_\mu \quad , \quad \mu = 0,1,2,3. \tag{3}$$

Here X^μ and P_μ are constant and satisfy the constraint

$$-P_0^2 + \mathbf{P}^2 + m^2 = 0.$$

In eqs. (2) and (3), we have excluded the arbitrary parameter τ on the solution curves, so that they determine just the solution *arc* in $\tilde{\Gamma}$. There is a unique solution arc through any point of $\tilde{\Gamma}$, whereas the parametrization by τ is arbitrary ("parametrized systems"), since it depends on the arbitrary function X.

Observe that we have "switched off" the usual condition $P_0 < 0$ that must be satisfied by our particle. This condition has no analogy in cosmology! It can, however, be "switched on" again later, if desired.

The set of all C^∞-functions on Γ forms a so-called Poisson algebra $\mathcal{P}$: the set is closed with respect to operation of linear combination and product of functions as well as of Poisson brackets (PB). Those elements of $\mathcal{P}$ that have weakly vanishing PB's with the constraint,

$$\{f, C\} = gC \quad , \quad g \in \mathcal{P},$$

form a Poisson subalgebra $\tilde{\mathcal{P}}$. An important subset of $\tilde{\mathcal{P}}$ is the following

$$p_\mu, \tag{4a}$$

$$J_l = \epsilon_{lmn} x^m p_n, \tag{4b}$$

$$K_l = x^0 p_l + x^l p_0. \tag{4c}$$

They generate, via PB's, infinitesimal Poincaré transformations in Γ.

The constraints themselves form an ideal $\mathcal{J}$ of $\tilde{\mathcal{P}}$:

$$\mathcal{J} := \{f \in \mathcal{P} \mid f = hC, h \in \mathcal{P}\}.$$

The classes of equivalence in $\tilde{\mathcal{P}}$ with respect to $\mathcal{J}$ are called *observables.* For example, p_0 is equivalent to a function $\Omega(p)$ satisfying

$$\Omega(p) = -\omega(\mathbf{p}) \quad p_0 < 0,$$

$$\Omega(p) = +\omega(\mathbf{p}) \quad p_0 > 0,$$

because

$$p_0 = \mp\omega(\mathbf{p}) \quad \text{on} \quad \tilde{\Gamma}_\pm.$$

We will denote the class of a function $o\epsilon\tilde{\mathcal{P}}$ by $\{o\}$. The classes form themselves a Poisson algebra, $\mathcal{P}_\sigma$; the operations for classes are defined by those of their representatives (see [10] for details). We can consider the functions (4) as a particular choice of representatives in the equivalence classes of Poincaré generators.

2.3 Dirac's Forms of Relativistic Dynamics

In this section, we explain some old ideas by Dirac [11] and use them to define the time evolution.

2.3.1 The General Theory

Everything starts with a choice of a particular hypersurface $\mathcal{S} \subset \Gamma$, defined by the equation $T = 0$, where T is a phase function. We assume that:

a) every solution arc γ has exactly one intersection with $\mathcal{S}$,

b) at this intersection, $\mathcal{S}$ is transversal to γ.

Consider the group $\mathcal{G}$ of symplectic transformations in Γ generated by a chosen set of observable representatives. $\mathcal{G}$ is a symmetry group of the system, because it leaves $\tilde{\Gamma}$ invariant. For our model, we choose for $\mathcal{G}$ the Poincaré group generated by (4).

Applying elements of $\mathcal{G}$ to $\mathcal{S}$, we obtain a family $\mathcal{T}$ of surfaces in Γ that satisfy the conditions a) and b). $\mathcal{T}$ forms a manifold diffeomorphic to $\mathcal{G}/\mathcal{G}_\mathcal{S}$, where $\mathcal{G}_\mathcal{S}$ is the stability group of $\mathcal{S}$. In general, $\mathcal{T}$ is not a foliation.

Each element $\mathcal{S}'$ of $\mathcal{T}$ is a "Heracleitian time instant" (see, e.g. [12]), because the classical system must be at exactly one point of $\mathcal{S}' \cap \tilde{\Gamma}$. Hence $\mathcal{S}' \cap \tilde{\Gamma}$ is a complete system of independent events. Probability distributions along $\mathcal{S}' \cap \tilde{\Gamma}$ are normalizable, etc.

Next, Dirac reduces the classical theory, defining what I will call "Dirac's representatives of observables". Let $\{o\}$ be an observable and $o \in \{o\}$. Let us define

$$o_D = o\,|_{\mathcal{S}\cap\tilde{\Gamma}},$$

Dirac's representative of $\{o\}$. It is a function on the submanifold $\mathcal{S} \cap \tilde{\Gamma}$ of Γ that is independent of the choice $o \in \{o\}$.

From the property b) of $\mathcal{S}$, it follows immediately that the pull-back of the symplectic form to $\mathcal{S} \cap \tilde{\Gamma}$ is a symplectic form on $\mathcal{S} \cap \tilde{\Gamma}$. Let us denote the corresponding PB on $\mathcal{S} \cap \tilde{\Gamma}$ by $\{.,.\}'$. Then, one can show the following theorem.

Theorem 1: Let $\{o_1\}$ and $\{o_2\}$ be two observables.

Then,

$$\{o_1, o_2\}_D = \{o_{1D}, o_{2D}\}'.$$

Calculating Dirac's representatives of the generators of $\mathcal{G}$, Dirac observed that some generators – namely the generators of $\mathcal{G}_\mathcal{S}$ – give "simple" expressions for their representatives, while the remaining generators lead to "complicated" Dirac representatives. Dirac calls the latter generators *"Hamiltonians"*, as they generate "motion" between different elements of $\mathcal{T}$.

In order to make the theory as simple as possible, Dirac proposes to choose $\mathcal{S}$ so that $\mathcal{G}_\mathcal{S}$ will be a maximal subgroup of $\mathcal{G}$ (there are other reasons to make $\mathcal{G}_\mathcal{S}$ as large as possible, see later). In this way, he finds just "three forms of relativistic dynamics" for our model:

1) instant form, $T = x^0$,
2) front form, $T = \frac{1}{\sqrt{2}}(x^0 - x^3)$,
3) point form, $T = (x^0)^2 - \mathbf{x}^2 - 1$.

Dirac's ideas suggest an approach to an old problem: that of time evolution. Indeed as all observables in a parametrized theory have to be constant along dynamical trajectories, it seems that the "dynamics is frozen". Trying to get some life into this rigid world, Rovelli introduced the so-called "evolving constants of motion" [13]. A different way, which will also lead just to the usual Heisenberg picture, is based on a choice of a particular Hamiltonian, H, say, and so define a "time direction" ξ_H (Hamiltonian vector field for the phase space function H) which is in general different from the "direction of motion" ξ_C. That is, we *define* the time derivative $\frac{\partial o}{\partial t}$ of any observable o by

$$\frac{\partial o}{\partial t} = \{o, H\}. \tag{5}$$

The geometrical meaning of $\frac{\partial o}{\partial t}$ can be clarified, if we introduce the one-dimensional subgroup $h(t)$ of $\mathcal{G}$ that is generated by H. Then, we have a one-dimensional family $h(t)\mathcal{S}$ of surfaces in $\mathcal{T}$. Let o be an observable; we can compare the values of o at different "times",

$$o\,|_{\mathcal{S}\cap\tilde{\Gamma}}\ ,\quad o\,|_{h(t)\mathcal{S}\cap\tilde{\Gamma}},$$

only if "corresponding" pair of points, $p \in \mathcal{S} \cap \tilde{\Gamma}$ and $q \in h(t)\mathcal{S} \cap \tilde{\Gamma}$ are given. If $q \in \gamma_p$, the dynamical orbit through p, then there is no change. What we do is just to choose

$$q = h(t)p.$$

Hence, "time" seems to comprise two independent structures: the time surface $\mathcal{S}$, and the time direction H.

Finally, we can use the time surface $\mathcal{S}$ to gain some important additional observables (like particle position, etc.). For that we have to "extend" the dynamics (quite arbitrarily) to a whole neighbourhood of $\tilde{\Gamma}$. This can at best be done by choosing a particular constraint function C (factor X in eq. (1)). Then, there is a whole neighbourhood U of $\tilde{\Gamma}$ in Γ such that $\mathcal{S}$ is transversal to the orbits of ξ_C in U and U is generated from $\mathcal{S} \cap U$ by the flow of ξ_C. Suppose that $f : \Gamma \to \mathbf{R}$ is an arbitrary phase function, not necessarily an observable. We define an observable $f_{\mathcal{S}} : U \to \mathbf{R}$ as follows. Let $p \in U$. Then, there is a unique orbit γ_p of ξ_C through p, intersecting $\mathcal{S}$, and

$$f_{\mathcal{S}}(p) := f(\mathcal{S} \cap \gamma_p).$$

Clearly, $f_{\mathcal{S}}$ has strongly vanishing PB with C and weakly vanishing PB with any function (1). Moreover, the restriction of $f_{\mathcal{S}}$ to $\tilde{\Gamma}$ does not depend on the choice of C. Thus, we do have a unique observable. There is a relation of these observables to the "evolving constants of motion": $f_{\mathcal{S}}$ is just an element of such a family of "constants" corresponding to time $\mathcal{S}$. The time evolution of $f_{\mathcal{S}}$ will be, however, obtained from eq. (5) in our theory. Still, we call $f_{\mathcal{S}}$ "Rovelli type observable".

We will implement all these ideas in quantum theory. Let us, however, first calculate how they work in simple examples.

2.3.2 The Instant Form

Here, $T = x^0$, and $\mathcal{G}_S$ has six generators, p_l and J_l, say, $l = 1,2,3$. We have four independent Hamiltonians, p_0 and $K_l, l = 1,2,3$, say.

Let us choose the functions $x^\mu, \mu = 0,1,2,3$, and $p_l, l = 1,2,3$, as coordinates on $\tilde{\Gamma}$. Then, Dirac's representatives of the Poincaré generators are given by

$$p_{lD} = p_l, J_{lD} = J_l$$

(see formulas (4)),

$$p_{0D} = \begin{pmatrix} -\omega(\mathbf{p}) \\ +\omega(\mathbf{p}) \end{pmatrix}$$

$$K_{lD} = \begin{pmatrix} -x^l\omega(\mathbf{p}) \\ +x^l\omega(\mathbf{p}) \end{pmatrix}.$$

We have pairs of functions, corresponding to the pair of components $\tilde{\Gamma}_\pm$ of $\tilde{\Gamma}$.

As canonical coordinates on $\mathcal{S} \cap \tilde{\Gamma}$, (x^l, p_l) can be chosen. One can check that the corresponding PB, $\{.,.\}'$, reproduce the Poincaré algebra.

Let us choose

$$H = -p_0$$

as time direction. Then the time derivative along $\tilde{\Gamma}$ is just $\frac{\partial}{\partial x^0}$ and we have for example in Γ

$$\frac{\partial K_l}{\partial t} = -p_l,$$

as $K_l = x^0 p_l + p_0 x^l$. For Dirac's representatives we obtain in $\tilde{\Gamma} \cap \mathcal{S}$

$$\frac{\partial K_{lD}}{\partial t} = \{K_{lD}, p_{0D}\}' = -p_{lD}$$

as expected. The equation

$$\frac{\partial o_D}{\partial t} + \{H_D, o_D\} = 0$$

leads to what can be considered as classical counterpart of the standard Heisenberg picture (for a similar, but different idea, see Kuchař [14]).

Finally, let us extend the dynamics by choosing $X = 1$ in eq. (1). $\mathcal{S}$ will be transversal to this "dynamics" only, if $p_0 \neq 0$. Thus,

$$U = \Gamma \setminus \{p_0 = 0\}.$$

Choose $f = x^l$, $l = 1, 2, 3$. Then, Rovelli's position x^l_S is the following phase function in U (use eq. (2)):

$$x^l_S = x^l + \frac{p_l}{p_0} x^0.$$

Its Dirac's representative is, of course,

$$x^l_{SD} = x^l,$$

and its time derivative

$$\frac{\partial x^l_S}{\partial t} = -\frac{p_l}{p_0}.$$

2.3.3 The Front Form

Here $T = \frac{1}{\sqrt{2}}(x^0 - x^3)$ and it is advantageous to pass to null coordinates on Γ:

$$x^{\pm} = \frac{1}{\sqrt{2}}(x^0 \pm x^3), \quad x^a,$$

$$p_{\pm} = \frac{1}{\sqrt{2}}(p_0 \pm p_3) \ , \ p_a \ , \quad a = 1, 2.$$

Let us denote by $\mathcal{N}$ the surface $T = 0$. $\mathcal{G}_{\mathcal{N}}$ is generated by seven functions

$$p_+ \ , \ p_a \ , J_3 \ ,$$

$$K_{a-} = p_a x^- + p_+ x^a = \frac{1}{\sqrt{2}}(K_a - \epsilon_{ak3} J_k),$$

$$K_{+-} = p_+ x^+ - p_- x^- = K_3,$$

$a = 1, 2$, and we have three independent Hamiltonians :

$$p_-,$$

$$K_{a+} = p_a x^+ + p_- x^a = \frac{1}{\sqrt{2}}(K_a + \epsilon_{ak3} J_k),$$

$a = 1, 2$. Let us choose the functions x^-, x^a, x^+, p_1, p_2, and p_+ as coordinates on $\tilde{\Gamma}$ (p_+ is negative along $\tilde{\Gamma}_+$ and positive along $\tilde{\Gamma}$!). Then, Dirac's representatives of $\mathcal{G}_{\mathcal{N}}$-generators are obtained from the above functions by setting $x^- = 0$, whereas the Hamiltonians become

$$p_{-D} = \frac{\eta^2}{2p_+} \ ,$$

$$K_{a+D} = p_a x^+ + \frac{\eta^2}{2p_+} x^a,$$

where

$$\eta^2 = p_1^2 + p_2^2 + m^2.$$

Canonical coordinates on $\mathcal{N} \cap \tilde{\Gamma}$ are (x^a, x^+, p_a, p_+).

Rovelli's positions, $x^1_{\mathcal{N}}$ and $x^+_{\mathcal{N}}$, are given by

$$x^a_{\mathcal{N}} = x^a + \frac{p_a}{p_+} x^-,$$

$$x^+_{\mathcal{N}} = x^+ - \frac{p_-}{p_+} x^-,$$

etc.

2.4 First Steps of Quantization

The algebraic method of quantization has been described by Ashtekar [4] and further developed [5], [6] and [7]. Basic for this methods are certain abstract algebras (of "operators"); the "states" are secondary, forming representation spaces of the algebras. We will apply this procedure in a short-cut version, to the relativistic particle.

Classical Poisson algebras have three independent operations. An algebra of operators on a Hilbert space has only two: the commutator is not independent of composition. That's why the Poisson algebra of all (or even all polynomial) phase functions cannot be isomorphic to an operator algebra (van Hove theorem, see, e.g. [15], p.435). One way out of this difficulty is to choose only a small subset of phase functions – the so-called *elementary variables* – and to represent only the algebra which they generate.

Let us choose, for our model, all complex C^∞ functions on Γ that are at most linear in momenta,

$$f(x,p) = v^\mu(x)p_\mu + u(x).$$

These functions form a complex linear space $\mathcal{L}$ which is closed with respect to PB's and complex conjugation. We can also multiply any element of $\mathcal{L}$ by any function $w(x) \in \mathcal{L}$ and obtain again an element of $\mathcal{L}$.

Step 2 is to generate from $\mathcal{L}$ an algebra $\mathcal{A}$ with involution "$\star$" by introducing a new operation, an associative (but non-commutative) product $\hat{a}\hat{b} \in \mathcal{A}$ for all $\hat{a}$, $\hat{b} \in \mathcal{A}$ as follows.

a) There is a linear injection

$$i : \mathcal{L} \to \mathcal{A}$$

with the notation $i(f) = \hat{f}$.

b) Every element of $\mathcal{A}$ equals a (finite) complex linear combination of (finite) products of elements of $i(\mathcal{L})$.

c) For any two elements $f, g \in \mathcal{L}$, we require:

(i) $\widehat{\{f,g\}} = i\hbar[\hat{f},\hat{g}]$, where

$$[\hat{f},\hat{g}] = \hat{f}\hat{g} - \hat{g}\hat{f}$$

(canonical commutation rules).

(ii) $\hat{f}^\star = \hat{\bar{f}}$,

where $\bar{f}$ is the complex conjugation of f.

(iii) $\widehat{fg} = \frac{1}{2}\hat{f}\hat{g} + \frac{1}{2}\hat{g}\hat{f}$,

if f or g are independent of momenta (that is, whenever $fg \in \mathcal{L}$).

Step 3 is to represent each element $\hat{a} \in \mathcal{A}$ by a linear operator $R(\hat{a})$ on a linear space $\mathcal{D}$.

For finite dimensional models, the following method has been worked out by Tate [5] (also described in [6]).

Let us choose

$$\mathcal{D} := C_0^\infty(\mathcal{C}, \mathbb{C}),$$

(that is the linear space of smooth complex functions with compact support) and let μ be an arbitrary C^∞ volume element on $\mathcal{C}$ (nowhere vanishing 4-form). Then, for any $\Psi \in \mathcal{D}$ and for any element

$$f = v^\mu(x)p_\mu + u(x)$$

of $\mathcal{L}$, we define

$$R(\hat{f})\Psi := -i\hbar(L_{\vec{v}} + \frac{1}{2}\mathrm{Div}_\mu \vec{v})\Psi + \\ + u\Psi,$$

where $L_{\vec{v}}$ is the Lie derivative with respect to the vector field $\vec{v}(x)$, and

$$(\mathrm{Div}_\mu \vec{v})\mu = L_{\vec{v}}\mu.$$

To represent the whole algebra $\mathcal{A}$ we define for any two elements $\hat{a}$ and $\hat{b}$ of $\mathcal{A}$ and $\alpha, \beta \in \mathbb{C}$:

$$R(\alpha\hat{a} + \beta\hat{b}) = \alpha(R(\hat{a})\Psi) + \beta(R(\hat{b})\Psi),$$
$$R(\hat{a}\hat{b})\Psi = R(\hat{a})(R(\hat{b})\Psi),$$

and

$$(\Psi, R(\hat{a})\Phi) = (R(\hat{a})^\dagger\Psi, \Phi),$$

where

$$(\Psi, \Phi) = \int_{\mathcal{C}} \mu\bar{\Psi}\Phi. \tag{6}$$

one can show that all "relations" (i), (ii) and (iii) are preserved by R, that is

$$R(\widehat{\{f,g\}})\Psi = i\hbar(R(\hat{f})R(\hat{g}) - R(\hat{g})R(\hat{f}))\Psi,$$
$$R(\widehat{fg})\Psi = \frac{1}{2}(R(\hat{f})R(\hat{g}) + R(\hat{g})R(\hat{f}))\Psi,$$
$$R(\hat{f}^*) = R(\hat{f})^\dagger.$$

In what follows, we will identify the algebra $\mathcal{A}$ with $R(\mathcal{A})$.

2.5 Observables

Our next taks is to make the observables and constraints to elements of $\mathcal{A}$.

The generators of the Poincaré group (4) are elementary variables, so we can write immediately:

$$\hat{p}_\rho = -i\hbar\left(\frac{\partial}{\partial x^\rho} + \frac{1}{2}\mathrm{Div}_\mu \vec{p}_\rho\right), \tag{7a}$$

$$\hat{J}_l = -i\hbar\left(\epsilon_{lmn} x^m \frac{\partial}{\partial x^n} + \frac{1}{2}\mathrm{Div}_\mu \vec{J}_l\right), \tag{7b}$$

$$\hat{K}_l = -i\hbar\left(x^0 \frac{\partial}{\partial x^l} + x^l \frac{\partial}{\partial x^0} + \frac{1}{2}\mathrm{Div}_\mu \vec{K}_l\right), \tag{7c}$$

where $\vec{p}_\rho$, $\vec{J}_l$, and $\vec{K}_l$ are the corresponding vector fields on $\mathcal{C}$, e.g.

$$\vec{p}_\rho = \frac{\partial}{\partial x^\rho}.$$

The choice of factor ordering (7) satisfies the conditions:

a) proper classical limit,

b) canonical commutation rules,

c) symmetry with respect to (.,.).

Similarly, we require for the constraint operator $\hat{C}$:

a) proper classical limit,

b) canonical commutation rules within the set $\{\hat{C}, \hat{p}_\rho, \hat{J}_l, \hat{K}_l\}$,

c) independence of coordinate choice in $\mathcal{C}$ (covariance, see [16]).

Then, it follows that

$$\hat{C} = -\frac{1}{\sqrt{\mu'}}(\partial_0^2 - \Delta + m'^2)\sqrt{\mu'}, \tag{8}$$

where μ' is defined by

$$\mu = \mu' dx^0 \wedge dx^1 \wedge dx^2 \wedge dx^3,$$

$$\Delta = \left(\frac{\partial}{\partial x^1}\right)^2 + \left(\frac{\partial}{\partial x^2}\right)^2 + \left(\frac{\partial}{\partial x^3}\right)^2$$

and m'^2 is a constant differing from m^2 by some powers of $\hbar$. Observe that a real m'^2 leads to a symmetric operator $\hat{C}$. Let us choose $m' = m$.

Finally, observe that different μ's lead to equivalent representations of $\mathcal{A}$ (cf. [5]). Thus, we just set

$$\mu = dx^0 \wedge dx^1 \wedge dx^2 \wedge dx^3. \tag{9}$$

2.6 The Physical Space

The next step is to solve the operator constraint equation

$$\hat{C}\Psi = 0. \tag{10}$$

This is called the Wheeler-DeWitt equation if $\hat{C}$, as in our model, generates reparametrizations. Within some quantization methods (see Chap. 3 of this paper), (10) becomes the dynamical equation in the constructed quantum mechanics. "The" space of solutions to (10) is called "physical space".

Such a space, however, can depend on the form in which $\hat{C}$ is written (see the discussion in [1]). This form, in turn, will depend on the choice of time. Our strategy will be to make the final quantum mechanics as independent of the choice of time as possible. Such a flexibility can be obtained if one requires that $\hat{C}$ be a self-adjoint operator on some Hilbert space (cf. [1]).

There may be problems with this requirements. First, equation (10), as eigenvalue equation, may happen to possess too few solutions to construct any reasonable Hilbert space from it. An example is given in [17] (system D). Our present model is not like that, neither are the models of [6], but some care is necessary, if one is going to generalize this method.

Another well known problem is that $\hat{C}$ is to be self-adjoint with respect to a scalar product on the representation space of $\mathcal{A}$ like (6). However, zero will lie, as a rule, in the continuum part of the spectrum of $\hat{C}$, and the corresponding solutions to (10) will not be normalizable. Then, (6) seems to be only some auxiliary construct which has nothing to do with the final, physical scalar product, and the condition for $\hat{C}$ to be self-adjoint with respect to (6) seems to be rather artificial.

For the last problem, a solution can be based on the following observations. First, the representation space $\mathcal{D}$ together with the quadratic form (6) determines uniquely a structure of a so-called rigged Hilbert space; within this space, eigenspaces of operators with continuous spectrum are meaningful. Second, Green's identity yields a natural relation between (6) and the physical scalar product.

In general, the rigged Hilbert space is the following structure (see, e.g. [18]).

1) A "countably Hilbert nuclear space" $\mathcal{H}_0$. We will not give a definition here, because it is rather complicated. Let us just mention that all so-called test-function spaces in the theory of distributions can be given this structure. The corresponding topology is called Fréchet topology. In our special case, $\mathcal{D}$ will play the role of $\mathcal{H}_0$.

2) A positive quadratic form (.,.) on $\mathcal{H}_0$ that is Fréchet-continuous. Let $\mathcal{H}_1$ be the Hilbert space resulting from $\mathcal{H}_0$ and (.,.) by Cauchy completion. Clearly,

$$\mathcal{H}_0 \in \mathcal{H}_1.$$

In our special case, the form (6) has the reqired property on $\mathcal{D}$. The Hilbert space $\mathcal{H}_1$ is called "auxiliary Hilbert space" and denoted by $\mathcal{H}_a$.

3) The topological dual $\mathcal{H}_2$ of $\mathcal{H}_0$. (This is the space of all Fréchet-continuous linear functionals on $\mathcal{H}_0$.) In our case, the space $\mathcal{D}'$ of distributions plays the role of $\mathcal{H}_2$. As (.,.) is continuous, any element $\Psi \in \mathcal{H}_1$ defines an element $\Psi \in \mathcal{H}_2$ by

$$\Psi(\varphi) = (\Psi, \varphi) \quad \forall \varphi \in \mathcal{H}_0.$$

In this sense

$$\mathcal{H}_1 \subset \mathcal{H}_2.$$

The triple $\mathcal{H}_0$, $\mathcal{H}_1$, $\mathcal{H}_2$ (often only $\mathcal{H}_0$) is called rigged Hilbert space. Its crucial property is described by the following theorem [18].

<u>Theorem 2</u>: Let $\hat{A}$ be a linear operator

a) with domain $\mathcal{H}_0$ and range in $\mathcal{H}_0$

b) essentially self-adjoint in $\mathcal{H}_1$.

Then, there is a complete system of *generalized eigenvectors* ξ_a of $\hat{A}$ in $\mathcal{H}_2$:

$$\xi_a(\hat{A}\varphi) = a\xi_a(\varphi) \quad \forall \varphi \in \mathcal{H}_0, \quad a \in \mathbf{R}.$$

In fact, the notion of generalized eigenvectors is not new to physicists. They are, e.g., the "scattering states normalized to δ-function". The rigged Hilbert space at our disposal is $\mathcal{D}, \mathcal{H}_a, \mathcal{D}'$; all observables (7) and the constraint (8) satisfy the conditions of Theorem 2. Similarly, the cosmological models of [6] exhibit the same structure and the physical space constructed in [6] is included in the following definition.

<u>Definition</u>: The physical space $\mathcal{H} \subset \mathcal{D}'$ is the generalized eigenspace of $\hat{C}$ to the generalized eigenvalue zero.

Thus, the physical space is well-defined. It is, however, no subspace of $\mathcal{H}_a$, so the scalar product in $\mathcal{H}_a$ cannot be restricted to $\mathcal{H}$. There is a hope that the algebraic method will supply a construction of such a scalar product.

2.7 The Physical Scalar Product

We start by making the observables to operators in $\mathcal{H}$. Let $\hat{o}$ be a linear Fréchet-continuous operator,

$$\hat{o} : \mathcal{D} \to \mathcal{D},$$

with domain $\mathcal{D}$. Then, there is a unique *dual* operator

$$\hat{o}' : \mathcal{D}' \to \mathcal{D}'$$

with domain $\mathcal{D}'$, defined as follows:

$$(\hat{o}'\xi)\,\varphi = \xi\,(\hat{o}\varphi) \quad \forall \xi \in \mathcal{D}'\,,\ \varphi \in \mathcal{D}.$$

With this notation,

$$\mathcal{H} = \left\{\xi \in \mathcal{D}' \mid \hat{C}'\xi = 0\right\}.$$

All operations on $\mathcal{A}$ clearly have their dual counterparts. In particular,

$$(\hat{o}')^\star = (\hat{o}^\star)'.$$

Suppose that

$$\hat{C}\hat{o}\varphi - \hat{o}\hat{C}\varphi = \hat{C}\hat{f}\varphi \quad \forall \varphi \in \mathcal{D},$$

where $\hat{f} : \mathcal{D} \to \mathcal{D}$ is a Fréchet-continuous linear operator. Then,

$$\hat{o}'\hat{C}'\xi - \hat{C}'\hat{o}'\xi = \hat{f}'\hat{C}'\xi \quad \forall \xi \in \mathcal{D}'.$$

In particular, if $\xi \in \mathcal{H}$, then

$$\hat{C}'\hat{o}'\xi = 0$$

and $\hat{o}'\xi \in \mathcal{H}$. Thus, $\hat{o}'$ restricted to $\mathcal{H}$ is an operator in $\mathcal{H}$.

All observables (7) can be made to operators on $\mathcal{H}$ in this way, as they, together with $\hat{C}$ given by (8), satisfy all above requirements (in fact, the corresponding $\hat{f}$'s are zero).

In general, the representation of the Lie algebra of observables by the operators $\hat{o}' \mid_{\mathcal{H}}$ on $\mathcal{H}$ will be reducible. Suppose that $\mathcal{H}_+ \subset \mathcal{H}$ is an invariant subspace on which the representation is irreducible. Let us restrict ourselves to $\mathcal{H}_+$. We ask for a scalar product $q(.,.)$ on $\mathcal{H}_+$ that satisfies the condition

$$q(\psi, \hat{o}'\varphi) = q(\hat{o}'^{*}\psi, \varphi)$$

for all $\psi, \varphi \in \mathcal{H}$ and all observables o. (This means that the scalar product is invariant with respect to infinitesimal transformations $1 + i\hat{o}'d\epsilon$, that is, with respect to the Poincaré group in our case.)

Thus, we have some nontrivial and physically reasonable requirements on the scalar product which can help to determine it. In many important examples, a product statisfying these requirement not only exists, but is even unique up to a positive factor (for general discussion and references, see Rendall, this volume and [19]). In the case that the method works, one question remains open: how the two products, (.,.) and $q(.,.)$ are related. There is still another way of construction $q(.,.)$ that answers this question and that works in all mentioned examples. It is based on DeWitt's generalization [20] of Green's identity.

Theorem 3: Let $(\mathcal{M}, \mu)$ be an n-manifold with a volume form μ, and $\hat{C}$ be a second-order differential operator on $C_o^2(\mathcal{M}, \mathbb{C})$ that is symmetric with respect to the scalar product (6).

Then, for any two functions $\psi, \varphi \in C_o^2(\mathcal{M}, \mathbb{C})$, there is a vector field

$$\vec{J}[\psi, \varphi; x]$$

on $\mathcal{M}$ such that

$$\overline{(\hat{C}\psi)}\varphi - \bar{\psi}(\hat{C}\varphi) = \mathrm{Div}_\mu \vec{J}.$$

For example, if $\hat{C}$ is given by eq. (8) and μ by eq. (9), then

$$J^\rho = \eta^{\rho\sigma}\left[(\partial_\sigma\bar{\psi})\varphi - \bar{\psi}(\partial_\sigma\varphi)\right]$$

is the well-known Klein-Gordon current vector. If $\hat{C}$ is of hyperbolic or parabolic type, $\vec{J}$ can be used to define a quadratic form on the space of solutions to eq. (10). This form is not positive on the whole physical space, but it is definite on its irreducible subspaces. We will find that this method will lead to the same scalar product on $\mathcal{H}_+$ as the algebraic one in our example. A coherent theory of such relations, however, does not exist as yet.

2.8 Forms of Dynamics in Quantum Theory

On our way to quantum theory, we have managed to convert the classical observables to essentially self-adjoint operators in $\mathcal{H}_a$; they will generate a unitary representation of the Poincaré group $\mathcal{G}$ on $\mathcal{H}_a$. (The general theory of such representation is given, e.g., in Chap. 11 of [21].) To implement Dirac's program, we have to identify the subgroup $\mathcal{G}_S$ of transformations inside of the submanifold $\mathcal{S} \cap \tilde{\Gamma}$. Let g_S be the Lie algebra of $\mathcal{G}_S$ and e_S its enveloping algebra (see [21]). We will find the generators of a maximal abelian subalgebra of e_S. If $\mathcal{G}_S$ is sufficiently large, these generators will define a complete set of commuting operators in $\mathcal{H}_+$. The common generalized eigenvectors of the complete set will form a basis in $\mathcal{H}_+$, whose elements will be labeled by the eigenvalues of the set. One has to calculate the matrix elements of all observables in this basis. We will find that the form of these "matrices" is closely related to classical Dirac's representatives.

Turning to our particular model, we quickly realize that the complete set can be chosen as $(\hat{p}_1, \hat{p}_2, \hat{p}_3)$ for the instant, $(\hat{p}_1, \hat{p}_2, \hat{p}_+)$ for the front, and $(\hat{J}_3, \hat{J}^2, -\hat{K}^2 + \hat{J}^2)$ for the point form. The physical states will be functions, $\psi(p_1, p_2, p_3)$, $\psi(p_1, p_1, p_+)$ or $\psi(m, j, \rho)$, respectively, and the quantum version of Dirac's representatives will act as operators on these functions. Let us work out the detail for the instant and front forms.

2.8.1 The Instant Form

Let us start by choosing a complete system of commuting operators for the whole auxiliary space $\mathcal{H}_a$ so that it contains the operators $\hat{p}_1, \hat{p}_2$ and $\hat{p}_3$. That is, we add $\hat{p}_0$. These four operators can be made simultaneously diagonal by the unitary transformation

$$U_{\text{inst}} : \mathcal{H}_a \to \mathcal{H}_{\text{inst}},$$

which is just the Fourier transformation from $L^2(\mathbf{R}^4, \mu) \to L^2(\mathbf{R}^4, \mu_{\text{inst}})$. U_{inst} will map the subspace $\mathcal{D} \subset \mathcal{H}_{\text{inst}}$ into $U_{\text{inst}}(\mathcal{D}) \subset \mathcal{H}_{\text{inst}}$. This is the well-known test function space $Z(\mathbf{R}^4)$, namely the space of "entire analytic slowly increasing functions on $\mathbf{R}^4$" (see, e.g. [22]). The volume form μ is mapped into μ_{inst},

$$\mu_{\text{inst}} = dp_0 \wedge dp_1 \wedge dp_2 \wedge dp_3.$$

The dual map U'_{inst} sends the distributions from Z' to $\mathcal{D}'$. U_{inst}, being unitary, has an inverse, U^{-1}_{inst}, and

$$(U^{-1}_{\text{inst}})' : \mathcal{D}' \to Z'.$$

The two rigged Hilbert spaces $(\mathcal{D}, \mathcal{H}_a, \mathcal{D}')$ and $(Z, \mathcal{H}_{\text{inst}}, Z')$ are called *isomorphic*. In this way, any unitary map of the auxiliary Hilbert space $\mathcal{H}_a$ can be extended to a map of the whole rigged Hilbert space, $(\mathcal{D}, \mathcal{H}_a, \mathcal{D}')$, that preserves the rigged Hilbert space structure. The image $U_{\text{inst}}\hat{o}U_{\text{inst}}^{-1}$ of any self-adjoint operator $\hat{o}$ in $\mathcal{H}_a$ will have the same spectrum as $\hat{o}$, and the corresponding eigenspaces are mapped into each other by $(U_{\text{inst}}^{-1})'$. It is exactly this invariance that motivated our definition of Wheeler-DeWitt equation as eigenvalue equation.

Let us calculate these images of all observables and constraints. We obtain setting $\hbar = 1$:

$$\hat{C} = -p_0^2 + \omega^2(\mathbf{p}),$$

$$\hat{p}_\mu = p_\mu,$$

$$\hat{J}_l = -i\epsilon_{lmn}p_m\frac{\partial}{\partial p_n},$$

$$\hat{K}_l = ip_l\frac{\partial}{\partial p_0} + ip_0\frac{\partial}{\partial p_l},$$

$$\hat{x}_S^l = i\frac{\partial}{\partial p_l} + i\frac{p_l}{p_0}\frac{\partial}{\partial p_o} - \frac{i}{2}\frac{p_l}{p_0^2}.$$

We observe that all these quantities are at most first order differential operators. This enables us, using Tate's [5] recipe, to find factor ordering that exactly reproduces all PB's. In particular, the term $\frac{i}{2}\text{Div}_{\mu_{\text{inst}}}\vec{v}$ is non-zero for the position operator.

The general solution to the eigenvalue equation $\hat{C}'\xi = 0$ can now easily be found. The above form of the operator $\hat{C}$ implies that any distributional solution to eq. (10) must be concentrated on $\tilde{\Gamma}$ (see , e.g. [22]). Substituting the general form of such a distribution as given in [22] into eq. (10), we obtain

$$\xi = \psi(p)\delta(-p_0^2 + \mathbf{p}^2 + m^2), \tag{11}$$

where $\psi \in Z(\mathbf{R}^4)$ is arbitrary. These solutions will form a dense subspace of the final physical space.

Within the algebraic method of quantizing constraint systems that we are describing, one avoids any sort of gauge fixing (in the case of parametrized systems, a time choice) as long as possible. In our special case, we have not yet made such a fixing. Now, we are going to represent the physical space as a space of certain unconstraint functions (see below). Such a representation is not unique, and, as it will turn out, it smuggles in quite a particular time choice. In this way, we arrive at a particular "form of dynamics" and a particular form of observables. The following factorization of the argument of the δ-function in (11),

$$-p_0^2 + \mathbf{p}^2 + m^2 = (p_0 + \omega(\mathbf{p}))(-p_0 + \omega(\mathbf{p})).$$

leads to a particular form of ξ:

$$\xi = \frac{\psi_+(\mathbf{p})}{2\omega(\mathbf{p})}\delta(p_0 + \omega(\mathbf{p})) + \frac{2\psi_-(\mathbf{p})}{\omega(\mathbf{p})}\delta(-p_0 + \omega(\mathbf{p})), \tag{12}$$

where

$$\psi_\pm(\mathbf{p}) = \psi(\mp\omega(\mathbf{p}), \mathbf{p})$$

are the promised functions of p_1, p_2, and p_3. They are just restrictions of ψ to $\tilde{\Gamma}_\pm$, written in particular coordinates on $\tilde{\Gamma}_\pm$. In this way, we have a linear isomorphism

$$\xi \leftrightarrow \begin{pmatrix} \psi_+ \\ \psi_- \end{pmatrix} \in Z(\mathbf{R}^3) \times Z(\mathbf{R}^3).$$

The action of observables on $Z(\mathbf{R}^3) \times Z(\mathbf{R}^3)$ is obtained in a straightforward way, just letting the observables act on the function (12) and rearranging the result again in the form (12). The result is:

$$\hat{p}_0 = -\omega(\mathbf{p})\sigma_3, \tag{13a}$$

$$\hat{p}_l = p_l\sigma_0, \tag{13b}$$

$$\hat{J}_l = -i\epsilon_{lmn}\sigma_0 p_m \frac{\partial}{\partial p_n}, \tag{13c}$$

$$\hat{K}_l = -i\omega(\mathbf{p})\sigma_3 \frac{\partial}{\partial p_l}, \tag{13d}$$

$$\hat{x}_S^l = i\sqrt{\omega(\mathbf{p})}\sigma_0 \frac{\partial}{\partial p_l} \frac{1}{\sqrt{\omega(\mathbf{p})}}. \tag{13e}$$

Here, σ_0 and σ_3 are the matrices,

$$\sigma_0 = \begin{pmatrix} 1 & 0 \\ 0 & 1 \end{pmatrix}, \sigma_3 = \begin{pmatrix} 1 & 0 \\ 0 & -1 \end{pmatrix}.$$

If one compares these operators with Dirac's representatives of Sec. 2.3.2, one observes a nice coincidence (up to factor order). Let us also remark that the above representation of the algebra of observables is reducible, and that only an irreducible subspace can be chosen as a physical Hilbert space.

A scalar product on $Z(\mathbf{R}^3) \times Z(\mathbf{R}^3)$ can be obtained either by requiring the operators (13) to be symmetric (up to a positive constant factor), or via Green's identity. The Fourier transform of Green's identity can be written in the form:

$$\mu_{\text{inst}} \left|_{\tilde{\Gamma}_\pm} = dC \wedge \tilde{\mu}_\pm,\right.$$

and we obtain:

$$\tilde{\mu}_\pm = \pm \frac{1}{2\omega(\mathbf{p})} dp_1 \wedge dp_2 \wedge dp_3,$$

which gives the "textbook" scalar product

$$\left(\begin{pmatrix} \psi_+ \\ \psi_- \end{pmatrix}, \begin{pmatrix} \varphi_+ \\ \varphi_- \end{pmatrix} \right)_{\text{inst}} = \int_{\mathrm{I\!R}^3} \frac{d^3p}{2\omega} \bar{\psi}_+ \varphi_+ + \int_{\mathrm{I\!R}^3} \frac{d^3p}{2\omega} \bar{\psi}_- \varphi_-.$$

(We have "rectified" the sign of $\tilde{\mu}_-$; this is a well-defined operation, as the two irreducible subspaces are unique). Now, define $\mathcal{H}^{\text{inst}}_{\text{phys}}$ as the completion of $Z(\mathbf{R}^3) \times Z(\mathbf{R}^3)$ with respect to $(.,.)_{\text{inst}}$.

Finally, we can construct *a* time evolution if we choose a particular time direction as defined by a particular Hamiltonian. Let

$$\hat{H}_{\text{inst}} = -\hat{p}_0.$$

In this way, $\hat{x}^0$ becomes t, and

$$\hat{H}_{\text{inst}} = \omega(\mathbf{p})\sigma_3.$$

So far, all operators and states are defined at $t = 0$. The evolution of an observable $\hat{o}$ can be obtained by solving the Heisenberg equation:

$$i\frac{\partial \hat{o}}{\partial t} + \left[\hat{H}_{\text{inst}}, \hat{o}\right] = 0.$$

All our observables happen to satisfy

$$\left[\hat{H}_{\text{inst}}, \left[\hat{H}_{\text{inst}}, \hat{o}\right]\right] = 0.$$

It follows that the general solution to Heisenberg equation is

$$\hat{o}(t) = \hat{o} + i\left[\hat{H}_{\text{inst}}, \hat{o}\right] t.$$

In this way, we obtain a non-trivial time dependence only for $\hat{K}_l$ and $\hat{x}^l_S$

$$\hat{K}_l(t) = -i\omega\sigma_3\frac{\partial}{\partial p_l} - p_l\sigma_0 t,$$

$$\hat{x}^l_S(t) = i\sqrt{\omega}\sigma_0\frac{\partial}{\partial p_l}\frac{1}{\sqrt{\omega}} + \frac{p_l}{\omega}\sigma_3 t.$$

Observe that the (Schrödinger picture) operator $\hat{x}^l_S(0)$ coincides with the Newton-Wigner operator.

2.8.2 The Front Form

We will proceed in a way completely analogous to previous sections. Thus, we can skip the details.

Let us make the operators p_-, p_1, p_2 and p_+ diagonal this time. The corresponding unitary map,

$$U_{\text{front}} : \mathcal{H}_a \to \mathcal{H}_{\text{front}}$$

is the Fourier transform between $L^2(\mathbf{R}^4, \mu)$ and $L^2(\mathbf{R}^4, \mu_{\text{front}})$, where

$$\mu_{\text{front}} = dp_- \wedge dp_1 \wedge dp_2 \wedge dp_+.$$

The constraint and the observables become

$$\hat{C} = 2p_+p_- + \eta^2,$$

$$\hat{p}_\pm = p_\pm,$$

$$\hat{p}_a = p_a,$$

$$\hat{J}_3 = -ip_1\frac{\partial}{\partial p_2} + ip_1\frac{\partial}{\partial p_1},$$

$$\hat{K}_{+-} = ip_+\frac{\partial}{\partial p_+} - ip_-\frac{\partial}{\partial p_-},$$

$$\hat{K}_{a\pm} = ip_\mp\frac{\partial}{\partial p_a} + ip_a\frac{\partial}{\partial p_\pm},$$

$$\hat{x}^a_N = i\frac{\partial}{\partial p_a} + i\frac{p_a}{p_+}\frac{\partial}{\partial p_-},$$

$$\hat{x}^+_N = i\frac{\partial}{\partial p_+} - i\frac{p_-}{p_+}\frac{\partial}{\partial p_-} - \frac{i}{2}\frac{1}{p_+}.$$

The general solution to the eigenvalue equation $C'\xi = 0$ has again the form (11). This time, however, we factorize as follows

$$-2p_+p_- + \eta^2 = -2p_+\left(p_- - \frac{\eta^2}{2p_+}\right)$$

and obtain the linear isomorphism

$$\xi \leftrightarrow \begin{pmatrix} \psi_+ \\ \psi_- \end{pmatrix} \in Z(\mathbf{R}_+ \times \mathbf{R}^2) \times Z(\mathbf{R}_+ \times \mathbf{R}^2)$$

where

$$\begin{aligned} \psi_\pm(q_1, q_2, q_+) &= \psi(p_-^\pm(q), p_+^\pm(q), p_1, p_2) \\ p_+^\pm &= \mp q_+, \\ p_-^\pm &= \mp\frac{\eta^2}{2q_+}, \\ p_a &= q_a. \end{aligned}$$

In this way, we achieve that

$$q_+ \in (0, \infty)$$

on both halves $\tilde{\Gamma}_\pm$ of the constraint hypersurface.

The form of the operators on $Z(\mathbf{R}_+ \times \mathbf{R}^2) \times Z(\mathbf{R}_+ \times \mathbf{R}^2)$ is

$$\begin{aligned} \hat{p}_+ &= -q_+\sigma_3 \\ \hat{p}_- &= -\frac{\eta^2}{2q_+}\sigma_3 \\ \hat{p}_a &= q_a \\ \hat{K}_{a+} &= -i\sigma_3\left(\frac{\eta^2}{2q_+}\frac{\partial}{\partial q_a} + q_a\frac{\partial}{\partial q_+}\right) \\ \hat{J}_3 &= -i\left(q_1\frac{\partial}{\partial q_2} - q_2\frac{\partial}{\partial q_1}\right) \\ \hat{K}_{a-} &= -iq_+\sigma_3\frac{\partial}{\partial q_a} \\ \hat{K}_{+-} &= iq_+\frac{\partial}{\partial q_+} \end{aligned}$$

$$\hat{x}^a_{\mathcal{N}} = i\frac{\partial}{\partial q_a}$$

$$\hat{x}^+_{\mathcal{N}} = -i\sigma_3\sqrt{2q_+}\frac{\partial}{\partial q_+}\frac{1}{\sqrt{2q_+}}.$$

They are all symmetric with respect to the scalar product

$$\left(\begin{pmatrix}\psi_+\\ \psi_-\end{pmatrix}, \begin{pmatrix}\varphi_+\\ \varphi_-\end{pmatrix}\right)_{\text{front}} = \int_{\mathrm{IR}_+\times\mathrm{IR}^2}\frac{d^2q\,dq_+}{2q_+}\bar{\psi}_+\varphi_+ + \int_{\mathrm{IR}_+\times\mathrm{IR}^2}\frac{d^2q\,dq_+}{2q_+}\bar{\psi}_-\varphi_-,$$

which has been obtained from the Fourier transformation of Green's identity (rectifying the sign). The corresponding Hilbert space is $\mathcal{H}^{\text{front}}_{\text{phys}}$. Finally, let us generate the time evolution by

$$\hat{H}_{\text{front}} = -\hat{p}_-.$$

Then, the time is x^-, and we obtain the Heisenberg operators

$$\hat{K}_{a-}(t) = -iq_+\sigma_3\frac{\partial}{\partial q_a} - q_a t,$$

$$\hat{K}_{+-}(t) = -iq_+\frac{\partial}{\partial q_+} - \frac{\eta^2}{2q_+}\sigma_3 t,$$

$$\hat{x}^a_{\mathcal{N}}(t) = i\frac{\partial}{\partial q_a} + \frac{q_a}{q_+}\sigma_3 t,$$

$$\hat{x}^+_{\mathcal{N}}(t) = -i\sigma_3\sqrt{2q_+}\frac{\partial}{\partial q_+}\frac{1}{\sqrt{2q_+}} + \frac{\eta^2}{2q_+}t,$$

all remaining observables being time independent.

2.9 Relation Between the Two Quantum Mechanics

We have chosen two different time surfaces, $x^0 = 0$ and $x^- = 0$, and two different time directions generated by $-p_0$ and $-p_-$. As a result, we have obtained two Hilbert spaces, $\mathcal{H}^{\text{inst}}_{\text{phys}}$ and $\mathcal{H}^{\text{front}}_{\text{phys}}$, two algebras, $\mathcal{A}_{\text{inst}}$ and $\mathcal{A}_{\text{front}}$, of observables represented as operators in the Hilbert spaces, and two Hamiltonians $\hat{H}_{\text{inst}}$ and $\hat{H}_{\text{front}}$. Let us look for relations between these quantum mechanics.

First, we find a simple relation between the states. Indeed, the composite map U defined by

$$U' = U'_{\text{inst}} \circ \left(U^{-1}_{\text{front}}\right)'$$

maps $\mathcal{H}^{\text{inst}}_{\text{phys}}$ onto $\mathcal{H}^{\text{front}}_{\text{phys}}$ and is invertible. (This is a completely general property of our construction.) We find that

$$U' \begin{pmatrix} \psi_+(\mathbf{p}) \\ \psi_-(\mathbf{p}) \end{pmatrix} = \begin{pmatrix} \varphi_+(\mathbf{p}) \\ \varphi_-(\mathbf{p}) \end{pmatrix}$$

where

$$\varphi_\pm(\mathbf{q}) = \psi_\pm\left(\mathbf{p}^\pm(\mathbf{q})\right),$$

and

$$p_a^\pm = q_a, \tag{14a}$$

$$p_3^\pm = \mp\frac{1}{\sqrt{2}}\left(q_+ - \frac{\eta^2}{2q_+}\right). \tag{14b}$$

Thus, U' is just a "transformation of coordinates". This is connected with the mutual dependence of two complete commuting systems, $\{\hat{p}_0, \hat{p}_1, \hat{p}_2, \hat{p}_3\}$, and $\{\hat{p}_-, \hat{p}_1, \hat{p}_2, \hat{p}_+\}$. The functions related by U' describe the *same physical* state: as the diagonal operators are time independent, their distribution does not depend on the time surface at which it is measured.

Moreover, the map U' is unitary. The Jacobian of the transformation (14) is

$$\frac{\partial(p_1, p_2, p_3)}{\partial(q_1, q_2, q_+)} = \mp\frac{\omega(\mathbf{p})}{q_+}$$

so that the volume forms are U'-related.

Next, let us consider the observables. As the two quantum mechanics describe the same physical system, the observable algebras $\mathcal{A}_{\text{inst}}$ and $\mathcal{A}_{\text{front}}$ must have elements which represent the same observables. These are clearly the generators of the Poincaré group. We can easily check that the operators representing the same generators of the Poincaré group are mapped into each other by U'. For example,

$$U'\hat{p}_0^{\text{inst}}U'^{-1} = \hat{p}_0^{\text{front}}, \tag{15}$$

where

$$\hat{p}_0^{\text{inst}} = -\omega(\mathbf{p})\sigma_3,$$

$$\hat{p}_0^{\text{front}} = -\frac{2q_+^2 + \eta^2}{2\sqrt{2}q_+}\sigma_3.$$

Hence: we have two *unitarily equivalent* representations of the Poincaré group. This result justifies further our claim that the U'-related functions are physically the same states: any measurement of a property related to any generator of Poincaré group will lead to the same result on U'-related states.

In spite of these facts, the two quantum mechanics are *not unitarily equivalent* in the usual sense; for that, one had to have that

$$\hat{H}_{\text{front}} = U'\hat{H}_{\text{inst}}(U')^{-1},$$

however, we obtain from eq. (15)

$$U'\hat{H}_{\text{inst}}(U')^{-1} = \frac{2q_+^2 + \eta^2}{2\sqrt{2}q_+}\sigma_3,$$

whereas

$$\hat{H}_{\text{front}} = -p^{\text{front}} = \frac{\eta^2}{2q_+}\sigma_3.$$

It is interesting to compare our pair of time choices with a pair of time choices that are related by a Poincaré transformation. For example, let T_1 and H_1 be as before, $T = x_0$, $H_1 = -p_0$, and choose

$$T_2 = x^0 \cosh\alpha + x^1 \sinh\alpha,$$

$$H_2 = -p_0 \cosh\alpha + p_1 \sinh\alpha.$$

It is easy to check that the two resulting quantum mechanics *will* be unitarily equivalent.

2.10 Conclusions

By implementing Dirac's program in quantum theory, we have found a way of how time choice enters the algebraic method. For the free relativistic particle, we have managed to choose times in such a way that the corresponding quantum mechanics are not unitarily equivalent.

As was explained in Sec. 2.9, the two quantum mechanics are, nevertheless, *physically* equivalent, because they realize unitarily equivalent representations of all common observables - generators of Poincaré group. To achieve this the definition of physical space as eigenspace played an important role.

Observe that the inequivalence of our two constructs has nothing to do with factor ordering. We have solved the factor ordering problem in an ideal way, all the relevant commutators being duely related to the corresponding Poisson brackets.

These considerations suggest that the notion of unitary equivalence may be too narrow for quantum gravity and that only after suitable generalizing it we could be able to construct a quantum theory "independent" of time choice.

3 Schrödinger Picture Approach to the Wheeler-DeWitt Equation

3.1 Introduction

In this chapter, we will approach the Wheeler-DeWitt equation in a different way: we will consider it as a Schrödinger equation. This is an old, but still very popular method, suitable especially for minisuperspace models. A recent relatively rigourous description is given in [7]; from the conceptual point of view, [8] is most useful.

Crucial for the method is the observation that the Wheeler-DeWitt equation for a wave function Ψ on a finite dimensional configuration space $\mathcal{C}$ of a given model can also be considered (at least formally) as a field equation for a quantum field Ψ on a curved spacetime $\mathcal{C}$. Then, some methods of the quantum field theory on a curved background can be used for the construction of quantum mechanics of the model. In particular, the space of states of the model will be analogous to the one-particle Hilbert space of the corresponding quantum field theory. The method will work only if the dynamical evolution will leave the one-particle Hilbert space invariant. A necessary condition for this is the existence of a particular kind of symmetry (see [7]., [8]). Such a symmetry may exist for some models, but it does not seem to exist for the full-fledged general relativity in ADM formulation (see [23]). The method can be useful, because it requires just *one* (although somewhat special) symmetry, whereas the algebraic method will work only if there is a sufficiently *large* number of (general) symmetries (generated by observables).

The model which we choose will demonstrate this point quite clearly. We just take a spherically symmetric thin shell made of incoherent matter (matter points interacting only by gravity). For such a system, a total rest of mass M is well-defined and a constant of motion. We will consider M as a defining parameter of the system (not as a dynamical variable). The classical dynamics of the shell exhibits gravitational collapse to black holes and singularities. Thus, we can hope to obtain some hints on quantum theory of collapse.

The most interesting feature of the model is the incompleteness of its classical dynamics connected with the existence of singularities. The configuration space has a boundary where the radius R of the shell vanishes. This boundary can be reached in a finite time interval not only by classical trajectories (incompleteness of classical dynamics), but also by wave packets solving the Wheeler-DeWitt equation (incompleteness of Wheeler-DeWitt dynamics).

The main body of papers dealing with quantum field theory on curved space times assumes that the background spacetimes are globally hyperbolic. Then, the dynamics of the field equation must be complete, and this completeness plays an important role in the construction of quantum field theory (cf. [7] and [24]). Thus, we cannot use this kind of construction. Still, there are exceptions. Thus, in [25], the background spacetime for a quantum field theory is the anti-DeSitter spacetime, which is not globally hyperbolic. A well-defined quantum dynamics is obtained in [25] from an additional requirement of global conservation. Due to the large symmetry of anti-DeSitter space, there is a number of quantities like, e.g., the total energy, which are conserved "locally", that is for small intervals of time. Global conservation means one for all times, so that nothing can go in or out through the boundary. A similar point of view was adopted in [26] and [27] for our system. We will review these papers as well as some newer results [9].

We will find in particular that the condition of global conservation of probability and energy leads to existence and uniqueness of a quantum mechanics of the shell, if the total rest mass M is smaller than Planck mass. In this case, which may have to do with elementary particle scattering rather than with star collapse, no black hole is forming and no singularity existing. This may (but need not) mean that the model has more than just a pedagogical value.

3.2 Classical Analysis of the Model

In general relativity, thin shells are considered as three-dimensional timelike surfaces Σ forming a common boundary to two spacetimes, $\mathcal{M}_+$ and $\mathcal{M}_-$, where certain boundary conditions are imposed on all fields (see, e.g. [28]). In our case, the spacetime $\mathcal{M}_+$ outside the shell will be a portion of Schwarzschild solution with mass parameter E, which, in the units of energy, indicates the total energy of the system. The internal spacetime $\mathcal{M}_-$

will be a portion of Minkowski spacetime. In $\mathcal{M}_-$, there will be an inertial system of coordinates $\{T, X^1, X^2, X^3\}$ where the mass center of the shell lies at $X^1 = X^2 = X^3 = 0$; it is determined up to shifts of T and rotations of $X^k, k = 1,2,3$. Let us denote the radius of the shell in $\mathcal{M}_-$ by R. Any dynamical trajectory of the shell can be described by a function $R(t)$.

Einstein's equations are, then, equivalent to the following equation for $R(T)$ (see, e.g. [26]):

$$E = \frac{M}{\sqrt{1-\dot{R}^2}} - V, \tag{15}$$

where

$$V = \frac{m^2}{2R}.$$

and

$$m^2 = GM^2 = \left(\frac{M}{M_{\text{Planck}}}\right)^2,$$

as we choose the units such that $c = \hbar = 1$.

There are three types of classical solutions:

a) Bound solutions with $E < M$, which start at the singularity $R = 0$, expand to maximal radius and recollapse to $R = 0$.

b) Ingoing scattering solutions $E \geq M$, starting at $R = +\infty$, contracting, forming a black hole at $R = 2GE$ and collapsing down to $R = 0$.

c) Outgoing scattering solutions: just a time-reversal of b.

We observe that eq. (15) coincides formally with the dynamical equation for radial motion of a relativistic particle of mass M, spin 0 and electric charge q in the Coulomb field of a central charge Q, if

$$\epsilon_o Qq = -\frac{m^2}{2},$$

where ϵ_o is the dielectrical constant of vacuum. Thus, we can import methods and results from relativistic scalar electrodynamics; the corresponding system, relativistic hydrogen atom was studied long ago ([29], [30]).

Starting this program, we have to rewrite eq. (15) in form of a super-Hamiltonian, to get rid of the square root:

$$C = -(P_T + V)^2 + P_R^2 + M^2, \tag{16}$$

where P_T and P_R are momenta conjugate to T and R, respectively. The extended configuration space is $\mathcal{C} \cong \mathbf{R}_+ \times \mathbf{R}$, spanned by $T \in \mathbf{R}$ and $R \in \mathbf{R}_+$, $\mathbf{R}_+ = (0, \infty)$. To obtain eq. (15), we observe that

$$\{P_T, C\} = 0,$$

so P_T is a constant of motion. If we set

$$P_T = -E,$$

calculate P_R from the equation

$$\dot{R} = \frac{\{R, C\}}{\{T, C\}},$$

and substitute for P_T and P_R in the constraint equation, we obtain (15).

3.3 The First Steps of Quantization

Consider the complex linear space of function $\Psi(T, R)$,

$$\Psi : \mathcal{C} \to \mathbb{C},$$

and represent the canonical commutation rules for T, R, P_T and P_R by the operators on this space:

$$\hat{T} = T \quad , \quad \hat{R} = R,$$

$$\hat{P}_T = -i\frac{\partial}{\partial T} \quad , \quad \hat{P}_R = -i\frac{\partial}{\partial R}.$$

Then, the simplest factor order in $\hat{C}$ leads to the following operator constraint equation - the Wheeler-DeWitt equation – for Ψ:

$$-\left(i\frac{\partial}{\partial T} + V\right)^2 \Psi - \frac{\partial^2 \Psi}{\partial R^2} + M^2 \Psi = 0. \tag{17}$$

We observe that the Monge cone of eq. (17) is the light cone of the metric

$$ds^2 = dT^2 + dR^2.$$

It is the so-called DeWitt metric; it makes the configuration space $\mathcal{C}$ to a spacetime. The boundary $R = 0$ of $\mathcal{C}$ is timelike with respect to DeWitt metric; hence, the spacetime $\mathcal{C}$ is not globally hyperbolic.

To proceed further, we rewrite Eq. (17) in form of Schrödinger equation. To this aim, we introduce a two-component notation (the bold-face letters with denote matrices):

$$\boldsymbol{\Psi} = \begin{pmatrix} \psi_1 \\ \psi_2 \end{pmatrix},$$

where $\psi_1 = \Psi$ and

$$\psi_2 = \frac{\partial \Psi}{\partial T}.$$

Then, eq. (17) is equivalent to

$$i\frac{\partial}{\partial T}\Psi = \mathbf{H}\Phi, \tag{18}$$

where

$$\mathbf{H} = \begin{pmatrix} 0 & , & i \\ -i\mathcal{E} & , & -2V \end{pmatrix},$$

and

$$\mathcal{E} = -\frac{\partial^2}{\partial R^2} + M^2 - V^2.$$

The operator $\mathbf{H}$ is well defined on the domain $\mathcal{D}$:

$$\mathcal{D} = C_0^\infty(\mathbf{R}_+ \times \mathbf{R}) \times C_0^\infty(\mathbf{R}_+ \times \mathbf{R});$$

it will be called *Hamiltonian.*

A scalar product for eq. (19) can be obtained as follows. Consider an ansatz

$$q(\Psi, \Phi) = \int_0^\infty dR \Psi^\dagger \mathbf{Q} \Phi,$$

where $\mathbf{Q}$ is an R-dependent 2×2 matrix. The condition that $\mathbf{H}$ be symmetric with respect to $q(.,.)$ reads

$$\mathbf{H}^\dagger \mathbf{Q} - \mathbf{Q}\mathbf{H} = 0.$$

All hermitean solutions to this equation are given by

$$\mathbf{Q} = c\begin{pmatrix} 2V , & i \\ -i , & 0 \end{pmatrix}, c \in \mathbf{R}.$$

Let us set $c = \frac{1}{2}$. Then,

$$q(\Psi, \Phi) = \frac{1}{2}\int_0^\infty dR\,(2V\bar{\psi}_1\varphi_1 + i\bar{\psi}_1\varphi_2 - i\bar{\psi}_2\varphi_1) \tag{19}$$

is just the "Klein-Gordon product" corresponding to eq. (17). We call $q(.,.)$ *"charge form"*.

Another important form is the so-called "energy form", e, defined by

$$e(\Psi, \Phi) = q(\Psi, \mathbf{H}\Phi),$$

or

$$e(\Psi, \Phi) = \frac{1}{2}\int_0^\infty dR\left(\bar{\psi}_1'\varphi_1' + M^2\bar{\psi}_1\varphi_1 - V^2\bar{\psi}_1\varphi_1 + \bar{\varphi}_2\varphi_2\right), \tag{20}$$

where the prime denotes the derivative with respect to R.

3.4 What Kind of Mathematical Problem We Are Facing

Studying the "Hamiltonian" **H** and the "scalar product" $q(.,.)$, one arrives at two well-known (physical) problems: first, **H** is not bounded from below and $q(.,.)$ is not positive definite. The solution can be given according to the usual pattern of quantum field theory. The second problem is more serious: the dynamics defined by eq. (4) on $\mathcal{D}$ is not complete; some wave packets from $\mathcal{D}$ will evolve a non-zero value at $R = 0$ reaching the boundary in finite time interval and so will run out of $\mathcal{D}$. To define what happens further is a *real* problem.

The mathematical machinery to deal with such problems is the theory of self-adjoint extensions of symmetric operators (see, e.g. [31]). However, if we want to apply these tools, we have to formulate our aims more precisely. Let us formulate it as the following mathematical problem (later referred to as "the Problem").

We seek

(i) a Hilbert space $(\mathcal{H}, (.,.))$ in which $\mathcal{D}$ is dense, and

(ii) a self-adjoint extension $\tilde{\mathbf{H}}$ of **H** in $\mathcal{H}$, such that

(a) $\mathcal{H}$ splits into a "positive" and a "negative" subspaces, $\mathcal{H}_+$ and $\mathcal{H}_-$,

$$\mathcal{H} = \mathcal{H}_+ \bigoplus_{\perp} \mathcal{H}_-,$$

with the projectors

$$P_\pm \mathcal{H} = \mathcal{H}_\pm,$$

being invariant with respect to $\tilde{\mathbf{H}}$ and satisfying

$$\inf\, \sigma\left(\mathbf{H}_{|\mathcal{H}_+}\right) = E_0^+ > 0, \tag{21a}$$

$$\sup\, \sigma\left(\mathbf{H}_{|\mathcal{H}_-}\right) = E_0^- < 0, \tag{21b}$$

where $\sigma(A)$ is the spectrum of an operator A;

(b) if we define the form $\tilde{q}(.,.)$ by

$$\tilde{q}(\Psi, \Phi) = (\Psi, (P_+ - P_-)\Phi) \quad \forall \Psi, \Phi \in \mathcal{H},$$

then

$$q(\Psi, \Phi) = \tilde{q}(\Psi, \Phi) \quad \forall \quad \Psi, \Phi \in \mathcal{D};$$

(c) if $\tilde{e}(.,.)$ denotes the closure of the energy form $e(.,.)$, then

$$\tilde{e}(\Psi, \Phi) = \tilde{q}(\Psi, \tilde{\mathbf{H}}\Phi) \quad \forall \Psi, \Phi \in \mathrm{Dom}\tilde{\mathbf{H}} \cap \mathcal{H}_e.$$

Some comments are in order.

A) In our point (a), the positive and negative subspace, $\mathcal{H}_{\pm}$ are defined, and the condition of "mass gap" is expressed. This condition is crucial for our proofs of existence and uniqueness. As yet, we do not understand systems without mass gap; results of [25] seem to suggest that uniquness could be lost for such systems.

B) The condition (b) expresses the expectation that the scalar product $(.,.)$ is obtained from the charge form $q(.,.)$, if we change its sign on $\mathcal{H}_{-}$. The condition (c) requires that the relation between the Hamiltonian and the two forms $q(.,.)$ and $e(.,.)$ remains valid for their extensions.

C) There is no straightforward method to solve the above Problem. Indeed, self-adjoint extensions $\tilde{\mathbf{H}}$ of $\mathbf{H}$ are well-defined only, if the scalar product $(.,.)$ is known. This is, in turn, only defined if the positive and negative subspaces of the self-adjoint operator $\tilde{\mathbf{H}}$ are known. One can find an indirect way, if one derives some consequences of the conditions (a)-(c). We formulate them as the following corollaries.

Corollary 1: The extended energy form $\tilde{e}(.,.)$ is positive definite,

$$\tilde{e}(\Psi, \Psi) \geq E_0(\Psi, \Psi)$$

for all $\Psi \in \text{Dom}\tilde{\mathbf{H}} \cap \mathcal{H}_e$, where

$$E_0 = \min(E^+, -E_0^-).$$

Corollary 2: The extended charge form $q(.,.)$ is bounded by $\tilde{e}(.,.)$,

$$|\, \tilde{q}(\Psi, \Phi)\, |^2 \leq \frac{1}{E_0^2} \tilde{e}(\Psi, \Psi) \tilde{e}(\Phi, \Phi)$$

for all $\Psi, \Phi \in \text{Dom}\tilde{\mathbf{H}} \cap \mathcal{H}_e$.

Both Corollaries 1 and 2 hold on $\mathcal{D}$ so they are a kind of necessary conditions for the Problem to have solutions. The Corollary 2 is closely related to the existence of mass gap.

Corollary 3: The evolution operator $e^{i\tilde{\mathbf{H}}t}$ leaves the subspace $\mathcal{H}_e$ and the forms $\tilde{q}(.,.)$ and $\tilde{e}(.,.)$ invariant.

This is a "global conservation law".

Corollary 4: The operator $\tilde{\mathbf{H}}\, |_{\mathcal{H}_e}$ is self-adjoint on the "auxiliary" Hilbert space $(\mathcal{H}_e, \tilde{e})$.

This Corollary yields the desired method of solution: we look for self-adjoint extensions of $\mathbf{H}$ in the auxiliary Hilbert space; this is, in turn, directly constructible as the completion of $\mathcal{D}$ with respect to $e(.,.)$.

A general proof of existence and uniqueness of such an extension then proceeds as follows. The Corollary 2 implies that $q(.,.)$ is bounded on $\mathcal{D}$, so there is a bounded extension $\bar{q}(.,.)$ to $\mathcal{H}_e$ of $q(.,.)$. $\bar{q}(.,.)$ defines a unique bounded self-adjoint operator $\mathbf{h}$ on $\mathcal{H}_e$ such that

$$\bar{q}(\Psi,\Phi) = \tilde{e}(\Psi,\mathbf{h}\Phi) \quad \forall\Psi,\Phi \in \mathcal{H}_e$$

(see, e.g. [31]). The operator $\mathbf{h}$ satisfies the equation

$$e(\Psi,\Phi) = q(\Psi,\mathbf{h}^{-1}\Phi) \quad \forall\Psi,\Phi \in \mathcal{D}.$$

As $q(.,.)$ is non-degenerate, the restriction of $\mathbf{h}^{-1}$ to $\mathcal{D}$ must coincide with $\mathbf{H}$. Thus, $\mathbf{h}^{-1}$ is a self-adjoint extension of $\mathbf{H}$ in $\mathcal{H}_e$.

That proves the existence. To prove the uniqueness, let us assume that $\bar{\mathbf{H}}$ is a self-adjoint extension of $\mathbf{H}$ in $\mathcal{H}_e$ that is different from $\mathbf{h}^{-1}$. One can show that two different self-adjoint extensions of the same symmetric operator cannot commute, hence,

$$e^{i\bar{\mathbf{H}}t}\mathbf{h} \neq \mathbf{h}e^{i\bar{\mathbf{H}}t}.$$

Then,

$$\bar{q}(e^{i\bar{\mathbf{H}}t}\Psi, e^{i\bar{\mathbf{H}}t}\Phi) = \tilde{e}(e^{i\bar{\mathbf{H}}t}\Psi, \mathbf{h}e^{i\bar{\mathbf{H}}t}\Phi) =$$
$$\tilde{e}(\Psi, e^{-i\bar{\mathbf{H}}t}\mathbf{h}e^{i\bar{\mathbf{H}}t}\Phi) \neq \tilde{e}(\Psi,\mathbf{h}\Phi)$$

at least for some $\Psi,\Phi \in \mathcal{H}_e$, so $\bar{q}(.,.)$ is not preserved by $e^{i\bar{\mathbf{H}}t}$, and $\bar{\mathbf{H}}$ does not solve the Problem according to the Corollary 3.

The construction of $(\mathcal{H},(.,.))$ and $\bar{\mathbf{H}}$ is now standard (see, e.g. [24]). Observe that we have used only the "abstract" properties as given in points (a)-(c) as yet. Thus, the described procedure is a general method of dealing with Wheeler-DeWitt equation. It will work, provided there is a mass gap.

3.5 Application to the shell

We start by checking the necessary conditions. As to the positivity of $e(.,.)$, eq. (20) shows immediately that there may be a problem due to the potential term. One can show the following theorem.

Theorem 4. Let $m \leq 1$. Then,

$$e(\Psi,\Psi) \geq \frac{M^2}{2}\int_0^\infty dR\Psi^\dagger\Psi \quad \forall\Psi \in \mathcal{D}.$$

Moreover, for $m > 1$, there is $\Psi \in \mathcal{D}$ such that $e(\Psi, \Psi) < 0$.

Similarly, the e-boundedness of $q(.,.)$ (the mass gap) can be shown only for small M:

Theorem 5. Let $m < 1$. Then,

$$| q(\Psi, \Phi) | \leq 2\frac{m^2 + \sqrt{1 - m^4}}{M^2\sqrt{1 - m^4}} \|\Psi\|_e \|\Phi\|_e$$

for all $\Psi, \Phi \in \mathcal{D}$. Moreover, if $m \geq 1$, then $q(.,.)$ is not bounded by $e(.,.)$.

As a consequence of theorems 4 and 5 we have to distinguish two regimes: the so-called "heavy" and "light" systems.

3.5.1 The Heavy Systems

These are defined by $M \geq M_{\text{Planck}}$. One can think of collapsing stars, for example. Our proof of existence and uniqueness of a unitary, energy-conserving quantum mechanics does not work in this case.

The situation is in some respect analogous to that in relativistic electrodynamics of the hydrogen atom. For a super-critial central charge

$$Z \gtrsim \frac{137}{2}$$

(this holds in the scalar electrodynamics, see, e.g. [32]), there is no one-particle Hilbert space, either. In the full-fledged, second quantized electrodynamics, the super-critical charge cannot exist longer time: it will produce pairs and polarize vacuum. A "second quantization" of our model would be, in principle, possible. The problem that such a procedure would mean a "third quantization" of gravity is absent, as the shell matter has been only first-quantized as yet. However, the analogy to electrodynamics ceases here, because the sign of the potential term cannot be inverted for gravity.

Thus, behaviour of the heavy systems can be quite different from what we know about electrodynamics. In particular, there may be room for black hole formation, Hawking's effect, etc. As yet, however, there are only speculations.

3.5.2 The Light Systems

These are defined by $M < M_{\text{Planck}}$; one could consider scattering of elementary particles, for example. In this case, our general method works and there is a unique unitary and energy preserving theory. We have even explicitly constructed this theory [9]. We have calculated the deficiency subspaces (see, e.g. [31]) and obtained a particular boundary condition for wave functions at $R = 0$. This coincides with Sommerfeld's boundary condition [29]. Thus, we can take over all old well-known bound state stationary wave functions, changing only some numerical values; one finds easily the scattering states [27].

The spectrum of the stationary bound states is given by

$$E_n = M\frac{2(\lambda+n)}{\sqrt{m^4+4(\lambda+n)^2}} \quad , \quad n = 0, 1, \ldots,$$

where

$$\lambda = \frac{1}{2} + \frac{1}{2}\sqrt{1-m^4}$$

The spectrum of the stationary scattering states is (M, ∞). All stationary wave functions are time-reversal symmetric. They vanish, together with the corresponding probability and energy currents, at $R = 0$.

These results suggest that there is no black hole formation and that the shell bounces at $R > 0$ so that no singularity forms, either. It is amusing to observe that the quantum system has no problem with "penetrating causal barriers" that are unsurmountable for classical systems. Technically, this is based on superposition principle: for example, the scattering wave function is a linear combination of ingoing and outgoing modes; the spacetime geometry implied by the ingoing mode contains a black hole and shrinking shell whereas that of the outgoing mode contains a white hole and expanding shell. The whole linear combination does not define any "sharp" spacetime.

These considerations throw some light on the semiclassical approximation, which is sometimes used in quantum model collapse calculations. Such an approximation is always associated with one particular spacetime which holds all the way down to the horizon and singularity. It seems, therefore, not to be an adequate approach, at least for light systems.

In this connection, one can mention some resent papers ([32], [33]) on two-dimensional models of gravity ("dilaton black holes"), in which a *non-perturbative* approach is attempted. These methods could confirm or disprove our suggestions (all fields in these models are massless, so they seem to be "light").

Acknowledgement. Useful discussions with A. Ashtekar, A. Higuchi and C. Isham are acknowledged.

References

[1] C.J. Isham: Conceptual and Geometrical Problems in Quantum Gravity. In: H. Mitter, H. Gausterer (Eds.) "Recent Aspects of Quantum Fields". Springer-Verlag, Berlin (1991) 123-230.

[2] K.V. Kuchař: Time and Interpretations of Quantum Gravity. In: Proceedings of the 4th Canadian Conference on General Relativity and Relativistic Astrophysics. World Scientific, Singapore (1992) 211-314.

[3] C.J. Isham: Canonical Quantum Gravity and the Problem of Time. In: "Integrable Systems, Quantum Groups, and Quantum Field Theories". Kluwer Academic Publishers, London (1993) 157-288.

[4] A. Ashtekar: Lectures on Non-Perturbative Canonical Gravity. World Science, Singapore, 1991.

[5] R.S. Tate: An Algebraic Approach to the Quantization of Constrained Systems: Finite Dimensional Examples. Ph.D. Thesis, Syracues University, 1992.

[6] A. Ashtekar, R. Tate, C. Uggla: Int.J.Mod.Phys. **D2**, 15 (1993).

[7] A. Ashtekar, R.S. Tage, C. Uggla: Minisuperspaces: Symmetries and Quantization. Preprint SU-CP-92/2-5 (1993).

[8] K.V. Kuchař, in *Conceptual Problems of Quantum Gravity* Eds. A. Ashtekar and J. Stachel. Birkhauser, Boston, 1991.

[9] P. Hájiček and B.S. Kay: Quantum Collapse of a Self-Gravitating Shell: Construction of a Quantum Mechanics. Preprint, Santa Barbara, 1993.

[10] P.A.M. Dirac: Lectures on Quantum Mechanics. New York, Yeshiva University, 1964.

[11] P.A.M. Dirac: Rev.Mod.Phys. **21**, 392 (1949).

[12] W.G. Unruh and R.M. Wald: Phys.Rev. **D40**, 2598 (1989).

[13] C. Rovelli: Phys.Rev. **D42**, 2638 (1990); Phys.Rev. **D43**, 442 (1991).

[14] K.V. Kuchař: Canonical Quantization of Generally Covariant Systems. Talk at International Conference on Gravitation and Cosmology, Goa, India, 14-19 Dec. 1987.

[15] R. Abraham and J.E. Marsden: Foundation of Mechanics. London, Benjamin, 1985.

[16] P. Hájiček and K.V. Kuchař: Phys.Rev. **D41**, 1091 (1990).

[17] P. Hájiček: Class. Quantum Gravity **7**, 871 (1990)

[18] I.M. Gel'fand and Ya. Vilenkin: Generalized Functions. Vol. IV. Academic Press, New York, 1965.

[19] A. Rendall: Unique Determination of an Inner Product by Adjointness Relation in the Algebra of Quantum Observables. Class. Quantum Grav. 10 (1993) 2261-2269.

[20] B.C. DeWitt: in *Relativity, Groups and Topology*. Eds. C. DeWitt and B.C. DeWitt. New York, Gordon & Breach, 1964.

[21] A.O. Barut and R. Raczka: Theory of Group Representations and Applications. Warshawa, Polish Scientific Publishers, 1977.

[22] I.M. Gel'fand and C.E. Shilov: Generalized Functions. Vol. I. Academic Press, New York, 1964.

[23] K.V. Kuchař: J.Math.Phys. **22**, 2640 (1981).

[24] B.S. Kay: Commun.Math. Phys. **62**, 55 (1978).

[25] S.J. Avis, C.J. Isham and D. Storey: Phys.Rev. **D18**, 3565 (1978).

[26] P. Hájiček: Commun.Math.Phys. **150**, 545 (1992).

[27] P. Hájiček, B.S. Kay and K.V. Kuchař: Phys.Rev. **D46**, 5439 (1992).

[28] W. Israel: Nuovo Cimento **44B**, 1 (1966);
K.V. Kuchař: Czech.J.Phys. **B18**, 435 (1968).

[29] A. Sommerfeld: Wave Mechanics. Dutton, New York, 1930.

[30] H. Bethe: Intermediate Quantum Mechanics. Benjamin, New York, 1965.

[31] M. Reed and B. Simon: Methods of Modern Mathematical Physics. Vol. I and II. Academic Press, New York, 1973.

[32] J.D. Bjorken and S.D. Drell: Relativistic Quantum Mechanics. McGraw-Hill, New York, 1964.

[33] E. Verlinde and H. Verlinde: A Unitary S-Matrix for $2d$ Black Hole Formation and Evaporation. Preprint, PUPT-1380, IASSNS-HEP-93/8 (1993).
E. Verlinde, H. Verlinde and D. Schoutens: Quantum Black Hole Evaporation. Preprint, PUPT-1395, IASSNS-HEP-93/25 (1993)

[34] A. Mikovic: Non-Perturbative Two-Dimensional Dilation Gravity. Preprint, Imperial-TP/92-93/44.

Canonical Quantum Gravity and the Question of Time

C.J. Isham

1 Summary of Canonical General Relativity

1.1 The ADM Formalism for Classical General Relativity

The purpose of this lecture is to provide a short introduction to the canonical quantisation of general relativity. I have elected to focus on the 'problem of time': partly because this is of considerable interest in its own right, and partly because it provides a useful way of organising the various approaches to canonical quantum gravity. What follows is very much a 'bare-bones' account covering just the basic ideas. For more details and comprehensive reference lists the reader should consult one of the recent reviews of the problem of time in quantum gravity [Kuc92a, Kuc92b, Ish93a].

Any discussion of canonical quantum gravity must begin with an account of the famous Arnowitt, Deser and Misner (ADM) analysis of the canonical structure of classical general relativity. Fortunately, this has been covered in depth in the lectures of Beig and Wipf, and here I shall state just those results that are most useful for our purposes.

1. The starting point is a four-dimensional spacetime manifold M that is diffeomorphic to $\Sigma \times \mathbb{R}$ where Σ is some compact three-manifold [1]. This topological property of M allows the choice of a reference foliation $\mathcal{F}^{\mathrm{ref}} : \Sigma \times \mathbb{R} \rightarrow M$ with respect to which the canonical decomposition will be performed.

2. Now let g be a Lorentzian metric on M and let $\mathcal{F}^{\mathrm{ref}}$ be such that each leaf $\mathcal{F}^{\mathrm{ref}}_t : \Sigma \rightarrow M, t \in \mathbb{R}$, is spacelike with respect to g. Thus, in a formal sense, the foliation parameter $t \in \mathbb{R}$ can serve as a definition of 'time'. However, such a parameter has no direct physical meaning: indeed, one of the main problems in the canonical formalism is to introduce a notion of time that is more physical; the reference foliation $\mathcal{F}^{\mathrm{ref}}$ is then part of the background structure, and the final physical results may well be required to be independent of its choice.

As explained in Beig's lectures, the Lorentzian metric is projected along, and normal to, the hypersurfaces of the foliation to give a canonical picture in which

[1] If Σ is non-compact, several of the expressions that follow must be augmented with surface terms.

the configuration variable is a Riemannian metric q on Σ whose conjugate p is related to the extrinsic curvature K by [2] $p^{ab} := -\frac{1}{\kappa^2}|q|^{\frac{1}{2}}(K^{ab} - q^{ab}K_c{}^c)$.

3. The canonical version of the field equations of general relativity can be derived from the action

$$S[q,p,N,\mathbf{N}] = \int dt \int_\Sigma d^3x \, (p^{ab}\dot{q}_{ab} - N\mathcal{H}_\perp - N^a\mathcal{H}_a) \tag{1}$$

where N is the lapse function, N^a is the shift vector, and $\mathcal{H}_a$ and $\mathcal{H}_\perp$ are defined by [3]

$$\mathcal{H}_a(x;q,p] := -2D_b\, p_a{}^b(x) \tag{2}$$

$$\mathcal{H}_\perp(x;q,p] := \kappa^2 \mathcal{G}_{ab\,cd}(x,q]\, p^{ab}(x)\, p^{cd}(x) - \frac{|q|^{\frac{1}{2}}(x)}{\kappa^2} R(x;q] \tag{3}$$

where

$$\mathcal{G}_{ab\,cd}(x;q] := \frac{1}{2}|q|^{-\frac{1}{2}}(x)\big(q_{ac}(x)q_{bd}(x) + q_{bc}(x)q_{ad}(x) - q_{ab}(x)q_{cd}(x)\big) \tag{4}$$

is the 'DeWitt supermetric' on the space of three-metrics [DeW67] and $\kappa^2 := 8\pi G/c^2$. The functions $\mathcal{H}_a$ and $\mathcal{H}_\perp$ of the canonical variables (q,p) play a key role in the theory and are known as the *supermomentum* and *super-Hamiltonian* respectively.

4. The functional

$$H[N,\mathbf{N}](t) := \int_\Sigma d^3x \, (N\mathcal{H}_\perp + N^a\mathcal{H}_a) \tag{5}$$

of the canonical variables (q,p) acts as the Hamiltonian of the system. Varying the action (1) with respect to q_{ab} and p^{ab} gives the dynamical equations

$$\dot{q}_{ab}(x,t) = \frac{\delta H[N,\mathbf{N}]}{\delta p^{ab}(x,t)}, \tag{6}$$

$$\dot{p}^{ab}(x,t) = -\frac{\delta H[N,\mathbf{N}]}{\delta q_{ab}(x,t)} \tag{7}$$

while varying N^a and N leads respectively to

$$\mathcal{H}_a(x;q,p] = 0 \tag{8}$$

and

$$\mathcal{H}_\perp(x;q,p] = 0 \tag{9}$$

which are *constraints* on the canonical variables (q,p).

---

[2] The Roman letters a,b refer to a coordinate system on the three-manifold Σ.

[3] The notation $F(x;f]$ means that F is a *function* of a point x in a finite-dimensional manifold and a '*functional*' of the function f (*i.e.*, f is a point in an infinite-dimensional manifold).

5. The classical canonical algebra of the system is expressed by the basic Poisson brackets

$$\{q_{ab}(x), q_{cd}(x')\} = 0 \tag{10}$$
$$\{p^{ab}(x), p^{cd}(x')\} = 0 \tag{11}$$
$$\{q_{ab}(x), p^{cd}(x')\} = \delta^c_{(a}\delta^d_{b)}\,\delta(x, x') \tag{12}$$

which can be used to cast the dynamical equations into a canonical form in which the right hand sides of (6) and (7) are $\{q_{ab}(x), H[N, \mathbf{N}]\}$ and $\{p^{ab}(x), H[N, \mathbf{N}]\}$ respectively.

A crucial property of the canonical formalism is the closure of the Poisson brackets of the super-Hamiltonian and supermomentum, computed using (10–12); *i.e.*, the set of constraints is *first class*. This is contained in the fundamental relations

$$\{\mathcal{H}_a(x), \mathcal{H}_b(x')\} = -\mathcal{H}_b(x)\,\partial^{x'}_a\delta(x, x') + \mathcal{H}_a(x')\,\partial^x_b\delta(x, x') \tag{13}$$
$$\{\mathcal{H}_a(x), \mathcal{H}_\perp(x')\} = \mathcal{H}_\perp(x)\,\partial^x_a\delta(x, x') \tag{14}$$
$$\{\mathcal{H}_\perp(x), \mathcal{H}_\perp(x')\} = q^{ab}(x)\,\mathcal{H}_a(x)\,\partial^{x'}_b\delta(x, x') - q^{ab}(x')\,\mathcal{H}_a(x')\,\partial^x_b\delta(x, x') \tag{15}$$

known as the *Dirac algebra*. Using the smeared variables[4] $H[N] := \int_\Sigma d^3x\, N(x)\,\mathcal{H}_\perp($ and $H[\mathbf{N}] := \int_\Sigma d^3x\, N^a(x)\,\mathcal{H}_a(x)$, where N and $\mathbf{N}$ are any scalar function and vector field on Σ, these equations can be written as

$$\{H[\mathbf{N}_1], H[\mathbf{N}_2]\} = H([\mathbf{N}_1, \mathbf{N}_2]) \tag{16}$$
$$\{H[\mathbf{N}], H[N]\} = H[L_{\mathbf{N}} N] \tag{17}$$
$$\{H[N_1], H[N_2]\} = H[\mathbf{N}] \tag{18}$$

where, in (18), $N^a(x) := q^{ab}(x)\big(N_1(x)N_2(x)_{,b} - N_2(x)N_1(x)_{,b}\big)$, and $L_{\mathbf{N}}$ denotes Lie differentiation along the vector field $\mathbf{N}$.

The geometrical interpretation of these expressions is as follows.

1. The Lie algebra of the diffeomorphism group Diff(Σ) is generated by the vector fields on Σ with minus the commutator of a pair of vector fields $[N_1, N_2]^a := N_1^b\, N_2^a{}_{,b} - N_2^b\, N_1^a{}_{,b}$ playing the role of the Lie bracket. Thus (16) shows that the map $\mathbf{N} \to -H[\mathbf{N}]$ is a homomorphism of the Lie algebra of Diff(Σ) into the Poisson bracket algebra of the theory.
2. Similarly, $H[N]$ can be interpreted as generating deformations of a hypersurface *normal* to itself as embedded in M. However, unlike the analogous statement for $H[\mathbf{N}]$, this interpretation applies only *after* the field equations have been solved.

Two important things should be noted about the 'gauge' algebra (16–18):

[4] The integrals are well-defined since both $\mathcal{H}_a$ and $\mathcal{H}_\perp$ are densities on Σ.

1. It is *not* the Lie algebra of Diff(M) even though this was the covariance group of the original theory.
2. The presence of the q^{ab} factor in the right hand side of (18) means it is *not* a genuine Lie algebra at all.

These two features are closely related since the Dirac algebra (16–18) is essentially the Lie algebra of Diff(M) *projected* along, and normal to, a spacelike hypersurface. The significance of this in the quantum theory will emerge later.

1.2 The Role of the Constraints

The constraints (8–9) are of major importance in both the classical and the quantum theories of gravity, and it is useful at this point to gather together various results concerning them.

1. The constraints are consistent with the equations of motion (6) and (7) in the sense of being automatically maintained in time. For example, $d\mathcal{H}_\perp(x)/dt = \{\mathcal{H}_\perp(x), H[N, \mathbf{N}]\}$, and the right hand side vanishes on the constraint surface in phase space by virtue of the closing nature of the algebra (13–15).
2. The constraints lead to a well-posed Cauchy problem for the dynamical equations once the undetermined quantities N and $\mathbf{N}$ have been fixed in some way.
3. If a Lorentzian metric g on the spacetime M satisfies the vacuum Einstein equations $G_{\alpha\beta}(X) = 0$ then the constraint equations (8–9) are satisfied on all spacelike hypersurfaces of M.
4. Conversely, let (M, g) be a Lorentzian spacetime with the property that the constraint equations (8–9) are satisfied on every spacelike hypersurface. Then g necessarily satisfies all ten Einstein field equations $G_{\alpha\beta}(X) = 0$.

This last result means that the dynamical aspects of the Einstein equations are already contained in the constraints alone. This plays a crucial role in the Dirac quantisation programme discussed in section 3.

1.3 Internal Time

The use of an 'internal time' is a significant feature of most approaches to canonical quantum gravity. The key idea is to introduce quantities $\mathcal{T}(x; q, p]$ and $\mathcal{Z}^a(x; q, p]$, $a = 1, 2, 3$, that can serve as the time and spatial coordinates for a spacetime event; the collection of four functions $(\mathcal{T}, \mathcal{Z}^a)$ will be written as $\mathcal{X}^A$, $A = 0, 1, 2, 3$. As we shall see, this is related to the problem of rewriting (1) as a true canonical action for just the physical modes of the gravitational field. Thus the discussion that follows is complementary to the treatment in Beig's lectures of the reduced formalism. The main steps are as follows.

1. Start with a *given* Lorentzian metric g on M and fix a reference foliation $\mathcal{F}^{\mathrm{ref}} : \Sigma \times \mathbb{R} \to M$ with spacelike leaves.

2. Pick functionals $\mathcal{T}(x;q,p]$ and $\mathcal{Z}^a(x;q,p]$ and locate an event X in M by assigning it the 'internal' space and time coordinates:

$$T(\mathcal{F}^{\text{ref}}(x,t)) := \mathcal{T}(x;q(t),p(t)] \tag{19}$$

$$Z^a(\mathcal{F}^{\text{ref}}(x,t)) := \mathcal{Z}^a(x;q(t),p(t)] \tag{20}$$

where $x \in \Sigma$ and $t \in \mathbb{R}$ are such that $X = \mathcal{F}^{\text{ref}}(x,t)$, and where $q(t)$ and $p(t)$ denote the metric and momentum induced from g on the hypersurface $\mathcal{F}_t^{\text{ref}}(\Sigma) := \{\mathcal{F}^{\text{ref}}(x,t) \mid x \in \Sigma\}$ of M.

3. Perform a canonical transformation

$$\left(q_{ab}(x), p^{cd}(x)\right) \mapsto \left(\mathcal{X}^A(x), \mathcal{P}_B(x); \phi^r(x), \pi_s(x)\right) \tag{21}$$

in which the $12 \times \infty^3$ variables $\left(q_{ab}(x), p^{cd}(x)\right)$ are mapped into

- the four functions $\mathcal{X}^A(x)$ specifying a particular choice of internal space and time coordinates;
- their four conjugate momenta $\mathcal{P}_B(x)$;
- the two modes $\phi^r(x)$, $r = 1,2$, which represent the physical degrees of freedom of the gravitational field;
- their conjugate momenta $\pi_s(x)$, $s = 1,2$.

The statement that $\mathcal{P}_B$ are the momenta conjugate to the four internal coordinate variables $\mathcal{X}^A$ means they satisfy the Poisson bracket relations (computed using the basic Poisson brackets (10–12))

$$\{\mathcal{X}^A(x), \mathcal{X}^B(x')\} = 0 \tag{22}$$

$$\{\mathcal{P}_A(x), \mathcal{P}_B(x')\} = 0 \tag{23}$$

$$\{\mathcal{X}^A(x), \mathcal{P}_B(x')\} = \delta^A_B\, \delta(x,x') \tag{24}$$

while the corresponding relations for the 'physical' variables (ϕ^r, π_s) are

$$\{\phi^r(x), \phi^s(x')\} = 0 \tag{25}$$

$$\{\pi_r(x), \pi_s(x')\} = 0 \tag{26}$$

$$\{\phi^r(x), \pi_s(x')\} = \delta^r_s\, \delta(x,x'). \tag{27}$$

It is also assumed that all cross brackets of $\mathcal{X}^A$ and $\mathcal{P}_B$ with ϕ^r and π_s vanish:

$$\begin{aligned} 0 = \{\phi^r(x), \mathcal{X}^A(x')\} = \{\phi^r(x), \mathcal{P}_B(x')\} = \\ \{\pi_s(x), \mathcal{X}^A(x')\} = \{\pi_s(x), \mathcal{P}_B(x')\}. \end{aligned} \tag{28}$$

Note that some or all of these relations may need to be generalised to take account of the global topological properties of the function spaces concerned. There may also be global obstructions to some of the steps.

4. Express the super-Hamiltonian and supermomentum as functionals of these new canonical variables and write the canonical action (1) as

$$S[\phi, \pi, N, \mathbf{N}, \mathcal{X}, \mathcal{P}] = \int dt \int_\Sigma d^3x \left(\mathcal{P}_A \dot{\mathcal{X}}^A + \pi_r \dot{\phi}^r - N\mathcal{H}_\perp - N^a \mathcal{H}_a\right) \tag{29}$$

where all fields are functions of x and t, and where $\mathcal{H}_\perp$ and $\mathcal{H}_a$ are rewritten as functionals of $\mathcal{X}_A$, $\mathcal{P}_B$, ϕ^r and π_s. Note that ϕ^r, π_s, $\mathcal{X}^A$, $\mathcal{P}_B$, N and $\mathbf{N}$ are all to be varied in (29) as independent functions.

5. Remove $4 \times \infty^3$ of the $8 \times \infty^3$ non-dynamical variables by solving [5] the constraints $\mathcal{H}_\perp(x) = 0$ and $\mathcal{H}_a(x) = 0$ for the variables $\mathcal{P}_A(x)$ in the form

$$\mathcal{P}_A(x) + h_A(x; \mathcal{X}, \phi, \pi] = 0. \tag{30}$$

6. Remove the remaining $4 \times \infty^3$ non-dynamical variables by 'deparametrising' the canonical action functional (29) by substituting into it the solution (30) of the initial value equations. This gives

$$S[\phi, \pi] = \int dt \int_\Sigma d^3x \left\{ \pi_r(x,t)\dot{\phi}^r(x,t) - h_A\big(x; \chi_t, \phi(t), \pi(t)\big] \dot{\chi}_t^A(x) \right\} \tag{31}$$

which yields the correct field equations for the physical fields ϕ^r and π_s. Note that, in (31), the four quantities $\mathcal{X}^A$ are no longer to be varied but have instead been set equal to some *prescribed* functions χ_t^A of x. This is valid since, after solving the constraints, the remaining dynamical equations of motion give no information about how the variables $\mathcal{X}^A$ evolve in parameter time t.

The lapse function and shift vector play no part in this reduced variational principle. However, it will be necessary to reintroduce them if one wishes to return to a genuine spacetime picture. This can be done by solving the additional set of Einstein equations that are missing from the set generated by the reduced action. These are elliptic partial-differential equations for N and $\mathbf{N}$.

The equations of motion derived from the reduced action (31) are those corresponding to the Hamiltonian

$$H_{\text{true}}(t) := \int_\Sigma d^3x\, \dot{\chi}_t^A(x)\, h_A\big(x; \chi_t, \phi(t), \pi(t)\big] \tag{32}$$

and can be written in the form

$$\frac{\partial \phi^r(x,t)}{\partial t} = \left\{ \phi^r(x,t), H_{\text{true}}(t) \right\}_{\text{red}} \tag{33}$$

$$\frac{\partial \pi_s(x,t)}{\partial t} = \left\{ \pi_s(x,t), H_{\text{true}}(t) \right\}_{\text{red}} \tag{34}$$

where $\{\, ,\}_{\text{red}}$ denotes the Poisson bracket evaluated using only the physical modes ϕ^r and π_s. Thus, at least formally, the system has been reduced to one that looks like a conventional canonical field theory.

[5] There may be global obstructions to this step.

1.4 The Multi-time Formalism

Although a particular time label t has been used in the above (corresponding to a reference foliation of M into a one-parameter family of hypersurfaces), this label is in fact quite arbitrary. This is reflected in the arbitrary choice of the functions χ_t^A in step 6. Indeed, the physical fields (ϕ^r, π_s) ought to depend only on the hypersurface in M on which they are evaluated, not on the way in which that hypersurface happens to be included in a particular foliation. Thus it should be possible to write the physical fields ϕ^r, π_s as functions of $x \in \Sigma$ and functionals of the internal coordinates which are now regarded as *arbitrary* functions χ^A of $x \in \Sigma$ rather than being some fixed set.

This is indeed the case. More precisely, the Hamiltonian equations of motion (33–34), which hold for all choices of the functions χ^A, imply that there exist Hamiltonian densities $h_A(x;\chi]$ such that $\phi^r(x;\chi]$ and $\pi_s(x;\chi]$ satisfy the functional differential equations

$$\frac{\delta\phi^r(x;\chi]}{\delta\chi^A(x')} = \{\phi^r(x;\chi], h_A(x';\chi]\}_{\text{red}} \tag{35}$$

$$\frac{\delta\pi_s(x;\chi]}{\delta\chi^A(x')} = \{\pi_s(x;\chi], h_A(x';\chi]\}_{\text{red}}. \tag{36}$$

This so-called 'bubble-time' or 'multi-time' canonical formalism will play an important role in our discussions of the problem of time in quantum gravity [Kuc72].

2 Time Chosen Before Quantisation

2.1 Preliminary Remarks

We must now consider how to quantise the classical canonical structure defined above. In the notation of my first lecture, we are concerned with a type-I scheme in which some quantum algorithm is to be applied to the given classical system. As emphasised in that lecture, there are a number of important *prima facie* questions that can be asked of any approach to quantum gravity; in the context of canonical quantum gravity, the following points are particularly relevant:

1. Canonical quantisation employs a fixed *background* three-manifold Σ but no background metric will be assumed.
2. The problem of *time* is a major issue and can be addressed in a variety of different ways. The following questions are especially important:
   - Is 'time' an exact concept, or one that is valid only in some semi-classical sense?
   - At what stage in the procedure is it necessary to identify a 'physical' time? In particular, must this be done before quantising, or can a quantum theory be constructed first and then a notion of time extracted afterwards? This is closely related to the question of whether the key constraints $\mathcal{H}_a = 0 = \mathcal{H}_\perp$ are to be imposed before, or after, quantising.

- How is the notion of time related to the definition of a physical 'observable'? And how is this reflected in the probabilistic interpretation of the quantum theory?

3. The role played by spacetime *diffeomorphisms* depends on how the constraints are handled. In particular, the Dirac algebra (13–15) is the canonical equivalent of the Lie algebra of the spacetime diffemorphism group Diff(M).

In what follows, these issues are discussed in terms of the standard metric-based approaches to canonical quantum gravity. As a consequence, severe operator-divergence and operator-ordering problems arise. The Ashtekar version of canonical general relativity is much better in this respect and is discussed in detail in several of the other contributions to this volume.

2.2 The Main Ideas

One major approach to canonical quantum gravity involves quantising the classical reduced theory discussed above. Thus the first step is to select internal coordinate functions $\mathcal{T}$ and $\mathcal{Z}^a$, and to perform the canonical transformation (21). The constraints are then solved classically (30) to yield the classical multi-time evolution equations (35–36).

The first stage in quantisation is to replace the Poisson brackets (25–27) of the physical gravitational modes with canonical commutation relations [6]

$$[\widehat{\phi}^r(x), \widehat{\phi}^s(x')] = 0 \tag{37}$$
$$[\widehat{\pi}_r(x), \widehat{\pi}_s(x')] = 0 \tag{38}$$
$$[\widehat{\phi}^r(x), \widehat{\pi}_s(x')] = i\hbar\delta^r_s\, \delta(x, x'). \tag{39}$$

The next step is to convert the classical Hamiltonian densities h_A in (30) into operator versions formed from the operators $\widehat{\phi}^r(x)$, $\widehat{\pi}_s(x)$, and then take the Heisenberg-picture analogue of the classical dynamical equations (35–36), or else use the operators $\widehat{h}_A$ to form the multi-time Schrödinger equation

$$i\hbar\frac{\delta\Psi_\chi}{\delta\chi^A(x)} = h_A(x; \chi, \widehat{\phi}, \widehat{\pi}]\Psi_\chi \tag{40}$$

for the state vector Ψ_χ.

The main advantages of this approach are as follows.

1. Time is external to the system being quantised and, in this sense, we are close to a standard type of quantum theory.
2. Modulo technical problems with operator divergences and ordering, there is an unambiguous Hilbert space structure provided by the representation of the canonical commutation relations (37–39). There also appears to be a natural probabilistic interpretation in which Ψ_χ is the probability distribution for physical modes on a spacetime surface labelled by the four functions $\chi^A(x)$ (but see below).

[6] In fact, the topological structure of the reduced space is quite complicated and (37–39) could not apply globally. However, we shall ignore such subtleties here.

3. A physical *observable* can be defined unambiguously as any function (possibly χ-dependent) of the physical modes (ϕ_r, π^s).

2.3 The Main Problems

Unfortunately, the positive factors of this approach to quantisation are offset by a number of major questions and problems. I shall list these in some detail as they illustrate many important issues, most of which have analogues in other approaches to canonical quantum gravity.

What Does the Formalism Mean? The Schrödinger equation (40) describes the 'evolution' of the state as a functional of the functions χ^A. However, in the classical theory, the choice of a specific set of such 'gauge' functions is related (via the internal coordinate functionals $\mathcal{X}^A$) to a particular hypersurface of M only *after* the full set of classical field equations has been solved (including the equations for the lapse function and shift vector). Correspondingly, the quantum equation (40) describes how functionals of certain pieces of the canonical data depend on the values of other pieces, but it has no immediate spacetime interpretation.

The Spacetime Problem. A related problem with spacetime interpretation arises from the fact that if a different reference foliation is used then the 'equal-t' hypersurface passing through a point X in M is likely to change, and then so is the 'time' value $T(X)$ allotted to the event X. Viewed from a spacetime perspective, this feature is pathological since coordinates on a manifold are local scalar functions. It is physically highly undesirable since it means that the results and the interpretation of the theory depend on the choice of the background foliation $\mathcal{F}^{\text{ref}}$.

This behaviour of the internal time $\mathcal{T}(x; q, p]$ can be avoided only if it has a vanishing Poisson bracket with the generator of 'tilts' or 'bends' of the hypersurface. In particular, we require $\{\mathcal{T}(x), H[N]\} = 0$ for all test functions N that vanish at a point $x \in \Sigma$ [Kuc91]. This requirement is rather strong and clearly excludes anything like $R(x; q]$ that is a local functional of the three-geometry alone. It also excludes many choices that are functionals of both q and p, including the mean extrinsic curvature function $\mathcal{T}(x; g] := q_{ab}(x)p^{ab}(x)/|q|(x)$ (see the lectures by Beig).

Functionals $\mathcal{T}(x; q, p]$ of the canonical data that *are* local spacetime scalars do exist; for example, take the square ${}^{(4)}R_{\alpha\beta\gamma\delta}(X; g]\, {}^{(4)}R^{\alpha\beta\gamma\delta}(X; g]$ of the Riemann curvature tensor of a vacuum Einstein spacetime and re-express it in terms of the canonical data on a spacelike hypersurface. But not all functionals $\mathcal{T}(x; q, p]$ of this type can serve as time functions. Two necessary conditions are:

1. $\{\mathcal{T}(x), \mathcal{T}(x')\} = 0$;
2. For any given Lorentzian metric g on M that satisfies the vacuum Einstein field equations, the hypersurfaces of equal $\mathcal{T}$ time must be *spacelike*. (These hypersurfaces can be evaluated using any convenient reference foliation $\mathcal{F}^{\text{ref}}$; the answer will be independent of the choice of $\mathcal{F}^{\text{ref}}$ for internal spacetime functionals that are compatible with the spacetime problem.)

There are no known examples of internal spacetime functionals that satisfy these conditions. Of course, another possibility is to construct scalars from *matter* fields in the theory—an option that has been discussed at length by Kuchař and collaborators [KT91a, KT91b, SK92, Kuc92b].

The Spatial Metric Reconstruction Question. At a classical level, the canonical transformation (21) can be inverted to express the metric $q_{ab}(x)$ as a functional $q_{ab}(x; \mathcal{X}, \mathcal{P}, \phi, \pi]$ of the coordinate variables $(\mathcal{X}^A, \mathcal{P}_B)$ and the physical degrees of freedom (ϕ^r, π_s). The question is whether something similar can be done at the quantum level. In particular:

- is it possible to make sense of an expression like $q_{ab}(x; \mathcal{X}, \widehat{\mathcal{P}}, \widehat{\phi}, \widehat{\pi}]$ with $\widehat{\mathcal{P}}_A$ being replaced by $-\widehat{h}_A$?;
- if so, is there any sense in which this operator looks like an operator version of a Riemannian *metric*?

In effect we are asking whether some aspect of the original three-dimensional geometric picture can be recovered in the quantum theory.

The Time-Dependent Hamiltonian Question. Solutions to the constraints are typically explicit functionals of the internal spacetime coordinate functions; in particular, the Hamiltonian is time-dependent. In a normal physical situation this means that energy can be fed into—or taken out of—the quantum system: a situation that arises whenever the system open, with the time dependence of the Hamiltonian being determined by the environment to which the system couples. However, a compact three-manifold Σ (the 'universe') has no external environment, and so the time-dependence does seem a little odd.

The Global Time Problem. This concerns the possibility of performing a canonical transformation (21) with the desired characteristics. This raises the following questions:

1. What properties must the functionals $\mathcal{X}^A(x; q, p]$ possess in order to serve as internal spacetime coordinates? For example, in the classical theory with a given Lorentzian metric g, a hypersurface of constant internal $\mathcal{T}$ time is required to be *spacelike*. But what does this mean in a quantum theory with no background causal structure?
2. Even classically, if such functions do exist, is it possible to perform a canonical transformation of the type (21) which is such that:
   (a) the constraints $\mathcal{H}_\perp(x; q, p] = 0$ and $\mathcal{H}_a(x; q, p] = 0$ can be solved globally (on the phase space) for $\mathcal{P}_A$ in the form (30); and
   (b) there is a unique such solution?

   If this is not possible at the classical level, what are the implications for the quantum theory?

The Definition of the Operators $\widehat{h}_A(x)$. Even apart from the question of ultraviolet divergences, many problems appear when trying to construct operator equivalents of the Hamiltonian/momentum densities $h_A(x; \mathcal{X}, \phi, \pi]$ that arise as the solutions (30) of the initial-value constraints for the conjugate variables $\mathcal{P}_A(x)$. For example:

1. As mentioned earlier, the solution for $\mathcal{P}_A(x)$ may exist only locally in phase space, and there may be more than one such solution. In the latter case it might be possible to select a particular solution on 'physical grounds' (such as the requirement that the physical Hamiltonian be a positive functional of the canonical variables) but the status of such a step is not clear since it means that certain classical solutions to the field equations are deliberately excluded.

2. Even if it does exist, the classical solution for $h_A(x;\mathcal{X},\phi,\pi]$ is likely to be a very complicated expression of the canonical variables. Indeed, it may exist only in some implicit sense: a good example is the solution of the elliptic partial-differential equation that determines h_A in the case of the mean extrinsic curvature time (see the lectures by Beig). The solution will also be a very non-local function of the canonical variables.

These properties of the classical theory pose various problems at the quantum level. For example:

- Operator ordering is likely to be a major difficulty. This is particularly relevant to the problem of functional evolution discussed below.
- The operator that represents a physical quantum Hamiltonian is required to be self-adjoint and positive. In a simple model, the positivity requirement may involve just selecting a particular solution to the constraints, but self-adjointness and positivity are very difficult to check in a situation in which even the classical expression is only an implicit function of the canonical variables.

The Multiple Choice Problem. Classically, all choices for the internal spacetime coordinates have an equal standing since they are related by canonical transformations. However, this poses a real problem at the quantum level because of the well-known Van-Hove phenomenon whereby most classical canonical transformations *cannot* be represented by unitary operators whilst maintaining the irreducibility of the canonical commutation relations. Hence there is no reason to suppose that the theories corresponding to different choices of time will agree. It should be emphasised that this phenomenon is not related to ultra-violet divergences but arises already in finite-dimensional approximations.

The Problem of Functional Evolution. The problem of functional evolution is concerned with the consistency of the dynamical evolution generated by the constraints. There is no problem at the classical level: the consistency of the constraints (8), (9) with the dynamical equations (6), (7) follows from the first-class nature of the constraints, *i.e.*, their Poisson brackets vanish on the constraint subspace by virtue of the Dirac algebra (13–15). Similarly, the consistency of the internal time dynamical equations (35–36) is ensured by the Poisson bracket relations between $h_A(x)$ and $h_B(x')$. This means that the classical evolution of the system from one initial hypersurface to another is independent of the family of hypersurfaces chosen to interpolate between them.

In the quantised theory, the analogous requirement on the quantum operators is

$$\frac{\delta \widehat{h}_A(x;\chi]}{\delta \chi^B(x')} - \frac{\delta \widehat{h}_B(x;\mathcal{X}]}{\delta \chi^A(x')} + \frac{1}{i\hbar}[\widehat{h}_A(x;\chi], \widehat{h}_B(x';\chi]] = 0. \qquad (41)$$

If this fails, the functional Schrödinger equation (40) breaks down. This is a non-trivial requirement since the existence of anomalous commutators is a notorious property of quantum field theories.

3 Time Identified After Quantisation

3.1 Dirac Constraint Quantisation

In these approaches to the problem of time, a quantum theory is constructed without solving the constraints, which are then imposed at the quantum level. The identification of 'time' is made only *after* this process, and is used to give the physical interpretation of the theory, particularly the probabilistic aspects. However, this final quantum structure may be related only loosely to that with which the construction begins. As we shall see, this leads to a picture of quantum gravity that is radically different from the one above.

The starting point is the operator version of the Poisson-bracket algebra (10–12) in the form of the canonical commutation relations

$$[\widehat{q}_{ab}(x), \widehat{q}_{cd}(x')] = 0 \qquad (42)$$

$$[\widehat{p}^{ab}(x), \widehat{p}^{cd}(x')] = 0 \qquad (43)$$

$$[\widehat{q}_{ab}(x), \widehat{p}^{cd}(x')] = i\hbar\, \delta^c_{(a}\delta^d_{b)}\, \delta(x, x') \qquad (44)$$

of operators defined on the three-manifold Σ. Several things should be said about this algebra:

- From a strict canonical perspective, operator status is being assigned to physical *and* non-physical modes of $q_{ab}(x)$.
- The use of all components of $q_{ab}(x)$ gives a more geometrical formalism although, strictly speaking, the commutation relations (42–44) are incompatible with the classical property that q is a Riemannian *metric*. This can be alleviated by changing the algebra somewhat [Ish84].
- In standard canonical quantum field theory, the analogue of (42) reflects the ideas of microcausality. However, the functional form of the gravitational constraints is independent of any foliation of spacetime, and therefore it is not clear what this 'microcausal' property means in terms of the usual ideas of an 'equal-time' hypersurface, or indeed how the notion of spacetime structure (as opposed to spatial structure) appears at all.

The key structural question in this approach to quantum gravity is how the constraint equations (8–9) are to be handled. The essence of the Dirac approach is to impose them as constraints on the physically allowed states in the form

$$\mathcal{H}_a(x; \widehat{q}, \widehat{p})\Psi = 0 \qquad (45)$$

$$\mathcal{H}_\perp(x; \widehat{q}, \widehat{p})\Psi = 0. \qquad (46)$$

We recall that, in the classical theory, the constraints are *equivalent* to the dynamical equations in the sense that if they are satisfied on all spatial hypersurfaces of a Lorentzian metric g, then g necessarily satisfies the full set of Einstein field equations. The analogue in the quantum theory is the assumption that the operator constraints (45–46) are the *sole* technical content of the theory, *i.e.*, dynamical evolution equations are not imposed as well.

However, this has a curious effect. According to the first-order action (1), the canonical Hamiltonian is (5) where N and $\mathbf{N}$ are external, c-number functions. But then, in the Schrödinger equation

$$i\hbar\frac{d}{dt}\Psi_t = \widehat{H}[N,\mathbf{N}](t)\Psi_t, \tag{47}$$

if the state Ψ_t satisfies the constraint equations (45–46), we see that it has no time dependence at all!

This so-called 'frozen formalism' caused much confusion when it was first discovered since it seems to imply that nothing happens in a quantum theory of gravity. These days, this is understood to reflect the absence of any external, preferred time parameter, and therefore—in particular—the need to discuss physical time using functionals of the internal variables.

3.2 Problems with the Dirac Approach

Many problems arise when attempting to implement the Dirac scheme. For example:

1. To what extent can, or should, the classical Poisson-bracket algebra (13–15) be maintained in the quantum theory? In answering this question we are lead inevitably to the problems of regularisation, renormalisation, operator ordering, and potential anomalies. This is how the functional evolution problem appears in this approach to quantisation.
2. It is not clear what properties to expect of the constraint operators $\widehat{\mathcal{H}}_\perp(x)$ and $\widehat{\mathcal{H}}_a(x)$; in particular, should they be self-adjoint? Since one presumably starts with self-adjoint representations of the canonical commutation representations (42–44), it is perhaps natural to require self-adjointness for the super-Hamiltonian and supermomentum operators. However, this has been challenged several times, partly on the grounds that the Hilbert space structure on the space that carries the representation of the canonical algebra may be related only distantly to the Hilbert space structure that ought to be imposed on the *physical* states (*i.e.*, those that satisfy the constraints).
3. More generally, what is the relation between these two Hilbert spaces? This question has important implications for the problem of time.

3.3 The Diff(Σ) Constraints

In attempting to find concrete representations of the canonical algebra (42–44) it is natural to try an analogue of the familiar representation of the canonical commutation relations in elementary wave mechanics. Thus the state vectors are taken to be functionals $\Psi[q]$ of Riemannian metrics q on Σ, and the canonical operators are defined as

$$\left(\widehat{q}_{ab}(x)\Psi\right)[q] := q_{ab}(x)\Psi[q] \tag{48}$$

$$\left(\widehat{p}^{cd}(x)\Psi\right)[q] := -i\hbar\frac{\delta\Psi[q]}{\delta q_{cd}(x)}. \tag{49}$$

These equations are not well-defined mathematically but, nevertheless, they have been widely used in the canonical approach to quantum gravity.

Let us consider the Dirac constraints (45–46) in this representation. From a physical perspective it is easy to see the need for them. Formally, the domain space of the state functionals is the space Riem(Σ) of all Riemannian metrics on Σ, and to specify a metric at a point $x \in \Sigma$ requires six numbers (the components $q_{ab}(x)$ of the metric in some coordinate system). However, the physical gravitational system should have only *two* degrees of freedom per spatial point, and therefore four of the six degrees of freedom need to be lost. This is precisely what is achieved by the imposition of the constraints (45–46). As we saw in section 2, the same counting argument applies if the system is reduced to true canonical form before quantising.

The easiest constraints to handle are those in the first set (45), *i.e.*,

$$\left(\widehat{H}[\mathrm{N}]\Psi\right)[q] = 0.$$

We saw earlier that the classical functions $H[\mathrm{N}]$ are the infinitesimal generators of the diffeomorphism group of Σ, and the same might be expected to apply here. Indeed, the implications of these constraints are a direct analogue of those in conventional Yang-Mills gauge theories. Thus the group Diff(Σ) acts as a group of transformations on the space Riem(Σ) of Riemannian metrics on Σ, with $f \in$ Diff(Σ) sending $q \in$ Riem(Σ) to f^*q. Apart from certain technical niceties, this leads to a picture in which Riem(Σ) is fibered by the orbits of the Diff(Σ) action. Then (45) implies that the state functional Ψ is constant (modulo possible θ-vacua effects) on the orbits of Diff(Σ), and therefore passes to a function on the *superspace* Riem(Σ)/Diff(Σ) of Diff(Σ) orbits.

3.4 The Wheeler-DeWitt Equation

We must now consider the final constraints $\widehat{\mathcal{H}}_\perp(x)\Psi = 0$, $x \in \Sigma$. Unlike the constraints $H[\mathrm{N}]\Psi = 0$, these have no simple group-theoretic interpretation since, as remarked earlier, the presence of the explicit $q^{ab}(x)$ factor on the right hand side of (15) means that (13–15) does not define a genuine Lie algebra. Thus the operator-ordering problem becomes significant. If, as a simple example, we choose the ordering in which the p^{cd} variables are all placed to the right of the q_{ab} variables, the constraint (46) becomes

$$-\hbar^2\kappa^2\mathcal{G}_{ab\,cd}(x;q)\frac{\delta^2\Psi[q]}{\delta q_{ab}(x)\,\delta q_{cd}(x)} - \frac{|q|^{\frac{1}{2}}(x)}{\kappa^2}R(x;q)\Psi[q] = 0 \tag{50}$$

where $\mathcal{G}_{abcd}$ is the DeWitt metric defined in (4).

Equation (50) is the famous Wheeler-DeWitt equation. It is the heart of the Dirac constraint quantisation approach to the canonical theory of quantum gravity, and everything must be extracted from it. Needless to say, there are a number of problems and questions that must be considered. For example:

- The ordering chosen in (50) is a simple one but there is no particular reason why it should be correct.
- The Wheeler-DeWitt equation contains products of functional differential operators evaluated at the same spatial point and therefore is likely to produce $\delta(0)$ singularities when acting on a wide variety of possible state functionals. Thus, despite its elegant appearance, the equation is not well defined. The Ashtekar version of this equation is much better in this respect.
- A major question is how to approach the problem of solving the Wheeler-DeWitt equation; in particular, what is the precise interpretation of the equations $\widehat{\mathcal{H}}_\perp(x)\Psi = 0$, $x \in \Sigma$? If they mean that Ψ is a simultaneous eigenvector of self-adjoint operators $\widehat{\mathcal{H}}_\perp(x)$, $x \in \Sigma$, with eigenvalue 0 then, as with eigenfunction problems for ordinary differential operators, boundary conditions need to be imposed on Ψ, but the theory is not too informative about what these might be. This problem is discussed in the article by Hájíček.
- One of the hardest problems is to decide what the Wheeler-DeWitt equation means physically. In particular, the notions of 'time' and 'time-evolution' must be extracted from the equation in some way.

The last problem is particularly important and we will now discuss briefly a few of the attempts to give meaning to the Wheeler-DeWitt equation. The central issue is to decide on the inner product that should be imposed on these solutions: a difficulty that is intimately connected with the problem of time.

3.5 The Simple Schrödinger Interpretation

The representation (48–49) implies that a basic Hilbert space in the technical construction of the theory is the set of all functionals of q with the inner product

$$< \Psi, \Phi > := \int_{\mathrm{Riem}(\Sigma)} \mathcal{D}q\, \Psi^*[q]\, \Phi[q] \tag{51}$$

in which the measure $\mathcal{D}q$ is defined (rather non-rigorously) on the space Riem(Σ) of Riemannian metrics on the three-manifold Σ. Indeed, this is the inner product with respect to which (48–49) define operators that are hermitian.

Perhaps the most natural interpretation of the Wheeler-DeWitt equation is the 'simple Schrödinger interpretation' which asserts that (51) is also the *physically* relevant inner product. In particular, the probability of 'finding' a hypersurface in M on which the three-metric q lies in a measurable subset B of Riem(Σ) is

$$\mathrm{Prob}(q \in B; \Psi) = \int_B \mathcal{D}q\, |\Psi[q]|^2. \tag{52}$$

This approach has several advantages:

- As mentioned already, (51) is the inner product with respect to which the operators (48–49) are formally self-adjoint.
- This is also the inner product that is naturally used in discussions about whether the constraint operators $\widehat{\mathcal{H}}_\perp(x)$ and $\widehat{\mathcal{H}}_a(x)$ are self-adjoint.
- This interpretation gives a simple 'wave-packet' picture of what a classical spacetime looks like in the quantum theory; for example, one might say that a certain state Ψ corresponds to a particular Lorentzian metric g if $\Psi[q] \neq 0$ only for those three-metrics q that appear as some spacelike slice of g.

Unfortunately, this simple interpretation has three, closely related, difficulties:

- Viewed canonically, the *non*-physical modes of the gravitational field have the same status as the physical modes.
- The variables integrated over in the inner product (51) include what might be identified as an internal time. It is as if, in normal wave mechanics, a solution $\psi(x,t)$ to the time-dependent Schrödinger equation was normalised using the inner product $\int dx\, dt |\psi(x,t)|^2$ rather than $\int dx |\psi(x,t)|^2$.
- The inner product (51) is defined on *arbitrary* square-integrable functionals of q, not just those that are solutions of the Wheeler-DeWitt equation.

3.6 The Klein-Gordon Interpretation

These objections to the simple Schrödinger interpretation make it clear that two key questions in understanding the physical meaning of the Wheeler-DeWitt equation are:

- What inner product should be imposed on the *solutions* to this equation (rather than on arbitrary functions of q)?
- How is this related to the notion of time?

One approach to this issue is based on the observation that the Wheeler-DeWitt equation looks rather like a Klein-Gordon equation. More precisely, the DeWitt metric (4) on Riem(Σ) has a hyperbolic character in which the conformal modes of the metric play the role of time-like directions, *i.e.*, the transformation $q_{ab}(x) \mapsto F(x)q_{ab}(x)$, $F(x) > 0$, is a 'time-like' displacement in Riem(Σ). This suggests that it may be possible to choose some internal time functional $\mathcal{T}(x;q]$ so that the Wheeler-DeWitt equation can be written in the form

$$-\hbar^2\kappa^2\Big(\frac{\delta^2}{\delta\mathcal{T}^2(x)} - \mathcal{F}^{R_1 R_2}(x;\mathcal{T},\sigma]\frac{\delta^2}{\delta\sigma^{R_1}(x)\,\delta\sigma^{R_2}(x)}\Big)\Psi[\mathcal{T},\sigma] - \frac{|q|^{\frac{1}{2}}(x;\mathcal{T},\sigma]}{\kappa^2}R(x;\mathcal{T},\sigma]\Psi[\mathcal{T},\sigma] = 0 \tag{53}$$

where $\sigma^R(x)$, $R = 1,\ldots,5$ denotes the $5 \times \infty^3$ modes of the metric variables $q_{ab}(x)$ that remain after identifying the $1 \times \infty^3$ internal time modes $\mathcal{T}(x)$.

The starting point is the formal pairing

$$< \Psi, \Phi >:= \tag{54}$$

$$i \prod_x \int_{\mathrm{Riem}(\Sigma)} d\Sigma^{ab}(x)\, \Psi^*[q] \Big(\mathcal{G}_{ab\,cd}(x;q] \frac{\overrightarrow{\delta}}{\delta q_{cd}(x)} - \frac{\overleftarrow{\delta}}{\delta q_{cd}(x)} \mathcal{G}_{ab\,cd}(x;q] \Big) \Phi[q]$$

between solutions Ψ and Φ of the Wheeler-DeWitt equation. The functional integral is over some surface in Riem(Σ) that is spacelike with respect to the DeWitt metric (4), and $d\Sigma^{ab}(x)$ is the directed surface-element in Riem(Σ) at the point $x \in \Sigma$. It has the important property of being invariant under deformations of the 'spatial' hypersurface in Riem(Σ). This is the quantum-gravity analogue of the requirement in the normal Klein-Gordon equation that the scalar product be time independent.

Unfortunately, the right hand side of (54) cannot serve as a genuine Hilbert space inner product because, as in the analogous case of the point particle, it is not positive definite. Guided by the point-particle example, one natural way of trying to resolve this problem is to look for a vector field on Riem(Σ) that is a Killing vector for the DeWitt metric (4) and that scales the potential term $|q|^{\frac{1}{2}}(x)\, R(x;q]$ in an appropriate way. Sadly, Kuchař has shown that Riem(Σ) admits no such vector [Kuc81, Kuc91], and hence there is no possibility of defining physical states as an analogue of the positive-frequency solutions of the normal Klein-Gordon equation

3.7 Other Interpretations

The problems with the quantised reduced formalism, and the failures of the simple Schrödinger and Klein-Gordon interpretations in the Dirac approach to canonical quantum gravity, have spawned a number of other suggestions that I can only list here; full discussions and references can be found in the review articles cited earlier.

1. *The WKB approximation.* This uses the Wheeler-DeWitt equation but it is solved only in a WKB-type limit. In particular, this gives rise to a concept of internal time that works only in a semi-classical sense. This particular scheme has been the subject of considerable investigation and is discussed at length in the lectures by Kiefer.
2. *Third quantisation.* In the case of a relativistic particle with an external potential, the impossibility of isolating positive-frequency solutions is connected with a breakdown of the one-particle interpretation of the theory, and the standard resolution is to second-quantise the system by turning the Klein-Gordon wave function into a quantum field. It has been suggested several times that a similar process might be needed in quantum gravity with $\Psi[q]$ becoming an operator $\widehat{\Psi}[q]$ on some new Hilbert space. This procedure is usually called 'third quantisation' since the original Wheeler-DeWitt equation is already the result of a quantum field theory. However, it is unclear what this means, or if the problem of time can really be solved in this way.

3. *The conditional probability interpretation.* This is a development of the simple Schrödinger interpretation in which, once again, the Hilbert space is the set of all functionals $\Psi[q]$ that are square-integrable with respect to the inner product (51). However, $|\Psi[q]|^2$ is no longer regarded as the absolute probability density of finding a three-metric q but is instead thought of as the probability of finding the $2 \times \infty^3$ physical modes of q *conditional* on the remaining $1 \times \infty^3$ variables—the internal-time part of q—being equal to some specific function (I am assuming that the constraints $\widehat{\mathcal{H}}_a(x)\Psi = 0$ have already been solved). The claim or hope is that this does not require any specific split of q into physical parts and an internal time function; indeed, the interpretation is supposed to be correct for any such choice. The main problem with this scheme is that it involves a significant departure from conventional quantum theory and it is by no means obvious that the resulting structure is self-consistent.
4. *Evolving constants of motion.* Rovelli has advocated an interesting approach to the problem of time which starts a classical observable being defined as any functional $A[q,p]$ of the canonical variables $(q_{ab}(x), p^{cd}(x))$ whose Poisson bracket with all the constraint functions vanishes:

$$\{A, \mathcal{H}_a(x)\} = 0 \tag{55}$$

$$\{A, \mathcal{H}_\perp(x)\} = 0. \tag{56}$$

The main idea is to find self-adjoint representations of algebras of such observables with particular emphasis on 'evolving constants of motion': one-parameter families of observables whose dependence on the parameter can be interpreted as evolution with respect to an internal time [Rov90, Rov91a, Rov91c, Rov91b]. This scheme fits in particularly well with the Ashtekar programme and is discussed in more detail in the lectures of Hájiček.

4 Conclusion

To date, none of the approaches to the problem of time in canonical quantisation has yielded a formalism that exactly replicates a conventional dynamical system, and it is quite possible that the final verdict on the normal notion of time will be that it can apply, at best, only in some semi-classical limit of the quantum theory of gravity.

This failure to recover the normal, exact, notion of time is not in itself a great disaster; indeed, it is probably a necessary prerequisite in any theory that aspires, for example, to describe the origin of the universe as a quantum event. However, it does throw some doubt on the validity of the whole canonical quantisation programme. In particular, if the notion of time is meaningful only in a semi-classical limit, why should not the same be true of the concept of space?

If upheld, such an objection would suggest dropping the canonical methods altogether and starting from a more spacetime-oriented approach, but in a way that makes no prior reference to space and time. This may sound like a self-contradiction but a possible way of implementing such a programme has been suggested recently by Hartle [Har93] in the context of the 'consistent histories' approach to quantum theory, and I have been working recently on this scheme with the aim of converting it into a fully-specified mathematical structure [Ish93b]. However, further discussion of such a heretical possibility seems out of place at a meeting devoted to the canonical formalism in all its pristine glory!

References

[DeW67] DeWitt, B.: Quantum theory of gravity. I. The canonical theory, *Phys. Rev.* **160** (1967) 1113–1148

[Har93] Hartle, J.: Spacetime quantum mechanics and the quantum mechanics of spacetime, *in* 'Proceedings on the 1992 Les Houches School, Gravitation and Quantisation' (1993)

[Ish84] Isham, C.: Topological and global aspects of quantum theory, *in* B. DeWitt & R. Stora, eds, 'Relativity, Groups and Topology II', North-Holland, Amsterdam (1984) 1062–1290

[Ish93a] Isham, C.: Canonical quantum gravity and the problem of time, *in* 'Integrable Systems, Quantum Groups, and Quantum Field Theories', Kluwer Academic Publishers, London (1993) 157–288

[Ish93b] Isham, C.: Quantum logic and the histories approach to quantum theory (1993)

[Kuc72] Kuchař, K.: A bubble-time canonical formalism for geometrodynamics, *J. Math. Phys.* **13** (1972) 768–781

[Kuc81] Kuchař, K.: Canonical methods of quantisation, *in* C. Isham, R. Penrose & D. Sciama, eds, 'Quantum Gravity 2: A Second Oxford Symposium', Clarendon Press, Oxford (1981) 329–374

[Kuc91] Kuchař, K.: The problem of time in canonical quantization, *in* A. Ashtekar & J. Stachel, eds, 'Conceptual Problems of Quantum Gravity', Birkhäuser, Boston (1991) 141–171

[Kuc92b] Kuchař, K.: Extrinsic curvature as a reference fluid in canonical gravity, *Phys. Rev.* **D45** (1992) 4443–4457

[Kuc92a] Kuchař, K.: Time and interpretations of quantum gravity, *in* 'Proceedings of the 4th Canadian Conference on General Relativity and Relativistic Astrophysics', World Scientific, Singapore (1992) 211–314

[KT91a] Kuchař, K. & Torre, C.: Gaussian reference fluid and the interpretation of geometrodynamics, *Phys. Rev.* **D43** (1991) 419–441

[KT91b] Kuchař, K. & Torre, C.: Harmonic gauge in canonical gravity, *Phys. Rev.* **D44** (1991) 3116–3123

[Rov90] Rovelli, C.: Quantum mechanics without time: A model, *Phys. Rev.* **D42** (1990) 2638–2646

[Rov91a] Rovelli, C.: Is there incompatibility between the ways time is treated in general relativity and in standard quantum theory?, *in* A. Ashtekar & J. Stachel, eds, 'Conceptual Problems of Quantum Gravity', Birkhäuser, Boston (1991) 126–140

[Rov91b] Rovelli, C.: Quantum evolving constants. Reply to 'Comments on time in quantum gravity: An hypothesis', *Phys. Rev.* **D44** (1991) 1339–1341

[Rov91c] Rovelli, C.: Time in quantum gravity: An hypothesis, *Phys. Rev.* **D43** (1991) 442–456

[SK92] Stone, C. & Kuchař, K.: Representation of spacetime diffeomorphisms in canonical geometrodynamics under harmonic coordinate conditions, *Class. Quan. Grav.* 9 (1992) 757–776

The Semiclassical Approximation to Quantum Gravity

Claus Kiefer

1 Introduction

Despite many efforts in the last sixty years there does not yet exist a consistent theory of quantum gravity. Why, then, should one address an issue like the *semiclassical approximation to quantum gravity* if there is no theory available which one could approximate? The answer is simple. According to the correspondence principle, the notion of a classical spacetime should emerge in an appropriate limit from quantum theory. Thus, by applying semiclassical considerations one hopes to get an insight into some of the structures of the full, elusive, theory.

Semiclassical ideas have been very important in the history of quantum theory. The Bohr-Sommerfeld quantization formula

$$\frac{1}{2\pi\hbar}\oint p dx = n + \frac{1}{2}, \tag{1.1}$$

for example, was used before the advent of quantum mechanics as an ad hoc assumption to explain atomic spectra. Formula (1.1) can, of course, be recovered from quantum mechanics by applying WKB methods to the Schrödinger equation and is even equal to the exact result in some special cases such as the harmonic oscillator or the Coulomb potential. Higher order terms in a WKB expansion then yield corrections to (1.1) of the order of λ/L, where λ is the de Broglie wavelength, and L is the typical dimension of the system under consideration.

A more impressive application of semiclassical ideas is Schrödinger's approach to his wave equation by using Hamiltonian methods. These methods had been developed in the 19th century to provide a joint formalism of optics and mechanics, which was then, however, merely regarded to give a *formal* analogy without physical significance, since there was no obvious relation between the velocity of light and the velocity of a mechanical particle. This changed only with de Broglie who came up with the idea that particles *are* waves. It then seemed natural for Schrödinger to look for a wave equation which yields the Hamilton-Jacobi equation in some appropriate limit. What he basically did in his pioneering work (Schrödinger, 1926a) was to take the Hamilton-Jacobi equation[1]

$$H\left(q_i, \frac{\partial S}{\partial q^i}\right) \equiv \frac{1}{2m}(\nabla S)^2 + V = E \tag{1.2}$$

and make the following ansatz for a wave function:

$$\psi \equiv \exp\left(iS/\hbar\right). \tag{1.3}$$

(The constant $\hbar$ was first left open and later determined from the spectrum of the hydrogen atom.) Calculating the second derivatives of ψ with respect to the q_i and neglecting the second derivatives of S compared to its first derivatives (this is motivated by geometrical optics where S varies very little with x) he found after insertion into (1.2) the famous Schrödinger equation

$$\left(-\frac{\hbar^2}{2m}\nabla^2 + V\right)\psi = E\psi. \tag{1.4}$$

In contrast to de Broglie, Schrödinger has only waves, but they are waves in *configuration space* and *not* in three-dimensional space like light waves. Semiclassical considerations of this kind will also be the general theme in this article, but instead of starting from the Hamilton-Jacobi equation we will accept a "model equation" for the wave functionals in quantum gravity and then discuss consequences from the semiclassical expansion of this equation.

To present a last example of the importance of semiclassical considerations in the history of quantum theory, I would like to write down the corrections to the classical Maxwell action through quantum fluctuations of electrons and positrons, which was found by Heisenberg and Euler (1936) before the development of QED. In the weak field approximation the corrected Maxwell Lagrangian reads

$$\mathcal{L} = \frac{1}{8\pi}(\mathbf{E}^2 - \mathbf{B}^2) + \frac{e^4\hbar}{360\pi^2 m^4 c^7}\left[(\mathbf{E}^2 - \mathbf{B}^2)^2 + 7(\mathbf{EB})^2\right]. \tag{1.5}$$

This result can of course be found from full QED by "integrating out" the fermions and expanding in powers of $\hbar$. Since this is equivalent to an expansion in the number of loops, (1.5) is also called the one-loop effective Lagrangian. We will encounter an analogous level of approximation in the case of quantum gravity below. The knowledge of (1.5) allows the computation of physical processes such as the scattering of light by external fields (Delbrück scattering), which has been experimentally observed.

[1]In this paper he treats the stationary case. The time-dependent Schrödinger equation was introduced only in his last paper on *quantization as an eigenvalue problem* (Schrödinger, 1926b).

Semiclassical methods have also become increasingly popular in the field of chaos where people have even coined the term "postmodern quantum mechanics" for these methods.

What is the situation in quantum gravity? It would be ideal to start with a "theory of everything" like superstring theory which encompasses all interactions including gravity in a single quantum framework. Unfortunately, string theory has not yet reached the stage where any genuine quantum gravity predictions such as black hole evaporation can be made. A great deal of work has therefore been done on the level of *quantum general relativity*, the topic of this volume, where formal quantization rules are applied to general relativity without invoking any scheme of unification of interactions. Whether quantum general relativity exists as a consistent theory and whether it can be derived from an underlying "theory of everything" in an appropriate low-energy limit is an open issue (it depends in particular on the ratio of, say, the string scale to the Planck scale). As long as the status of a possible fundamental theory is unknown, it seems perfectly justified to investigate whether quantum general relativity can serve as a consistent theory of quantum gravity. The perturbative non-renormalisability of this approach may thus not necessarily present a fundamental problem. (In the following we will always use the term "quantum gravity" in this sense.) While most of the other contributions in this volume are concerned with the fundamental level itself, this article will basically address the following questions. Firstly, how can the framework of quantum field theory in a given background spacetime be recovered, at least formally, from quantum gravity and secondly, can one calculate corrections to this framework which come from quantum gravity and may even be testable? Within the framework of quantum field theory in a given spacetime one can derive concrete results like the Hawking temperature for black holes. It would thus be interesting to know the corrections to such results from quantum gravity.

The organisation of this article is as follows. In the next section I show how the functional Schrödinger equation for matter fields can formally be recovered from quantum gravity. I also present the nonrelativistic limit of the Klein-Gordon equation as a useful formal analogy.

Section 3 is a major part of this contribution and is devoted to the derivation of correction terms to the Schrödinger equation from quantum gravity. Applications include the corrections to energy expectation values of scalar fields in de Sitter space and non-unitary contributions to the evaporation of black holes. I also make a comparison with an analogous approximation scheme in QED.

In section 4 I address the issues of decoherence and back reaction, which are important concepts in the semiclassical expansion. I briefly discuss examples from quantum mechanics for illustration and then turn to the analogous situation in quantum gravity. The modification of semiclassical time through back reaction is carefully examined.

The last section gives a brief summary and a brief account of the topics which have not been addressed in this article. This includes the relevance of semiclassical considerations for the problem of time in quantum gravity as well as a comparison with the effective action approach and standard perturbation theory.

2 Derivation of the Schrödinger equation from quantum gravity

The central equation of canonical quantum gravity is the constraint equation $H\Psi = 0$. This was discussed in detail in other contributions to this volume. In the following we will mainly use the geometrodynamical language, since semiclassical discussions are most transparent in this picture. We will comment, however, on the situation using Ashtekar's variables at the end of this section.

Starting point is the full Wheeler-DeWitt equation

$$\mathcal{H}\Psi[h_{ab},\phi] \equiv \left(-\frac{16\pi G\hbar^2}{c^2}G_{abcd}\frac{\delta^2}{\delta h_{ab}\delta h_{cd}} - \frac{c^4}{16\pi G}\sqrt{h}(R-2\Lambda) + \mathcal{H}_m\right)\Psi = 0, \tag{2.1}$$

where h_{ab} is the three-metric, R the three-dimensional Ricci scalar, Λ the cosmological constant, and $\mathcal{H}_m$ is the Hamiltonian density for non-gravitational fields, denoted symbolically below by ϕ. In the remaining part of this article we will deal for simplicity exclusively with scalar fields. There is of course the usual factor ordering ambiguity in (2.1), but for definiteness we use the naive factor ordering with no first metric derivatives in the following discussion and comment on the general case briefly at the end.

One now introduces the parameter

$$M \equiv \frac{c^2}{32\pi G}, \tag{2.2}$$

with respect to which the semiclassical expansion will be performed. The Wheeler-DeWitt equation then reads

$$\left(-\frac{\hbar^2}{2M}G_{ab}\frac{\delta^2}{\delta h_a\delta h_b} + MV + \mathcal{H}_m\right)\Psi = 0, \tag{2.3}$$

where V stands for $-2c^2\sqrt{h}(R-2\Lambda)$, and we have introduced a condensed notation, labeling three-metric coefficients by h_a and components of the DeWitt-metric by G_{ab}, since contractions always involve index pairs.

An expansion with respect to the large parameter M should lead to sensible results if the relvant mass scales of non-gravitational fields are much smaller than the Planck mass. If there were no non-gravitational fields, an M expansion would be fully equivalent to an $\hbar$ expansion and thus to the usual WKB expansion for the gravitational field (this can be seen from (2.1) after multiplication with G). In the presence of non-gravitational fields, the M expansion is analogous to a Born-Oppenheimer expansion, where the large mass of the nuclei is replaced by M, and the small electron mass is replaced by the mass-scale of the non-gravitational field.

The first subsection is thus devoted to a brief review of the Born-Oppenheimer approximation in molecular physics. Most of the steps in that scheme will simply be extrapolated, in a formal way, to quantum gravity. We then present another helpful analogy – the nonrelativistic expansion of the Klein-Gordon equation. Subsection 3 will then present a detailed discussion of the derivation of the Schrödinger equation from the Wheeler-DeWitt equation, while the last section will give a brief review of the analogous situation in the Ashtekar framework.

2.1 Born-Oppenheimer approximation

Consider the following quantum mechanical Hamiltonian

$$H = \frac{P^2}{2M} + \frac{p^2}{2m} + V(R,r) \equiv \frac{P^2}{2M} + h, \tag{2.4}$$

where $M \gg m$. We seek an approximate solution to the stationary Schrödinger equation

$$H\Psi = E\Psi. \tag{2.5}$$

We assume that the spectrum of the "light" particle is known for each configuration R of the "heavy" particle, i.e.

$$h|n;R\rangle = \epsilon(R)|n;R\rangle.$$

The full state Ψ is then expanded into these eigenfunctions,

$$\Psi = \sum_k \psi_k(R)|k;R\rangle. \tag{2.6}$$

If one inserts this ansatz into (2.5) one finds, after multiplication with $\langle n;R|$ from the left and using the fact that these states are orthonormalised, an equation for the wave functions ψ_n,

$$\left(-\frac{\hbar^2}{2M}\nabla_R^2 + \epsilon_n(R) - E\right)\psi_n(R) =$$
$$\frac{\hbar^2}{M}\sum_k \langle n;R|\nabla_R k;R\rangle \nabla_R \psi_k(R) + \frac{\hbar^2}{2M}\sum_k \langle n;R|\nabla_R^2 k;R\rangle\psi_k(R). \tag{2.7}$$

Note that this is still an exact formula. The Born-Oppenheimer approximation now consists in the *neglection of off-diagonal terms* in (2.7).The result is an autonomous equation for the ψ_n, which is conveniently written as follows

$$\left(\frac{1}{2M}(-i\hbar\nabla - \hbar A(R))^2 - \frac{\hbar^2 A^2}{2M} - \frac{\hbar^2}{2M}\langle n; R|\nabla_R^2 n; R\rangle + \epsilon_n(R)\right)\psi_n(R) = E\psi_n(R), \tag{2.8}$$

where we have introduced the "Berry connection"

$$A(R) = -i\langle n; R|\nabla_R n; R\rangle. \tag{2.9}$$

Note that the momentum of the "slow" particle has been shifted according to

$$P \to P - \hbar A. \tag{2.10}$$

In most textbook treatments this connection is not taken into account, since it is assumed that the "fast" eigenfunctions can be chosen to be real, in which case the connection A vanishes. There are, however, situations where this cannot be done and A has to be taken into account, see for example Jackiw (1988a), and Wudka (1990). As we will see below, the complex nature of the "fast" wave functions is essential in quantum gravity.

We emphasise that the essential approximation in this scheme consists in the neglection of the off-diagonal terms in (2.7) and thus the reduction of the superposition (2.6) to a single product state. As I will briefly review in section 4 below, this can be justified by taking the unavoidable interaction with environmental degrees of freedom into account – the various components in (2.6) *decohere* from each other. Moreover, this interaction leads to decoherence *within* one component. The position of heavy molecules, for example, is "measured" by radiation (Joos and Zeh, 1985). In this sense one can therefore justify the substitution of a *fixed* position R_0 of the "heavy" particle into the "fast" states $\langle n; R|$. It is only in this limit that molecules with a well-defined shape emerge.

2.2 Non-relativistic limit of the Klein-Gordon equation

The Klein-Gordon equation reads, in the case of vanishing external field,

$$\left(\frac{\hbar^2}{c^2}\frac{\partial^2}{\partial t^2} - \hbar^2\nabla^2 + m^2c^2\right)\varphi(\mathbf{x}, t) = 0. \tag{2.11}$$

On comparison with (2.3) we recognise that the limit $M \to \infty$ is equivalent to the limit $c \to \infty$ in the Klein-Gordon case. It is, however, emphasised that this analogy is only a partial one. The Klein-Gordon equation (2.11) is an equation for a one-particle wave function, whereas the Wheeler-DeWitt equation (2.3) is already a "second-quantized" equation, i.e. the corresponding state is a wave *functional* on a configuration space, whose points are *field* configurations. Nevertheless, both equations have the structure of a wave equation and exhibit many formal similarities.

The non-relativistic expansion starts by writing the wave function $\varphi(\mathbf{x},t)$ as

$$\varphi(\mathbf{x},t) = \exp\left(iS(\mathbf{x},t)/\hbar\right) \tag{2.12}$$

and expanding

$$S = c^2 S_0 + S_1 + c^{-2} S_2 + \ldots. \tag{2.13}$$

Inserting this into (2.11) and comparing equal powers of the expansion parameter c^2 yields at order c^4

$$(\nabla S_0)^2 = 0, \tag{2.14}$$

so that S_0 depends on t only. The next order, c^2, yields

$$-\left(\frac{\partial S_0}{\partial t}\right)^2 + m^2 = 0. \tag{2.15}$$

This is a Hamilton-Jacobi type of equation, which gives real solutions if $m^2 \geq 0$, i.e. if there are no tachyons. The solutions can then immediately be written down:

$$S_0 = \pm mt + \textit{ constant}. \tag{2.16}$$

At this order of approximation we thus have for the wave function

$$\varphi_{\mp}(\mathbf{x},t) \approx \exp\left(\pm imc^2 t/\hbar\right). \tag{2.17}$$

Its interpretation is obvious: It describes a particle at rest with positive energy (lower sign) or with negative energy (upper sign). To neglect superpositions like $\varphi_+ + \varphi_-$ reflects the fact that field theoretic effects like pair creation are assumed to be negligible. Note that both φ_+ and φ_- are complex, despite the real character of the Klein-Gordon equation (2.11). We assume in the following the use of the positive energy wave function.

The next order (c^0) yields an equation for S_1:

$$2m\dot{S}_1 + (\nabla S_1)^2 - i\hbar\nabla^2 S_1 = 0. \tag{2.18}$$

This can be simplified by defining the wave function

$$\chi \equiv \exp\left(iS_1/\hbar\right), \tag{2.19}$$

which leads, for non-vanishing mass, to

$$i\hbar\dot{\chi} = -\frac{\hbar^2}{2m}\nabla^2\chi. \tag{2.20}$$

At this order one has thus recovered the non-relativistic Schrödinger equation. If one proceeds beyond this order, one finds, at order c^{-2}, the first relativistic corrections. The equation involving S_2 reads

$$i\hbar\ddot{S}_1 - i\hbar\nabla^2 S_2 - \dot{S}_1^{\,2} + 2m\dot{S}_2 + 2\nabla S_1 \nabla S_2 = 0, \tag{2.21}$$

which can be simplified if one rewrites S_1 in terms of χ and introduces the definition

$$\psi \equiv \chi \exp\left(iS_2/\hbar c^2\right). \tag{2.22}$$

The wave function ψ then obeys the modified Schrödinger equation

$$i\hbar\dot{\psi} = -\frac{\hbar^2}{2m}\nabla^2\psi - \frac{\hbar^4}{8m^3c^2}\nabla^2\nabla^2\psi. \tag{2.23}$$

The last term on the right-hand side of this equation is, of course, just the first relativistic correction to the kinetic energy, which can easily be obtained by expanding the square root $\sqrt{m^2c^4 + p^2c^2}$. The present expansion scheme has, however, the merit of systematically producing correction terms also in the presence of external fields. In the case of coupling to an external gravitational field, some of these correction terms are imaginary and thus explicitly violate unitarity in the matter sector (Lämmerzahl, 1993). We will encounter similar terms in the case of the Wheeler-DeWitt equation, see section 3.

I emphasise that (2.23) does not make sense physically as a differential equation (the effective Hamiltonian, for example, is unbounded from below). It only makes sense if the additional term is used to calculate small corrections to results found on the previous level of approximation. One can, for example, calculate the relativistic corrections to the energy eigenvalues of hydrogen-like pionic atoms:

$$\begin{aligned}\Delta E_{rel} &= -\int d^3x\,\psi^*_{nlm}\left(\frac{\hbar^4}{8m^3c^2}\nabla^2\nabla^2\right)\psi_{nlm} \\ &= -\frac{mc^2}{2}(Z\alpha)^4\left(\frac{1}{n^3(l+\frac{1}{2})} - \frac{3}{4n^4}\right).\end{aligned} \tag{2.24}$$

We now turn to the interesting case of the Wheeler-DeWitt equation.

2.3 Derivation of the functional Schrödinger equation from the Wheeler-DeWitt equation

We now make an expansion of the Wheeler-DeWitt equation (2.3) with respect to M, which is formally similar to the non-relativistic expansion of the Klein-Gordon equation presented above. To be specific, we take $\mathcal{H}_m$ to be the Hamiltonian density of a minimally coupled scalar field, i.e.,

$$\mathcal{H}_m = \frac{1}{2}\left(-\frac{\hbar^2}{\sqrt{h}}\frac{\delta^2}{\delta\phi^2} + \sqrt{h}h^{ab}\phi_{,a}\phi_{,b} + \sqrt{h}(m^2\phi^2 + U(\phi))\right), \tag{2.25}$$

where $U(\phi)$ can be an arbitrary potential. In analogy to (2.12) we make the ansatz

$$\Psi \equiv \exp(iS/\hbar), \tag{2.26}$$

and expand

$$S = MS_0 + S_1 + M^{-1}S_2 + \dots . \tag{2.27}$$

As remarked above, this expansion scheme will distinguish *all* gravitational degrees of freedom (as being "semiclassical") from the remaining ones (which are "fully quantum"). One would, of course, expect that not all degrees of the gravitational field are semiclassical, such as the "gravitons." It is straightforward to generalise this scheme to incorporate the more general situation. Vilenkin (1989), for example, makes the ansatz

$$h_{ab} = h_{ab}^{(0)} + \sqrt{8\pi G} q_{ab}.$$

Inserting this into the Wheeler-DeWitt equation, one recognises that the q_{ab} appear on the same footing as ϕ and are therefore "fully quantum." The expansion is thus, of course, no longer an expansion with respect to G. Generally speaking, the assumption behind this is that the quantum states of the "fast variables" depend on the "semiclassical variables" only adiabatically, see e. g. Halliwell and Hawking (1985) or Kiefer (1987).

In complete analogy to (2.14)ff. one finds after inserting (2.26) and (2.27) into (2.3) at the highest order, M^2,

$$\left(\frac{\delta S_0}{\delta \phi}\right)^2 = 0, \tag{2.28}$$

i.e. that S_0 depends only on the three-metric. If many matter fields are present one can draw this conclusion only if their kinetic energies are positive (provided, of course, S_0 is assumed to be either real or pure imaginary). Since the gravitational kinetic energy is indefinite, one is thus not able to derive the opposite limit, where all gravitational degrees of freedom are quantum and the non-gravitational ones semiclassical.

The next order, M^1, yields

$$\frac{1}{2} G_{ab} \frac{\delta S_0}{\delta h_a} \frac{\delta S_0}{\delta h_b} + V(h_a) = 0. \tag{2.29}$$

This is the Hamilton-Jacobi equation for the gravitational field (Peres, 1962). It is equivalent to all ten of Einstein's field equations (DeWitt, 1967; Gerlach, 1969; DeWitt, 1970). I want to briefly review how this equivalence can be shown, i. e. how the "dynamical laws follow from the laws of the instant" (Kuchař, 1992).

Suppose one is given a solution S_0 to (2.29). From this one can immediately read off the geometrodynamical field momentum

$$\pi^{ab} = M \frac{\delta S_0}{\delta h_{ab}}. \tag{2.30}$$

One half of Hamilton's equations of motion then give the connection between the field momenta and the "velocities" $\dot{h}_{ab}$,

$$\dot{h}_{ab} = -2NK_{ab} + N_{a|b} + N_{b|a}, \tag{2.31}$$

where K_{ab} are the components of the extrinsic curvature of three-space, whose relation to the momenta is

$$K_{ab} = -\frac{16\pi G}{c^2} G_{abcd}\pi^{cd}, \tag{2.32}$$

and N and N_a are respectively the lapse function and the shift vector, for both of which a choice must be made in order for $\dot{h}_{ab}$ to be specified. After this has been done, and an "initial" three-geometry has been chosen, (2.31) can be integrated to yield a whole *spacetime* with a definite foliation and a definite choice of coordinates on each member of this foliation. One can thus combine all the trajectories in superspace which describe the same spacetime to a "sheaf" (DeWitt, 1970). Lapse and shift can be chosen in such a way that these curves comprise a "sheaf of geodesics." Superposing many WKB solutions of the gravitational field, Gerlach (1969) has demonstrated how *one* specific trajectory in configuration space, i. e. one specific spacetime, can be traced out by a wave packet.

Since S_0 satisfies, in addition to the Hamilton-Jacobi equation (2.29), the three equations

$$\left(\frac{\delta S_0}{\delta h_{ab}}\right)_{|b} = 0, \tag{2.33}$$

the "complete solution" depends on two arbitrary functions on three-space. The relations (2.33) follow from an expansion of the momentum constraints in powers of M. As was shown by Moncrief and Teitelboim (1972), these equations follow from the validity of the Hamilton-Jacobi equation (2.29) itself, if one takes into account the Poisson bracket relations between the constraints $\mathcal{H}$ and $\mathcal{H}_a$. While (2.29) and (2.33) are equivalent to the G_{00} and G_{0i} part of the Einstein equations, the remaining six field equations can be found by differentiating (2.31) with respect to t and eliminating S_0 by making use of (2.31) and (2.29). Basically, this is the same procedure as can already be done in Newtonian mechanics.

The Hamilton-Jacobi equation (2.29), as it stands, is actually an infinity of equations, one at each space point. An alternative viewpoint is also useful (Giulini, 1993). It interprets (2.29) as being multiplied with a "test function" and thus represents *one* equation for *each* choice of lapse, i.e. for each choice of foliation. The advantage of this viewpoint is that, although it may be impossible to solve the Hamilton-Jacobi equation for each choice of lapse, it may be possible to do so for specific choices. This may be enough in certain situations to draw interesting physical conclusions. One way or the other, it will be assumed in the following that a specific solution of (2.29) has been chosen and a specific spacetime with a specific foliation has been constructed from it.

It has been remarked (Kuchař, 1992) that the sandwich problem may present some obstacle in solving the Hamilton-Jacobi equation. This is because a complete solution to (2.29) can be interpreted as generating a canonical transformation between two three-geometries which are assumed to be connected by a single spacetime (which is in a sense "sandwiched" between these three-geometries). The uniqueness of the interpolating four-geometry was only proven in special cases, for example if $2\rho - R > 0$ (Belasco and Ohanian, 1969). It may thus happen that the complete solution to (2.29) cannot generate a canonical transformation between all three-geometries. In the semiclassical approximation, however, we use a *special* solution to (2.29) and can always construct a spacetime from a given three-geometry using the construction presented above.

Proceeding with the semiclassical expansion, one finds at order M^0 an equation involving also S_1:

$$G_{ab}\frac{\delta S_0}{\delta h_a}\frac{\delta S_1}{\delta h_b} - \frac{i\hbar}{2}G_{ab}\frac{\delta^2 S_0}{\delta h_a \delta h_b} + \mathcal{H}_m = 0. \tag{2.34}$$

As in the Klein-Gordon case, one can simplify this equation by introducing the wave functional

$$\chi = D[h_a]\exp\left(iS_1/\hbar\right). \tag{2.35}$$

We will *choose* D in such a way that the equations become simple. This is achieved by demanding

$$G_{ab}\frac{\delta S_0}{\delta h_a}\frac{\delta D}{\delta h_b} - \frac{1}{2}G_{ab}\frac{\delta^2 S_0}{\delta h_a \delta h_b}D = 0, \tag{2.36}$$

which can also be written in the form of a "conservation law"

$$G_{ab}\frac{\delta}{\delta h_a}\left(\frac{1}{D^2}\frac{\delta S_0}{\delta h_b}\right) = 0. \tag{2.37}$$

The interpretation of this equation becomes immediately clear if one writes down its pendant in one-dimensional quantum mechanics:

$$\frac{\partial}{\partial x}\left(\frac{1}{D^2}\frac{\partial S_0}{\partial x}\right) = 0.$$

This is nothing but the continuity equation for time-independent states. The solution of this simple equation gives the well known expression for the WKB prefactor $D \propto \sqrt{p}$.

Using (2.37) one finds from (2.34)

$$i\hbar G_{ab}\frac{\delta S_0}{\delta h_a}\frac{\delta \chi}{\delta h_b} \equiv i\hbar\frac{\delta \chi}{\delta \tau} = \mathcal{H}_m\chi. \tag{2.38}$$

This is the functional Schrödinger equation in its local form (the Tomonaga-Schwinger equation) for quantum fields propagating on the classical spacetimes described by S_0. Note that the Schrödinger equation cannot be recovered if S_0 turns out to be a constant.

The above approach to semiclassical gravity has been discussed from various points of view by several authors including DeWitt (1967), Lapchinsky and Rubakov (1979), Banks (1985), Halliwell and Hawking (1985), Hartle (1986), and Kiefer (1987).

In (2.38) one has introduced a time functional $\tau(x; h_a]$ according to

$$G_{ab}(x)\frac{\delta S_0}{\delta h_a(x)}\frac{\delta\tau(x;h_a]}{\delta h_b(y)} = \delta(x-y). \tag{2.39}$$

This "many-fingered" time (which is also called *WKB time*) is defined on configuration space, but also yields a time parameter in each of the spacetimes described by S_0. This is clear from the above construction, since each spacetime is constructed from a specific foliation, which is labelled by τ. It is for this reason that the "spacetime problem" (Isham, 1992; Kuchař, 1992) does not come into play in the semiclassical approximation. This problem refers to the fact that τ is not a spacetime scalar, i. e. that it depends on the embedding which is chosen. While this may in fact present a problem for approaches which attempt to use functions like τ as fundamental time functions in quantum gravity, there is no such ambiguity in the semiclassical approximation, where the spacetimes emerge with a specific foliation.

We also note that the hypersurfaces $\tau = constant$ in configuration space do not coincide with the hypersurfaces $S_0 = constant$. This can be immediately seen by comparing the definition of τ (2.39) with the Hamilton-Jacobi equation (2.29).

Another problem discussed in this context is the "global time problem." As simple minisuperspace models indicate, τ may not exist globally in the space of three-geometries. This is, however, not a real obstacle for the semiclassical approximation in that one can simply restrict the attention to regions in superspace where it exists. Since we will be interested in the following to quantum corrections only in the neighbourhood of a given spacetime this does not really present a restriction.

Barbour (1992) has stressed that the above introduced WKB time is very similar to the notion of *ephemeris time* as it is used by astronomers. This time is determined in retrospect from actual observations of the celestial bodies, exploiting the dynamical equations of the gravitational field. The above WKB time thus plays a double role: From one point of view one may simply interpret it as a coordinate on spacetime, but on the other side it is constructed from the geometry of a given three-dimensional space and is thus a quantity which in principle can be determined from astronomical observations. That ultimately the whole three-geometry is implemeted into the construction of this notion of time can very clearly be seen in the motion of the binary pulsar *PSR* 1913+16:

If one attempts to determine ephemeris time from this system (exploiting the period decrease due to the radiation of gravitational waves) one reaches consistency only if the whole motion of the Galaxy and its "pull" on the binary pulsar is quantitatively taken into account (Damour and Taylor, 1991).

Up to this order, the total wave functional thus reads

$$\Psi \approx \frac{1}{D} \exp\left(iMS_0[h_{ab}]/\hbar\right) \chi[h_{ab}, \phi], \tag{2.40}$$

where χ obeys the Schrödinger equation (2.38).

2.4 Semiclassical gravity in the connection representation

Instead of using the three-metric and is conjugate momentum as the basic variables one can perform a canonical transformation at the classical level to a set of new canonical variables consisting of the densitised triad and a complex $SO(3)$ connection. Details are presented in various other contributions to this volume. The fundamental equation is, of course, again of the form $H\Psi = 0$, but the Hamiltonian reads differently, see Ashtekar (1991):

$$\begin{aligned} H &= \frac{G\hbar^2}{4\pi}\epsilon^{ijk}F^k_{ab}\frac{\delta^2}{\delta A^i_a \delta A^j_b} + \hbar^2\frac{\delta^2}{\delta\phi^2} + 2G^2\hbar^2\partial_a\phi\partial_b\phi\frac{\delta^2}{\delta A^i_a\delta A^i_b} \\ &\quad + \frac{\sqrt{2}\hbar^3}{3}(G^3m^2\phi^2 + G^2\Lambda)\eta_{abc}\epsilon^{ijk}\frac{\delta^3}{\delta A^i_a\delta A^j_b\delta A^k_c}, \end{aligned} \tag{2.41}$$

where$F^k_{ab} = \partial_a A^k_b - \partial_b A^k_a - \epsilon^{klm}A^l_a A^m_b$ is the field strength tensor associated with the complex connection $\mathbf{A}$, and η_{abc} is the metric-independent totally skew-symmetric density of weight -1. I have used here a *rescaled* connection $A^i_a \equiv G\bar{A}^i_a$, where $\bar{A}^i_a$ denotes the connection as it is used in Ashtekar (1991). Such a rescaling is convenient for performing the semiclassical expansion and has also been proven useful in the investigation of the strong gravity limit, see Ashtekar (1988). The distinguished property of (2.41) is the fact that the potential term proportional to R has been "swallowed" by the new variables. In addition, it contains only terms with functional derivatives (even third order derivatives) and that it is intrinsically complex since $\mathbf{A}$ is complex. Note also that a factor of G is associated with each functional derivative with respect to A^i_a which comes with a term containing ϕ.

The semiclassical expansion now proceeds analogously to the geometrodynamical case of the previous section (the parameter M is now just the inverse of G). The functional S_0 does not depend on non-gravitational fields, and it obeys the Hamilton-Jacobi equation

$$\frac{\epsilon^{ijk}}{4\pi}F^k_{ab}\frac{\delta S_0}{\delta A^i_a}\frac{\delta S_0}{\delta A^j_b} + \frac{i\sqrt{2}}{3}\Lambda\eta_{abc}\epsilon_{ijk}\frac{\delta S_0}{\delta A^i_a}\frac{\delta S_0}{\delta A^j_b}\frac{\delta S_0}{\delta A^k_c} = 0. \tag{2.42}$$

Note that since the momentum conjugate to A_a^i, $\tilde{E}_i^a$, is replaced by $\delta/\delta A_a^i$ (without an i) in the Schrödinger representation, the momentum is given by $\tilde{E}_i^a = i\delta S_0/\delta A_a^i$. The triad $\tilde{E}_i^a$ is not necessarily real, but it can be made real since it satisfies Gauss's law. Note that $S_0 = constant$ is always a solution of this equation.

The next order (G^0) yields the functional Schrödinger equation for the wave functional $\chi \equiv De^{iS_1}$,

$$i\hbar\frac{\epsilon^{ijk}}{4\pi}F_{ab}^k\frac{\delta S_0}{\delta A_a^i}\frac{\delta\chi}{\delta A_b^j} \equiv i\hbar\frac{\delta\chi}{\delta\tau} = \tilde{H}_m\chi, \tag{2.43}$$

where the Hamiltonian density $\tilde{H}_m$ is now given by the expression

$$\begin{aligned}\tilde{H}_m &= -\frac{\hbar^2}{2}\frac{\delta^2}{\delta\phi^2} + \frac{\delta S_0}{\delta A_a^i}\frac{\delta S_0}{\delta A_b^i}\partial_a\phi\partial_b\phi + \frac{i}{3\sqrt{2}}m^2\phi^2\eta_{abc}\epsilon_{ijk}\frac{\delta S_0}{\delta A_a^i}\frac{\delta S_0}{\delta A_b^j}\frac{\delta S_0}{\delta A_c^k}\\ &\quad +\sqrt{2}\Lambda\hbar\eta_{abc}\epsilon_{ijk}\frac{\delta S_0}{\delta A_a^i}\left(\frac{\delta S_0}{\delta A_b^j}\frac{\delta}{\delta A_c^k} - \frac{1}{D}\frac{\delta S_0}{\delta A_b^j}\frac{\delta D}{\delta A_c^k} + \frac{\delta^2 S_0}{\delta A_b^j\delta A_c^k}\right).\end{aligned} \tag{2.44}$$

A comparison with (2.41) exhibits that $\tilde{H}_m$ is equal to H as evaluated on the classical gravitational background determined by the Hamilton-Jacobi equation *except* for the last three Λ- dependent terms in (2.44) which arise due to the presence of the third functional derivatives in (2.41). These terms can be absorbed by a redefinition of the wave functional, but their interpretation is not yet clear.

It seems that the traditional semiclassical approximation is tied to the presence of only second order derivatives (i.e. quadratic momenta) in the Hamiltonian. Consider, for example, the anharmonic oscillator in quantum mechanics, which is defined by the Hamiltonian

$$H = \frac{p^2}{2} + \lambda x^4.$$

If one wished to use the momentum representation instead of the position representation for the wave function in the Schrödinger equation, one would have to substitute fourth order derivatives in the momentum for the second term in H. Performing a WKB expansion one finds terms analogous to the ones in (2.41), which one would have to absorb through a redefinition of the wave function to get a transparent interpretation. This problem reflects of course the unsolved issue of the Hilbert space structure of the full theory (which would allow much less states than the number of solutions to $H\Psi = 0$).

A second major difference to the geometrodynamical case is the complexity of the connection, as mentioned above. One has been able to write (2.43) as a functional Schrödinger equation after, in an appropriate region of configuration space, a time functional $\tau(\mathbf{x}; \mathbf{A}]$ was introduced, which satisfies

$$\delta(\mathbf{x}-\mathbf{y}) = \frac{\epsilon^{ijk}}{4\pi}F_{ab}^k(\mathbf{y})\frac{\delta S_0}{\delta A_a^i(\mathbf{y})}\frac{\delta\tau(\mathbf{x};\mathbf{A}]}{\delta A_b^j(\mathbf{y})}. \tag{2.45}$$

This time functional, however, is complex and thus not viable as a semiclassical time parameter. Since one generally assumes that the functionals in the connection representation are holomorphic, one can take the *real part* of τ as a candidate for WKB time.

Since one wants to recover *real* quantum gravity at a certain stage, one must implement appropriate reality conditions (Ashtekar, 1991). The assumption, which is motivated by a comparison with the Bargmann representation for the harmonic oscillator, is that they are used to determine the inner product which has to be imposed on the physical states. The present semiclassical approximation, however, only works on the level of the differential equation and therefore does not "know" about the Hilbert structure (see the remarks above). It may even be possible that both concepts are in conflict with each other, i. e. that the semiclassical approximation "runs out" of the Hilbert space (Louko, 1993). It is definitely not possible to implement the reality conditions on the wave functional in the same way as the constraint equation. One may therefore be able to decide the compatibility only after the appropriate Hilbert space structure of the full theory has been determined.

The above derivation has to be contrasted with the derivation of the Schrödinger equation by Ashtekar (1991). He expands the *classical* Hamiltonian constraint up to second order around some *given* classical configuration, which is chosen to be a flat triad on a non compact manifold. The important feature of the truncated constraint is its *linearity* in the triad, i. e. in the canonical momentum. As in the full theory, quantisation proceeds by implementing the truncated constraint as a condition on wave functionals. The result is

$$\begin{aligned} &-\hbar\frac{1}{G}(\Delta\frac{\delta}{\delta A^T(x)})\Psi(A^{TT},A^L,A^T,A^A) \\ &= GA^{TTab}(x)\left(A^{TT}_{ab}(x)\right)^*\Psi(A^{TT},A^L,A^T,A^A). \end{aligned} \tag{2.46}$$

The internal indices have here all been converted to space indices by the use of the given background triad, and we have decomposed the resulting "connection tensor" into its transverse-traceless, longitudinal, trace, and antisymmetric parts.

The important step is now to identify the term $G(\Delta)^{-1}A^T(x)$ as a "many-fingered" parameter $\tau(x)$, which has the dimension of time. (This is a *nonlocal* expression in the connection.) As in the above semiclassical expansion, it is, however, complex and thus not suitable as a physical time. But, again, assuming the holomorphicity of the wave functional one can identify its *imaginary part* as physical time, since then the left-hand side of (2.46) reads

$$-\hbar\frac{\delta}{\delta\tau(x)}\Psi = i\hbar\frac{\delta}{\delta(\mathrm{Im}\tau)(x)}\Psi, \tag{2.47}$$

while the right-hand side is the "physical" Hamiltonian in this order, i. e. it generates the dynamics of the transverse traceless degrees of freedom (the "gravitons"). One has thus recovered, in the weak field approximation, the Schrödinger equation from quantum gravity.

It was important in the above derivation that the classical constraint has been expanded up to quadratic order in the deviation of the fields from their flat space values, since the left-hand side of (2.46) would have been zero in the first order approximation. This has already been recognised by Kuchař (1970) in his analysis of the weak field limit in the geometrodynamical language: The linear gravitons do not contain "information about time." This information shows first up in the gravitational field which they produce in the next order. Ashtekar (1991) has related an *intrinsic* degree of freedom in configuration space, the imaginary part of the trace part of the connection, to physical time. In the geometrodynamical language, this imaginary part is basically given by the trace of the extrinsic curvature and is thus an "extrinsic time." Not surprisingly, this was the time parameter chosen by Kuchař (1970), who gave a lucid geometric interpretation: Consider a hypersurface in flat spacetime by specifying t as a function of x^i. If this hypersurface is attained from a flat hypersurface by a small deformation, Kuchař showed that the intrinsic geometry is only sensitive to the second order in this deformation ($h_{ij} \approx \delta_{ij} + t_{,i}\, t_{,j}$), while the extrinsic geometry is already sensitive to the first order ($K_{ik} \approx t_{,ik}$) and thus seems to be more appropriate in defining a time variable.

What are the main differences of this approach to the semiclassical expansion, which was presented above? One difference is that, whereas above the whole gravitational field was semiclassical in the first order, in the Ashtekar approach some of the degrees of freedom are treated as fully quantum, i. e. they appear in the Hamiltonian in the Schrödinger equation. This is, however, not a major difference since, as has been remarked in the previous section, one could easily generalise the scheme to incorporate such a situation by expanding with respect to a parameter different from G. It is thus not surprising that Ashtekar's definition of time contains the gravitational constant.

What is more important is that in the second approach a classical background has been chosen *ab initio* and only a truncated constraint, being linear in the canonical momentum, has been quantised. It is far from clear whether the same results can be recovered from the full theory in an appropriate limit.

Whereas time in the semiclassical approximation is constructed from the phase of the WKB state and thus essentially state-dependent, the weak field approach constructs time always from the same part of the configuration variables and is thus state-independent. Moreover, the semiclassical limit seems to be inappropriate for studying the weak field limit, since a flat background spacetime in its standard foliation corresponds to $S_0 = constant$ ("no change of configuration" $\leftrightarrow$ "no WKB time") and thus does not allow the recovery of a Schrödinger equation in this case. This difficulty of incorporating "flat spacetime" into the picture of quantum geometrodynamics will again show up in the next order of this approximation scheme.

3 Corrections to the Schrödinger equation from quantum gravity

3.1 General derivation and discussion

We continue with the semiclassical expansion scheme to the order M^{-1}, which leads to the calculation of *correction terms* to the functional Schrödinger equation (2.38). This order yields an equation involving S_2 (compare (2.27)) and reads explicitly

$$G_{ab}\frac{\delta S_0}{\delta h_a}\frac{\delta S_2}{\delta h_b}+\frac{1}{2}G_{ab}\frac{\delta S_1}{\delta h_a}\frac{\delta S_1}{\delta h_b}-\frac{i\hbar}{2}G_{ab}\frac{\delta^2 S_1}{\delta h_a\delta h_b}$$
$$+\frac{1}{\sqrt{h}}\left(\frac{\delta S_1}{\delta\phi}\frac{\delta S_2}{\delta\phi}-\frac{i\hbar}{2}\frac{\delta^2 S_2}{\delta\phi^2}\right)=0. \quad (3.1)$$

One first rewrites this equation by substituting S_1 in favour of χ, see (2.35), and then makes the following ansatz for S_2:

$$S_2[h_a,\phi]=\sigma_2[h_a]+\eta[h_a,\phi]. \quad (3.2)$$

Eq. (3.1) will be simplified, if one demands that σ_2 obeys:

$$G_{ab}\frac{\delta S_0}{\delta h_a}\frac{\delta\sigma_2}{\delta h_b}-\frac{\hbar^2}{D^2}G_{ab}\frac{\delta D}{\delta h_a}\frac{\delta D}{\delta h_b}+\frac{\hbar^2}{2D}G_{ab}\frac{\delta^2 D}{\delta h_a\delta h_b}=0. \quad (3.3)$$

The interpretation of this equation becomes immediately clear, if one writes out its analogue in one-dimensional quantum mechanics:

$$\sigma_2'=-\frac{\hbar^2}{4}\frac{p''}{p^2}+\frac{3\hbar^2}{8}\frac{p'^2}{p^3}, \quad (3.4)$$

where, again, $p=\partial S/\partial x\equiv S'$. Eq. (3.4) is nothing but the equation for the second-order WKB correction, see e. g. Landau and Lifshitz (1975). (Recall that for pure gravity an expansion in G is fully equivalent to an expansion in $\hbar$.) The separation of σ_2 from S_2 thus serves to separate the pure gravitational WKB(2) factor (which in this context is uninteresting) from the relevant part, which satisfies the corrected Schrödinger equation. We then find from (3.1) an equation involving η only:

$$G_{ab}\frac{\delta S_0}{\delta h_a}\frac{\delta\eta}{\delta h_b} = \frac{\hbar^2}{2\chi}\left(-\frac{2}{D}G_{ab}\frac{\delta\chi}{\delta h_a}\frac{\delta D}{\delta h_b}+G_{ab}\frac{\delta^2\chi}{\delta h_a\delta h_b}\right)$$
$$+\frac{i\hbar}{\sqrt{h}\chi}\frac{\delta\eta}{\delta\phi}\frac{\delta\chi}{\delta\phi}+\frac{i\hbar}{2\sqrt{h}}\frac{\delta^2\eta}{\delta\phi^2}. \quad (3.5)$$

Up to this order, the total wave functional is thus of the form

$$\Psi=\frac{1}{D}\exp\left(\frac{i}{\hbar}(MS_0+\sigma_2M^{-1})\right)\chi\exp\left(\frac{i\eta}{M\hbar}\right). \quad (3.6)$$

Note that the relative phases in this wave functional have been fixed by the semiclassical expansion. The wave functional

$$\psi \equiv \chi \exp\left(\frac{i\eta}{M\hbar}\right) \tag{3.7}$$

then obeys the "corrected Schrödinger equation" (Kiefer and Singh, 1991)

$$i\hbar\frac{\delta\psi}{\delta\tau} = \mathcal{H}_m\psi + \frac{\hbar^2}{M\chi}G_{ab}\left(\frac{1}{D}\frac{\delta D}{\delta h_a}\frac{\delta\chi}{\delta h_b} - \frac{1}{2}\frac{\delta^2\chi}{\delta h_a \delta h_b}\right)\psi. \tag{3.8}$$

Thus, if one knows the solutions to the previous order equations, one can evaluate the correction term on the right-hand side of (3.8). Note that it is *not* sufficient to know the solution, χ, of the Schrödinger equation (2.38), but that the knowledge of the gravitational prefactor, D, is also required.

As in the case of the uncorrected Schrödinger equation, Eq. (3.8) is equivalent to its integrated form, since a specific slicing of a specific spacetime has been chosen upon giving a solution to the Hamilton-Jacobi equation (2.29).

To write (3.8) in a more transparent way, we decompose the vector fields on the right-hand side of (3.8) into their respective components tangential and normal to the integral curves of the vector field $G_{ab}\delta S_0/\delta h_a$. This will enable us to use the Schrödinger equation (2.38) for the tangential part. We first decompose the first derivative according to

$$G_{ab}\frac{\delta\chi}{\delta h_a} = \alpha\, G_{ab}\frac{\delta S_0}{\delta h_a} + a_b, \tag{3.9}$$

where

$$\frac{\delta S_0}{\delta h_a}a_a = 0.$$

The coefficient α is determined by multiplying each side of (3.9) with $\delta S_0/\delta h_b$, summing over b, and making use of the Hamilton-Jacobi equation (2.29) as well as the Schrödinger equation (2.38). One finds

$$\alpha = \frac{i}{2\hbar V}\mathcal{H}_m\chi. \tag{3.10}$$

This, of course, holds only if the potential V is non-vanishing, i. e. in regions where the vector field $G_{ab}\delta S_0/\delta h_a$ is not "lightlike." (Note that in the general case V is nonzero, since it also includes contributions from "macroscopic" matter sources.) In the "lightlike" case α cannot be unambiguously determined, since the addition of a "lightlike" vector on the right-hand side of (3.9) does not change the scalar product of the left-hand side with $\delta S_0/\delta h_b$. But it is only this scalar product which is determined by the previous order equation (2.38).

I emphasise that the component a_b is *not* determined by the previous order equations, but only by the boundary conditions which are imposed on the *full* wave functional Ψ. As can be explicitly seen from (3.8), the second derivatives with respect to the three-metric now come into play. We will assume in the following that these component can be neglected, i. e. we will choose the boundary conditions such that χ is "peaked" around the considered worldline in configuration space, i. e. along the particular spacetime we have chosen (Cosandey, 1993).

The second derivative term on the right-hand side of (3.8) is now also decomposed into its tangential and normal components, and use of the previous order equations as well as of (3.10) is made. After some calculation one can then put (3.8) in the form (Kiefer and Singh, 1991)

$$i\hbar\frac{\delta\psi}{\delta\tau} = \mathcal{H}_m\psi + \frac{4\pi G}{c^4\sqrt{h}(R-2\Lambda)}\mathcal{H}_m^2\psi + i\hbar\frac{4\pi G}{c^4}\frac{\delta}{\delta\tau}\left(\frac{\mathcal{H}_m}{\sqrt{h}(R-2\Lambda)}\right)\psi. \quad (3.11)$$

This is the central equation of this section. Several comments are in order.

(1) The WKB(1) prefactor D has disappeared from the corrected Schrödinger equation after use of (2.36) was made. This is fortunate, since we do not have to deal with the pure gravitational part of the full wave functional.

(2) As remarked above, this form of the corrected Schrödinger equation only holds for non-vanishing gravitational potential. In the general case one must resort to the form (3.8).

(3) The correction terms in (3.11) are, of course, only formal as long as no regularisation scheme is brought into play to cure the divergences which come together with the functional derivatives. We will show in the section on applications how concrete physical prediction can nevertheless be extracted from these terms in special cases.

(4) The terms in (3.11) are independent of the factor ordering which is chosen for the gravitational kinetic term in (2.1).There may, however, be an ambiguity in the factor ordering of the gravity-matter coupling.

(5) If "macroscopic" matter sources are present, one has to replace $\sqrt{h}(R-2\Lambda)$ in the denominator of (3.11) by $\sqrt{h}(R-2\Lambda)-\epsilon_m/2Mc^2$, where ϵ_m is an effective matter energy density. That $M \propto G^{-1}$ comes explicitly into play in this expression, does not present a problem as long as the corresponding correction terms are small. This will become important in the example of black hole evaporation discussed below.

(6) In addition to the Wheeler-DeWitt equation (2.1) one must also expand the momentum constraints. This is straightforward and leads to the result that the wave functional remains unchanged under a spatial diffeomorphism *at each order* of this approximation scheme.

(7) The second correction term in (3.11) is pure imaginary and thus leads to a *violation of unitarity*. The occurrence of such a term is not surprising, since we have attempted to write down an effective equation for the non-gravitational fields alone. This can also be recognised from the point of view of the conservation laws for the wave functional. The full Wheeler-DeWitt

equation (2.1) is of the Klein-Gordon type and thus obeys the following conservation law:

$$\frac{1}{M}G_{ab}\frac{\delta}{\delta h_a}\left(\Psi^* \frac{\overleftrightarrow{\delta}}{\delta h_b} \Psi\right)+\frac{1}{\sqrt{h}}\frac{\delta}{\delta\phi}\left(\Psi^* \frac{\overleftrightarrow{\delta}}{\delta\phi} \Psi\right)=0. \tag{3.12}$$

Applying the M-expansion for the wave functional to this conservation law, one finds at the level of the corrected Schrödinger equation

$$\begin{aligned}&\frac{\delta}{\delta\tau}(\psi^*\psi)+\frac{\hbar}{2i\sqrt{h}}\frac{\delta}{\delta\phi}\left(\psi^* \frac{\overleftrightarrow{\delta}}{\delta\phi} \psi\right)\\ &-\frac{i\hbar}{2M}G_{ab}\left[\frac{\delta}{\delta h_a}\left(\psi^* \frac{\overleftrightarrow{\delta}}{\delta h_b} \psi\right)-\frac{1}{D}\frac{\delta D}{\delta h_a}\psi^*\frac{\delta}{\delta h_b}\psi\right]=0.\end{aligned} \tag{3.13}$$

The first two terms together are the "Schrödinger current" connected with (2.38), while the remaining terms yield correction terms proportional to M^{-1}. Upon functionally integrating this equation over the field ϕ and making the standard assumption that ψ falls off for large field configurations, one finds from (3.13)

$$\frac{d}{dt}\int\mathcal{D}\phi\,\psi^*\psi=\frac{8\pi G}{c^4}\int\mathcal{D}\phi\int d^3x\,\psi^*\frac{\delta}{\delta\tau}\left(\frac{\mathcal{H}_m}{\sqrt{h}(R-2\Lambda)}\right)\psi. \tag{3.14}$$

A comparison with (3.11) immediately exhibits that the violation of the Schrödinger conservation law produces the imaginary correction term in (3.11). I note again that a similar term emerges in the nonrelativistic expansion of the Klein-Gordon equation in the presence of an external Newtonian potential (Lämmerzahl, 1993).

Corrections to the Schrödinger equation have been discussed, in the context of one-dimensional minisuperspace models, by a number of authors including Brout and Venturi (1989), Singh (1990), Kowalski-Glikman and Vink (1990), and Bertolami (1991) (see also Singh (1993) for a review). The above, systematic, expansion in the context of the full Wheeler-DeWitt equation has been discussed by Kiefer and Singh (1991).

The above semiclassical expansion scheme can also be applied to standard quantum field theories like QED, see Kiefer, Padmanabhan, and Singh (1991), and Kiefer (1992a). The purpose of such an application lies primarily in the test of its viability within a well-understood framework. It provides, however, also a useful insight into the semiclassical expansion of QED itself. It demonstrates, for example, that one can start from the *stationary* equation for full QED and derive, as in the gravitational case, a *time-dependent* Schrödinger equation for the matter fields in an external electromagnetic background. Again, the "fast" matter degrees of freedom follow the "slow" electromagnetical degrees of freedom, which can thus serve as an "intrinsic clock."

Formally, the expansion is performed with respect to the electric charge (after a suitable rescaling of the vector potential has been made). The expression for the full wave functional up to order e^0 then reads, in analogy to the gravitational expression (2.40),

$$\Psi \approx \frac{1}{D} \exp\left(\frac{iS_0[\mathbf{A}]}{\hbar e^2}\right) \psi[\mathbf{A}, \phi, \phi^\dagger], \tag{3.15}$$

where $\mathbf{A}$ is the vector potential, and ϕ is a charged scalar field. The functional ψ obeys a functional Schrödinger equation, in which time is defined by S_0. Corrections to this Schrödinger equation can be found in order e^2 and look very similar to the gravitational terms in (3.11). In QED they can be easily interpreted in terms of Feynman diagrams with an internal photon line, see Kiefer (1992a). This in turn provides a bridge to the standard expansion of the effective action.

I want to conclude this section with a brief remark on the correction terms in the Ashtekar variables. Basically, the expansion proceeds as in the geometrodynamical case. The main difference is that there is *no* R term in the analogue of (3.11), since the potential V in the Ashtekar case is given by the cosmological constant and the macroscopic matter contribution only.

3.2 Applications

3.2.1 A minisuperspace example

The simplest example is to neglect all inhomogeneous degrees of freedom and keep only homogeneous ones like the scale factor a of a (closed) Friedmann Universe and a homogeneous scalar field ϕ. The analogue of (3.11) is then a quantum mechanical Schrödinger equation for a wave function $\psi(a, \phi)$ augmented by two correction terms:

$$i\hbar\dot{\psi} = H_m\psi + \frac{G}{3\pi c^4 a}\left(H_m^2 + i\hbar H_0\left[-\frac{\partial H_m}{a\partial a} + H_m\right]\right)\psi. \tag{3.16}$$

Here, $H_0 \equiv \dot{a}/a$ denotes the Hubble parameter of the Friedmann Universe, and H_m is the Hamiltonian for the scalar field. One recognises that the correction terms only become relevant for small scales ($a \to 0$), i. e. in the early Universe.

3.2.2 Scalar field in de Sitter space

As a specific example we treat the case of a minimally coupled scalar field, whose Hamiltonian density is given by (2.25) with $U \equiv 0$. The case of nonminmally fields involves extra subtleties and is treated elsewhere, see Kiefer (1993b). It is convenient to write the three-metric in the following form:

$$h_{ab} = h^{1/3}\tilde{h}_{ab}, \tag{3.17}$$

where h is the determinant of the three-metric. (The "conformal part" $\tilde{h}_{ab}$ thus has unit determinant.) The Hamilton-Jacobi equation (2.29) then reads

$$-\frac{3\sqrt{h}}{16}\left(\frac{\delta S_0}{\delta\sqrt{h}}\right)^2+\frac{\tilde{h}_{ac}\tilde{h}_{bd}}{2\sqrt{h}}\frac{\delta S_0}{\delta\tilde{h}_{ab}}\frac{\delta S_0}{\delta\tilde{h}_{cd}}-2\sqrt{h}(R-2\Lambda)=0. \tag{3.18}$$

We want to look for a solution of this equation which, after integrating (2.30), describes de Sitter spacetime in a flat foliation. This is easily achieved by setting $R=0$ in (3.18) and looking for a solution of the form $S_0=S_0(\sqrt{h})$. The desired solutions read

$$S_0=\pm 8\sqrt{\frac{\Lambda}{3}}\int\sqrt{h}d^3x\equiv\pm 8H_0\int\sqrt{h}d^3x, \tag{3.19}$$

where, again, H_0 is the Hubble parameter (which is a constant in this case). According to (2.38), S_0 defines a local time parameter in configuration space,

$$\frac{\delta}{\delta\tau}=-\frac{3\sqrt{h}}{8}\frac{\delta S_0}{\delta\sqrt{h}}\frac{\delta}{\delta\sqrt{h}}=\sqrt{3h\Lambda}\frac{\delta}{\delta\sqrt{h}},$$

from where the well kown expansion law for the scale factor can be found,

$$\frac{\partial a^3}{\partial t}=\int d^3y\frac{\delta\sqrt{h}(\mathbf{x})}{\delta\tau(\mathbf{y})}=3H_0a^3\ \rightarrow\ a(t)=e^{H_0t}. \tag{3.20}$$

Equation (2.36) for the prefactor can be immediately solved and yields $\delta D/\delta\tau=0$, i. e. D is constant on the whole de Sitter spacetime. The functional Schrödinger equation (2.38) then reads in its integrated form (setting $\hbar=1$ in the following)

$$i\dot{\psi}=\int d^3x\left(-\frac{1}{2a^3}\frac{\delta^2}{\delta\phi^2}+\frac{a}{2}(\nabla\phi)^2+\frac{a^3}{2}m^2\phi^2\right)\psi. \tag{3.21}$$

Since de Sitter space is homogeneous, it is adequate to use a momentum representation, i. e.

$$\phi(\mathbf{x})=\int\frac{d^3k}{(2\pi)^3}\chi(\mathbf{k})e^{i\mathbf{kx}}\equiv\int d\tilde{\mathbf{k}}\chi_k e^{i\mathbf{kx}}, \tag{3.22}$$

and thus

$$\frac{\delta}{\delta\phi(\mathbf{x})}=\int d\tilde{k}e^{i\mathbf{kx}}\frac{\delta}{\delta\chi_k}\ ;\ \frac{\delta\chi_k}{\delta\chi_{k'}}=(2\pi)^3\delta^{(3)}(\mathbf{k}+\mathbf{k}').$$

We want to solve (3.21) by a Gaussian ansatz, i. e.

$$\psi=N(t)\exp\left(-\frac{1}{2}\int d\tilde{k}\Omega(\mathbf{k},t)\chi_k\chi_{-k}\right). \tag{3.23}$$

Gaussian states are used to describe generalised vacuum states in the Schrödinger picture, see e. g. Jackiw (1988b). This special form is of course tied to quadratic Hamiltonians like (2.25) with $U = 0$. According to (3.21), the Gaussian form (3.23) is preserved in time. Inserting this ansatz into (3.21) yields two equations for N and for Ω:

$$i\frac{\dot{N}}{N} = \frac{1}{2a^3}\mathrm{Tr}\Omega \equiv \frac{V}{2a^3}\int d\tilde{k}\ \Omega(\mathbf{k},t) \tag{3.24}$$

$$i\dot{\Omega} = \frac{\Omega^2}{a^3} - a^3(m^2 + \frac{k^2}{a^2}). \tag{3.25}$$

Eq. (3.25) can be simplified by writing

$$\Omega = -ia^3\frac{\dot{y}}{y}, \tag{3.26}$$

which leads to a linear second-order equation for y,

$$\ddot{y} + 3H_0\dot{y} + \left(m^2 + \frac{k^2}{a^2}\right)y = 0. \tag{3.27}$$

It is useful to introduce a conformal time coordinate η,

$$dt = ad\eta \ \rightarrow\ a(\eta) = -\frac{1}{H_0\eta},\ \eta \in (0, -\infty), \tag{3.28}$$

which yields

$$y'' + 2\frac{a'}{a}y' + (m^2a^2 + k^2)y = 0, \tag{3.29}$$

where $'$ denotes a derivative with respect to η. Inserting the evolution law (3.28) for the scale factor, this equation can be explicitly solved in terms of Hankel functions, see Guven, Lieberman, and Hill (1989). The question then arises what boundary conditions one has to choose to select a particular solution of (3.29). We will follow Guven *et al* and choose the *Bunch-Davies vacuum.* This is de Sitter invariant, i. e. invariant under $SO(4,1)$, and reduces to the Minkowski vacuum at early times, i. e. Ω tends to $\sqrt{k^2+m^2}$ as $t \rightarrow -\infty$, where the metric is essentially static and one can put $\dot{\Omega} = 0$ and $a = 1$ in (3.25). The demand for de Sitter invariance is the analogue of the demand for Poincaré invariance in Minkowski space. In $2+1$ dimensions it has been shown that the Bunch-Davies vacuum is the *only* de Sitter invariant vacuum state, see Jackiw (1988b).

We are now interested to calculate the expectation value of the Hamiltonian density with respect to (3.23). Using the solution for Ω which corresponds to the Bunch-Davies vacuum, one finds for $\langle\phi^2\rangle$,

$$\langle\phi^2\rangle = \int\frac{d\tilde{k}}{2\mathrm{Re}\Omega} \propto \frac{H_0^2}{4\pi^2}\int_0^\infty dyy,$$

i. e. a divergent result! The occurrence of ultraviolet divergencies is of course not surprising, and we need a regularisation scheme to extract physical predictions from (3.23). Guven *et al* (1989) employ a dimensional regularisation scheme: They evaluate all expressions in d space dimensions, where they are finite, and subtract terms which diverge in the limit $d \to 3$. After some lengthy calculation they find a finite result for the expectation value of the Hamiltonian which agrees with the result found by using the effective action. In the massless case, for example, one has (re-inserting $\hbar$ and c)

$$\langle \mathcal{H}_m \rangle = \frac{29\hbar H_0^4 a^3}{960\pi^2 c^3}. \tag{3.30}$$

This result is obtained on the level of the functional Schrödinger equation (3.21), which is the level of quantum field theory on a classical spacetime.[2] As we have seen in the last section, the next order of the semiclassical approximation scheme yields corrections to the Schrödinger equation. In the present case, Eq. (3.11) reads

$$i\hbar\frac{\delta\psi}{\delta\tau} = \mathcal{H}_m\psi - \frac{2\pi G}{c^4\sqrt{\hbar}\Lambda}\mathcal{H}_m^2\psi - i\hbar\frac{2\pi G}{c^4\Lambda}\frac{\delta}{\delta\tau}\left(\frac{\mathcal{H}_m}{\sqrt{\hbar}}\right)\psi. \tag{3.31}$$

The first correction term in (3.31) leads to a shift in the above expectation value, while the second term is a source of non-unitarity. I want to focus first on the real term. It causes a shift

$$\langle \mathcal{H}_m \rangle \to \langle \mathcal{H}_m \rangle - \frac{2\pi G}{3c^2a^3H_0^2}\langle \mathcal{H}_m^2 \rangle. \tag{3.32}$$

The last term is of course utterly divergent and it is not clear how to regularise it. On the covariant level it corresponds to a term like $\langle T_{\mu\nu}^2(x)\rangle$, i. e. to a four-point correlation function, whose regularisation in the presence of gravitational fields has, to my knowledge, not been addressed. Nevertheless, one can evaluate (3.32) in the present case, at least in a heuristic sense. The reason is that the Bunch-Davies vacuum state is an adiabatic vacuum state. The fluctuation of such a state would be zero in quantum mechanics. I assume that this holds also for the present field theoretic case, i. e. that

$$\langle \mathcal{H}_m^2 \rangle \approx \langle \mathcal{H}_m \rangle^2. \tag{3.33}$$

The covariant version of (3.33) is frequently employed as a criterion for semiclassical behaviour (Kuo and Ford, 1993). This will be discussed in more detail in the next section.

One thus gets the following prediction for the shift in the expectation value (3.30):

$$\langle \mathcal{H}_m \rangle \to \langle \mathcal{H}_m \rangle - \frac{841}{1382400\pi^3}\frac{G\hbar H_0^6 a^3}{c^8}. \tag{3.34}$$

Although found by using heuristic arguments, this is a definite prediction of the Wheeler-DeWitt equation! That the correction term is proportional to

[2] Note that $\langle \mathcal{H}_m \rangle = a^3\langle T_{00}\rangle$, where T_{00} is the 00- component of the energy-momentum tensor.

GH_0^6 is of course clear from dimensional arguments. We remark already at this point that there will be a second contribution to the energy shift which arises from the modification of semiclassical time through the back reaction of the matter fields. Surprisingly, this modification will just lead to a change of sign in the energy shift (3.34). The details will be presented in the next section.

What about the unitarity violating term? Again, its regularisation is unclear, but one can make the following rough estimate. It produces an imaginary contribution to the energy density of the order of magnitude

$$\begin{aligned} |\mathrm{Im}\epsilon| &= \frac{2\pi G\hbar}{3c^2 H_0^2} \frac{\langle \mathcal{H}_m \rangle}{\tau V a^3} \\ &\approx \frac{2\pi G\hbar}{c^2 H_0 V} \frac{\langle \mathcal{H}_m \rangle}{a^3} = \frac{29 G\hbar^2 H_0^3}{480\pi V c^5}. \end{aligned} \tag{3.35}$$

The presence of an energy density with an imaginary part is of course an indication of a possible instability of the system. The associated time scale t^* on which this "quantum gravitational instability of de Sitter space" may become relevant is given by

$$t^* = (2\mathrm{Im}\epsilon V)^{-1}\hbar = \frac{240\pi c^5}{29 G\hbar H_0^3} \sim \left(\frac{H_0^{-1}}{t_{Pl}}\right)^3 t_{Pl}, \tag{3.36}$$

where t_{Pl} is the Planck length. This time scale is thus only relevant if the "horizon scale" H_0^{-1} is of the order of the Planck length.

3.2.3 Evaporation of black holes

We now turn to an example where the second, imaginary, correction term in (3.11) becomes important. This will be the case if the mass of a black hole approaches the Planck regime through the emission of Hawking radiation (Kiefer, Müller, and Singh, 1993).

A black hole is produced by the collapse of matter, which we can assume for simplicity to be in a pure state initially. The presence of quantum fields on the black hole spacetime leads to Hawking radiation, i. e. the black hole emits thermal radiation and shrinks. What happens at the final stage of this evaporation is an open issue. If the evaporation is complete, only the Hawking radiation is left behind. Since thermal radiation cannot, however, be described by a pure state, there seems to be an evolution from pure states into mixed states. This was taken as a hint that quantum gravity may violate unitarity (Hawking, 1976) and that information is lost at a fundamental level. There are, however, suggestions how this "information loss paradox" can be resolved (a recent review of this issue is Page (1993)). One possibility is that the prescription through Hawking radiation is incomplete and that the true time evolution is unitary. This would mean that information must be encoded in the black hole radiation, for example by stimulated emission or correlations. A second possibility is that the black hole evaporation is not complete but ceases at a mass of the order of the Planck mass, leaving behind a Planck mass remnant. The main problem with this approach is that only little energy is available to store a huge amount of information.

The main drawback of all these approaches is that, although the problem is a genuine problem of quantum gravity, only the semiclassical theory has been considered. All scenarios put forward so far thus only rest on speculations. To address these issues properly, one should start from a specific approach to quantum gravity such as the Wheeler-DeWitt equation (2.1). While it is not yet clear how to apply this equation to the black hole case one can at least apply the corrected Schrödinger equation, see (3.11), and hope that the results obtained indicate some genuine features of quantum gravity.

As discussed above, the functional Schrödinger equation (2.38) describes the level of quantum field theory on a classical gravitational background, i. e. the level where the Hawking effect is calculated, see, e. g., Freese, Hill, and Mueller (1985) for a discussion in the functional Schrödinger picture. We choose a slicing of the background spacetime, which is obtained from S_0, such that $R = 0$. How can the correction terms in (3.11) then be evaluated? An important observation is that one deals here with an *asymptotically flat* space, in contrast to the compact case of the general discussion. The Wheeler-DeWitt equation (2.1), if integrated over space, is thus not correct as it stands but has to be supplemented by a surface term from infinity. This is nothing but the ADM mass, $\mathcal{M}$. One thus has to replace the term $\sqrt{h}R$ in the denominators of (3.11) by $-16\pi G\mathcal{M}/c^2$. This is already clear from dimensional arguments, since $\sqrt{h}R$ has dimensions of a length and the only length scale in the black hole case is the Schwarzschild radius.

One might wonder whether the functional Schrödinger equation (2.38) can be recovered at all in this case, since the flat slicing of Schwarzschild spacetime corresponds to a solution S_0 = *constant* of the Hamilton-Jacobi equation, and the whole spacetime would thus only be a point in the configuration space. This would certainly be the case for a compact space, but in the asymptotically flat case a distinguished time parameter is associated with the Poincaré group at infinity. This time parameter is the one that shows up in the Schrödinger equation (2.38). It is also the time parameter with respect to which the question of unitarity is discussed in the case of black hole evaporation.

The ratio of the first correction term in (3.11) to the dominant term, which is $\mathcal{H}_m$, is given by the ratio of the energy of the scalar field to $\mathcal{M}c^2$, which is small even if the mass of the hole approaches the Planck regime. This is not the case for the second term! It contributes the following term to the

Hamiltonian:

$$\Delta H_m \equiv -\frac{4\pi i G\hbar}{c^4}\int d^3x \mathcal{H}_m \frac{\partial}{\partial t}\left(\frac{c^2}{16\pi G\mathcal{M}(t)}\right). \tag{3.37}$$

We have here taken into account only the time derivative of the ADM mass, since it is the change of the background geometry which is the relevant contribution in the final stage of the evaporation (the decrease of the ADM mass due to Hawking evaporation). The time dependence of the mass of the black hole can be obtained by using the expression for the Hawking temperature and Stefan-Boltzmann's law, see e. g. DeWitt (1975). This leads to

$$\mathcal{M}(t) = \left(\mathcal{M}_0^3 - \frac{a\hbar c^4}{G^2}t\right)^{1/3}, \tag{3.38}$$

where $\mathcal{M}_0$ is the initial mass of the hole, and a is a numerical factor that depends on the details of the model (the number of particle species considered, etc.) and whose precise value is not important for the present qualitative discussion. Eq. (3.37) can of course only be used if one assumes that the black hole does not settle to a mass which is much bigger than the Planck mass.

The term (3.37) becomes important if its of the same order of magnitude than the Hamiltonian H_m of the scalar field, i. e. if

$$\frac{\hbar}{4M^2c^2}\frac{\partial\mathcal{M}}{\partial t} \approx \left(\frac{m_{Pl}}{\mathcal{M}}\right)^4 \approx 1. \tag{3.39}$$

The non-unitary term thus becomes important if the mass of the evaporating black hole approaches the Planck mass. In the above scenario, this happens after a time

$$\bar{t} \approx \left(\frac{\mathcal{M}_0}{m_{Pl}}\right)^3 t_{Pl}. \tag{3.40}$$

After the mass of the hole has entered the Planck regime, the semiclassical expansion breaks down and one would have to deal with the full Wheeler-DeWitt equation. Nevertheless, the correction term (3.37) may be a good approximation if, for example, the black hole settles to a remnant of the order of the Planck mass but such that the numerical value of (3.37) is still small compared to H_m. It may also happen that the final evaporation time is of the order of $\mathcal{M}_0^4$ (Carlitz and Willey, 1987) which would mean that (3.37) is an excellent approximation over a long period of time.

A non-hermitean Hamiltonian does not transform a pure state into a mixed state. It does, however, change the "degree of purity" for a density matrix. In the present case its sign is such that it *increases* the degree of purity, if the mass of the hole decreases, see Kiefer, Müller, and Singh (1993). Whether this can be interpreted as a hint for the unitarity of quantum gravity is an open issue.

Although the above considerations are of a heuristic nature, they nevertheless demonstrate that the correction terms to the Schrödinger equation can be applied to concrete physical examples.

4 Decoherence and back reaction

In the derivation of the Schrödinger equation from quantum gravity, as it was presented in section 2, a crucial assumption was that the gravitational part of the wave functional is of the very special form $D^{-1}\exp(iMS_0/\hbar)$. The use of such a special, complex, solution to the real Wheeler-DeWitt equation (2.1) has been criticised in particular by Barbour (1993), who regarded this choice as being equivalent to putting in the desired result, namely the time-dependent Schrödinger equation, by hand. The purpose of this section is to demonstrate that the use of such a single WKB state can be justified in a very natural, physical, way.

Consider, for example, a solution to (2.1) at order M^0 which is a superposition of the state (2.40) with its complex conjugate,

$$\Psi = \frac{1}{D}\exp(iMS_0/\hbar)\chi[h_{ab},\phi] + \frac{1}{D}\exp(-iMS_0/\hbar)\chi^*[h_{ab},\phi]. \tag{4.1}$$

Both χ and χ^* obey a time-dependent Schrödinger equation, where time is defined respectively by S_0 and $-S_0$. The superposition (4.1), however, does *not* describe a classical world, since there are *interference terms* between both components of (4.1). Note that the superposition (4.1) is analogous to the superposition (2.6) in the derivation of the Born-Oppenheimer approximation.

How can one get rid of these "unwanted" superpositions? The basis idea is the fact that states like (4.1) are highly correlated in the high-dimensional configuration space of gravitational and non-gravitational degrees of freedom, but that only very few of them are actually accessible to a localised observer. Instead of the full state (4.1) such an observer has thus to discuss the reduced density matrix, which is obtained from (4.1) by integrating out the (huge number of) irrelevant degrees of freedom. States of the accessible part, which are then correlated with orthogonal states of the "rest of the world" do not exhibit any interference terms locally. Information about the corresponding phase relations has been *delocalised* in the full configuration space. This mechanism is referred to as *decoherence*. In the derivation of the Born-Oppenheimer approximation (2.8), decoherence lies at the heart of the neglection of interference terms.

In the following I will first briefly discuss two examples from quantum mechanics and then show how decoherence can justify the use of only one of the components in (4.1). For a general discussion on decoherence I refer to the extensive literature on this subject - see, for example, Zeh (1992, 1993a) or Zurek (1991), and the references therein.

4.1 Decoherence in quantum mechanics: Two examples

The first example is the *localisation* of macroscopic objects through interaction with their natural environment, for example through scattering by photons or air molecules (Joos and Zeh, 1985). Be $|\mathbf{x}\rangle$ the centre of mass position eigenstate of the macroscopic body, and $|\chi\rangle$ an initial state of the environment such that the total state at, say, $t = 0$ is a product of these states. The interaction be such that this state evolves according to

$$|\mathbf{x}\rangle|\chi\rangle \stackrel{t}{\rightarrow} |\mathbf{x}\rangle S_{\mathbf{x}}|\chi\rangle, \tag{4.2}$$

where $S_{\mathbf{x}}$ denotes the scattering matrix. The state of the environment is thus correlated with the state of the body. If one starts now from an entangled, nonclassical, state for the body, it will envolve, according to the superposition principle, into an entangled state of the body with the environment, i. e.

$$\left(\int d^3x\varphi(\mathbf{x})|\mathbf{x}\rangle\right)|\chi\rangle \stackrel{t}{\rightarrow} \int d^3x\varphi(\mathbf{x})|\mathbf{x}\rangle S_{\mathbf{x}}|\chi\rangle. \tag{4.3}$$

There is no sign of any localisation for the body in this state! Since, however, the huge number of degrees of freedom of the environment is inaccessible to a local observer, the relevant object is the density matrix of the position of the scattering centre after scattering:

$$\begin{aligned}\rho(\mathbf{x},\mathbf{x}') &= \varphi(\mathbf{x})\varphi^*(\mathbf{x}')\langle\chi|S^\dagger_{\mathbf{x}'}S_{\mathbf{x}}|\chi\rangle \\ &\approx \varphi(\mathbf{x})\varphi^*(\mathbf{x}')\exp\left(-\Lambda t(\mathbf{x}-\mathbf{x}')^2\right).\end{aligned} \tag{4.4}$$

The Gaussian factor in (4.4) leads to the suppression of interference effects between different positions of the scattered object (Λ contains the details of the scattering process) – the object becomes localised, if the mass of the body is sufficiently large, and wave packets are prevented from spreading. The density matrix in (4.4) obeys a non-unitary master equation instead of a unitary von Neumann equation. As shown by Joos and Zeh (1985), even the scattering by the photons of the microwave background leads to the localisation of a dust grain in interstellar space.

Two important aspects of decoherence should be stressed (Zeh, 1993b). The first aspect follows from the very formalism of decoherence and is independent of any interpretation. It states that the phase relations of an entangled state like (4.3) can never be seen by a local observer according to the *probabilistic rules* used in conventional quantum theory. This does *not* mean, however, that only one component of this state survives – all components are still "there." The second aspect addresses the issue of why only one component is "observed." This involves the interpretation of the quantum formalism. Basically, there are two options: Either one has to assume that the decohered components are still co-existing, but are dynamically independent from each other (this dynamical independence is what decoherence achieves) or one has to invoke explicit collapse models (and thereby has to *change* conventional quantum theory) to single out a specific component, which is then the "observed one." For all practical purposes, this is equivalent to the branching into dynamically independent components in the first interpretation, where decoherence defines an ensemble of wave functions, the "preferred basis" (Zurek, 1993).

The second example is taken from nuclear physics, where deformed nuclei exhibiting a definite orientation can be found even if the total quantum state is an angular momentum eigenstate (Zeh 1967, 1993b). Consider an N-nucleon system which is described by a *stationary* Schrödinger equation. An *approximate* solution, as it is found through an appropriate variation principle, is given by the antisymmetrised produced, ϕ, of one-particle nucleon wave functions (Slater determinants). Since a superposition of different such ϕ is *not* a determinant again (due to the nonlinearity of the ansatz), these solutions need not be eigenstates of an operator like angular momentum which commutes with the Hamiltonian. They may thus describe a deformed nucleus, which has a definite orientation. Performing a superposition of different "deformed solutions,"

$$\Psi = \int d\Omega f(\Omega) U(\Omega)\phi, \tag{4.5}$$

where $U(\Omega)$ denotes a unitary representation of the rotation group and Ω may stand, e. g., for the Euler angles, one obtains, of course, again an approximate solution to the stationary Schrödinger equation. The amplitude $f(\Omega)$ may be chosen such that Ψ is an approximate eigenstate of angular momentum.

If many nucleons are present in the nucleus, a certain nucleon is correlated with a corresponding deformed potential generated by the remaining nucleons in each component of the superposition (4.5) ("strong coupling"). The matrix element between different orientations reads[3]

$$\begin{aligned} \langle U(\Omega)\phi | U(\Omega')\phi\rangle &= \prod_{i=1}^{N} \langle U(\Omega)\phi_i | U(\Omega')\phi_i\rangle \\ &\propto \exp\left(-\lambda(\Omega - \Omega')^2\right). \end{aligned} \tag{4.6}$$

For large N this is a narrow Gaussian peaked around $\Omega = \Omega'$ – the different orientations in (4.5) have decohered from one another. This is an example of a spontaneous symmetry breaking.[4] As can be seen from the existence of a rotational spectrum, these different orientations even seem to rotate slowly, in spite of the stationary nature of the total state! This is the analogue to the recovery of an approximate time-dependent Schrödinger equation for intrinsic degrees of freedom from the stationary Wheeler-DeWitt equation (2.1). Since $\langle U(\Omega')\phi | U(\Omega)\phi\rangle \approx \delta(\Omega - \Omega')$, one has $\langle \phi(\Omega) | \psi\rangle \approx f(\Omega)$ and can

[3] We assume here for simplicity that $\phi = \prod \phi_i$ (no antisymmetrisation).

[4] Other important examples are chiral molecules and the Higgs vacuum, see Zeh (1992).

thus assign an approximate wave function $f(\Omega)$ to the rotational degrees of freedom, describing a fictitious "rigid top."

If the nucleus contained an intrinsic quantum observer, such an observer would see the nucleus with a definite orientation. In the case of quantum cosmology, where the Wheeler-DeWitt equation (2.1) is applied to the whole Universe, all observers are intrinsic observers, and decoherence is the reason why they observe a classical world. This is discussed in the next subsection.

4.2 Decoherence of different WKB branches in quantum gravity

The analogue of the rotational symmetry of the nuclear Hamiltonian is here the (discrete) symmetry of the Wheeler-DeWitt Hamiltonian in (2.1) under complex conjugation. The semiclassical expansion presented in the previous sections shows that, to order M^0, the functionals $\chi[h_{ab}, \phi]$ of the non-gravitational fields obey an approximate time-dependent Schrödinger equation, which is a *complex* equation! Consequently, χ is in general complex and χ and χ^* thus couple drastically different to the components $\exp(iMS_0/\hbar)$ and $\exp(-iMS_0/\hbar)$. Since a huge number of degrees of freedom is involved, this leads to decoherence between the two components in (4.1), if these degrees of freedom are integrated out to get the reduced density matrix

$$\rho[h_{ab}, h'_{ab}] \equiv \mathrm{Tr}_{\phi}\Psi[h_{ab}, \phi]\Psi^*[h'_{ab}, \phi] \tag{4.7}$$

for the gravitational field. This general expectation is verified in concrete models (see Kiefer (1992c) for a review). Typically, global degrees of freedom like the scale factor a of a Friedmann Universe are chosen to represent the semiclassical gravitational degrees of freedom, while "small fluctuations" like density perturbations or gravitational waves play the role of the irrelevant degrees of freedom, which are integrated out. The functional χ is written in these examples as

$$\chi = \prod_{n=1}^{N} \chi_n(a; x_n), \tag{4.8}$$

where the x_n may be, for example, coefficients in the expansion of the scalar field ϕ into its harmonics. The result is that ρ becomes a narrow Gaussian in $a - a'$ (this is the analogue of the quantum mechanical example on localisation and may here be described as a "localisation within one WKB component" or "measurement of the three-geometry") and contains in addition a large suppression factor for the interference term between the WKB components in (4.1) (this is the analogue of the deformed nucleus or the chiral molecule). There are, however, situations where χ is approximately real and the two components in (4.1) interfere. This happens, for example, in regions which would correspond in the classical theory to the turning point of a recollapsing universe (Kiefer, 1992b), where $\partial S_0/\partial a$ becomes small. Decoherence thus justifies the selection of a definite WKB component in the semiclassical approximation as well as the localisation within one component. This is, however, not yet sufficient to justify the notion of a classical space*time*. It requires in addition the persistence of decoherence along a whole WKB trajectory and is related to the issue of back reaction which will be discussed in the next subsection.

I want to remark finally that it is of course not necessary to deal only with the whole Universe in quantum gravity. A quantum black hole, for example, may be viewed from outside and is thus a more direct analogue of the deformed nucleus.

4.3 Back reaction

In this section I want to address the question to what extent the presence of the non-gravitational matter fields lead to a modification of the Hamilton-Jacobi equation (2.29) at order M^{-1}, i. e. the question of back reaction. In particular, I want to discuss the conditions under which this back reaction is given by the *expectation value* of the Hamiltonian (2.25) with respect to the state χ satisfying the approximate Schrödinger equation. This is equivalent to study the range of validity of the semiclassical Einstein equations

$$G_{\mu\nu} = \frac{8\pi G}{c^4}\langle T_{\mu\nu}\rangle. \tag{4.9}$$

At order M, the value of the geometrodynamical momentum when applied to the WKB state of the gravitational field is given by the familiar expression (2.30). To find out how this expression may be changed at the next order, we first decompose χ into its absolute value and its phase,

$$\chi \equiv R\exp(i\theta/\hbar), \tag{4.10}$$

and then rewrite the Hamilton-Jacobi equation (2.29) as follows:

$$\frac{1}{2M}G_{ab}\left(M\frac{\delta S_0}{\delta h_a}+\frac{\delta\theta}{\delta h_a}\right)\left(M\frac{\delta S_0}{\delta h_b}+\frac{\delta\theta}{\delta h_b}\right)+MV-\frac{\delta\theta}{\delta\tau}+\mathcal{O}(M^{-1})=0. \tag{4.11}$$

Performing the expectation value of this equation with respect to χ and introducing the quantity

$$\Pi^a \equiv M\frac{\delta S_0}{\delta h_a}+\langle\chi|\frac{\delta\theta}{\delta h_a}\chi\rangle, \tag{4.12}$$

this can be written as

$$\frac{1}{2M}G_{ab}\Pi^a\Pi^b+MV-\langle\chi|\frac{\delta\theta}{\delta\tau}\chi\rangle+\mathcal{O}(M^{-1})=0. \tag{4.13}$$

The crucial assumption made at this stage is that Π^a is the geometrodynamical momentum at this order of approximation. We will specify later the conditions under which this assumption is valid. Taking this for granted, Eq. (4.13) is the desired "back reaction corrected" Hamilton-Jacobi equation. It can be conveniently rewritten by noting that

$$\langle\chi|\frac{\delta\theta}{\delta\tau}\chi\rangle = -\langle\chi|\mathcal{H}_m\chi\rangle, \tag{4.14}$$

which follows immediately from the fact that θ is the phase of a solution to the time-dependent Schrödinger equation (2.38). One can thus write (4.13) as

$$\frac{1}{2M}G_{ab}\Pi^a\Pi^b + MV + \langle\chi|\mathcal{H}_m\chi\rangle + \mathcal{O}(M^{-1}) = 0. \tag{4.15}$$

Making use of (2.30) and (2.38) one can write

$$\langle\chi|\mathcal{H}_m\chi\rangle = -\frac{\hbar}{M}G_{ab}\Pi^a A^b, \tag{4.16}$$

where we have introduced in analogy to (2.9) a "super Berry connection" (see Balbinot, Barletta, and Venturi, 1990; Datta, 1993; Kiefer, 1993a)

$$A^b = -i\langle\chi|\frac{\delta\chi}{\delta h_b}\rangle \tag{4.17}$$

which is defined on the space of three-metrics.

An equivalent way to write (4.13) is then, up to terms of the order M^{-1},

$$\begin{aligned}\Bigg\{\frac{1}{2M}G_{ab}\left(-i\hbar\frac{\delta}{\delta h_a} - \hbar A^a\right)\left(-i\hbar\frac{\delta}{\delta h_b} - \hbar A^b\right) + MV \\ -\frac{\hbar^2}{2M}G_{ab}A^aA^b + \mathcal{O}(M^{-1})\Bigg\}\psi_G = 0,\end{aligned} \tag{4.18}$$

where ψ_G is an effective gravitational wave function at this order of approximation. I emphasize that this equation is very similar to the equation (2.8) for the nuclear wave function in the Born-Oppenheimer approximation. The term quadratic in the connection, which is already of order M^{-1}, has only been retained in (4.18) because of this comparison.

An important consequence of the back reaction is that the notion of WKB time is modified since an additional term proportional to M^{-1} arises from (4.12). This term is thus of the same order as the corrections (3.11) to the Schrödinger equation. Using the definition (4.12), one can rewrite the time derivative introduced in (2.38) as follows:

$$\begin{aligned} i\hbar\frac{\delta\psi}{\delta\tau} &\equiv i\hbar G_{ab}\frac{\delta S_0}{\delta h_a}\frac{\delta\psi}{\delta h_b} = i\hbar G_{ab}\left(\frac{\Pi^a}{M} - \frac{1}{M}\langle\chi|\frac{\delta\theta}{\delta h_a}\chi\rangle\right)\frac{\delta\psi}{\delta h_b} \\ &\equiv i\hbar\frac{\delta\psi}{\delta\tilde{\tau}} - \frac{i\hbar}{M}G_{ab}\langle\chi|\frac{\delta\theta}{\delta h_a}\chi\rangle\frac{\delta\psi}{\delta h_b}. \end{aligned} \tag{4.19}$$

At order M^{-1} the "back reaction corrected" WKB time $\tilde{\tau}$ has to be used instead of the old τ. To find the induced contribution of this change in the definition of time to the corrected Schrödinger equation (3.11), we first decompose the vector field $G_{ab}\delta\psi/\delta h_a$ as in (3.9) into its components tangential and orthogonal to the flow lines of the vector field $G_{ab}\delta S_0/\delta h_a$ and assume, again, that the orthogonal component is much smaller than the tangential component. Thus,

$$G_{ab}\frac{\delta\psi}{\delta h_a} = \frac{i}{2\hbar V}\mathcal{H}_m\psi G_{ab}\frac{\delta S_0}{\delta h_a} + \mathcal{O}(M^{-1}). \tag{4.20}$$

With the help of (4.14) one can then write (4.19) in the form

$$i\hbar\frac{\delta\psi}{\delta\tau} = i\hbar\frac{\delta\psi}{\delta\tilde{\tau}} - \frac{1}{2MV}\langle\chi|\mathcal{H}_m\chi\rangle\mathcal{H}_m\psi, \tag{4.21}$$

so that the corrected Schrödinger equation (3.11) can be written as an equation with respect to the new time $\tilde{\tau}$:

$$\begin{aligned} i\hbar\frac{\delta\psi}{\delta\tilde{\tau}} &= \mathcal{H}_m\psi + \frac{4\pi G}{c^4\sqrt{h}(R-2\Lambda)}(\mathcal{H}_m^2 - 2\langle\mathcal{H}_m\rangle\mathcal{H}_m)\psi \\ &\quad + i\hbar\frac{4\pi G}{c^4}\frac{\delta}{\delta\tau}\left(\frac{\mathcal{H}_m}{\sqrt{h}(R-2\Lambda)}\right)\psi. \end{aligned} \tag{4.22}$$

The modification in the definition of time thus leads to an additional contribution to the shift of energies induced by quantum gravity. In our example of the scalar field in de Sitter space (section 3.2.2) this additional term leads to a *change of sign* in the shift of the expectation value for the Hamiltonian density. Instead of (3.34) one has

$$\langle\mathcal{H}_m\rangle \rightarrow \langle\mathcal{H}_m\rangle + \frac{841}{1382400\pi^3}\frac{G\hbar H_0^6 a^3}{c^8}. \tag{4.23}$$

From the corrected Hamilton-Jacobi equation (4.15) one finds in this example, using (3.30), an effective *increase* of the Hubble parameter

$$H_0^2 \rightarrow H_0^2\left(1 + \frac{29}{360\pi}\frac{GH_0^2\hbar}{c^5}\right). \tag{4.24}$$

The increase in H_0 is thus essentially determined by the square of the ratio of the Planck length to the horizon length. If our present Universe were in a de Sitter phase this increase in H_0 would be of the order of 10^{-120}.

Analogus to the Hamilton-Jacobi equation one gets a correction to the momentum constraints at this order of approximation, i. e.

$$D_a\Pi^{ab} = \langle\chi|T_0^b\chi\rangle. \tag{4.25}$$

One has thus recovered the semiclassical Einstein equations (4.9) at order M^{-1}.

What are the exact conditions under which the semiclassical Hamilton-Jacobi equation is valid? The first, obvious, condition is that the expectation value of $\mathcal{H}$ with respect to the state satisfying the approximate Schrödinger equation be *small* compared to the other terms in (4.15). This came out in a natural way through the expansion scheme with respect to the Planck mass. The second condition, as has been discussed in section 4.1, is the smallness of interference terms between different WKB components of the total state. This is the analogy of the smallness of interferences between the different components of (2.6) and is a necessary condition for the validity of a Born-Oppenheimer approximation. As shown in section 4.1, these various components decohere in realistic cases and thereby justify the use of a single WKB component.

The third, and most subtle, condition is the possibility to replace the geometrodynamical momentum at order M^{-1} by the expression (4.12). This can be done if there is a strong correlation in the gravitational sector between the momentum and the expression on the right-hand side of (4.12). There is no simple general condition for this to be the case. The usual approach is to employ the *Wigner function* for the reduced density matrix (4.7) obtained by integrating out non-gravitational degrees of freedom within a *single* WKB component. For a given density matrix in the position representation $\rho(q,q')$, the Wigner function is defined as

$$F_W(\bar{q},p) = \int d\Delta e^{-2ip\Delta}\rho(\bar{q}+\Delta,\bar{q}-\Delta), \tag{4.26}$$

where $\bar{q}=(q+q')/2$ and $\Delta=(q-q')/2$. The Wigner function is the closest one can get to a classical phase space distribution; because of the quantum uncertainty, the Wigner function is in general not positive definite. The Wigner function has been used in quantum gravity by several authors including Halliwell (1987), Padmanabhan and Singh (1990), Habib and Laflamme (1990), and Paz and Sinha (1991). To evaluate (4.26) one has to first calculate the reduced density matrix by using a definite solution of the Schrödinger equation satisfied by χ. In most examples discussed so far a Gaussian solution such as (3.23) has been chosen. The Wigner function (4.26) can then be evaluated and is in these examples typically of the form

$$\begin{aligned} F_W[h_a,\Pi^a] \;\approx\; & D^{-2}\sqrt{\frac{\pi}{\sigma^2}}\exp\left[-\frac{1}{\sigma^2}G_{ab}\left(\Pi^a - M\frac{\delta S_0}{\delta h_a} - \langle\chi|\frac{\delta\theta}{\delta h_a}\chi\rangle\right)\right. \\ & \left.\left(\Pi^b - M\frac{\delta S_0}{\delta h_b} - \langle\chi|\frac{\delta\theta}{\delta h_b}\chi\rangle\right)\right], \end{aligned} \tag{4.27}$$

where

$$\sigma^2 = \mathrm{Tr}\frac{|\delta\Omega/\delta\tau|^2}{4\mathrm{Re}\Omega^2} \tag{4.28}$$

turns out to be the inverse of the coherence width associated with the reduced density matrix itself. (Ω is the covariance of the Gaussian, see (3.23).) The conditions for the validity of the semiclassical Hamilton-Jacobi equation arising from these considerations are thus a) that the width (4.28) of the Gaussian (4.27) is much smaller than the peak value of the Gaussian ("strong correlation condition") and b) that the inverse width is much smaller than the components of the three-metric which are "measured" by the huge number of irrelevant degrees of freedom ("strong decoherence condition"). Typically, these conditions are fulfilled if the *fluctuation of the Hamiltonian density is small* in the state χ satisfying the approximate Schrödinger equation, i. e. if (3.33) holds. Under this condition, therefore, one would expect the expressions derived within this framework to hold very accurately. It is, however, still unclear what are the *most general conditions* which are compatible with the back reaction-corrected Hamilton-Jacobi equation (4.15).

5 Discussion

It is my purpose in this contribution to present, as I hope convincing, arguments that it is possible to extract physical insight from semiclassical gravity even if the underlying exact theory is not yet available. The formal framework is based on the canonical approach to quantum gravity and its central equation (2.1). Although the fundamental Hamiltonian may not be exactly of the form (2.1), general principles like the reparametrisation invariance of the classical theory suggest that a constraint equation of the form $H\Psi = 0$ will still play a central role. Semiclassical considerations may thus well be, at least to a large extent, independent of the concrete form of the fundamental equation. Moreover, the necessary requirement that the, well-tested, Schrödinger equation can be recovered as an approximate equation puts an important restriction on the fundamental theory. Corrections to the Schrödinger equation, on the other side, may turn out to be different from one theory to the other and thus allow the possibility to distinguish between alternative approaches. I emphasise that such correction terms lead to definite physical predictions such as the energy shift in (4.23) which can, at least in principle, be checked and thus confirmed or falsified.

I want to finally give a brief outlook on other important topics in semiclassical gravity which have not been addressed in this contribution. The first topic has to do with the relevance of semiclassical considerations for the problem of time in quantum gravity (Isham, 1992; Kuchař, 1992). Briefly speaking, the problem of time in quantum gravity is concerned with the "timeless nature" of the Wheeler-DeWitt equation (2.1) (by which one means the absence of the classical time parameter t) and the question whether a physical concept of time can be introduced at the most fundamental level or only in the semiclassical regime. I have presented an approximation scheme where the

Schrödinger equation together with its classical time parameter emerges in an appropriate limit. The expansion scheme is only valid if the higher order terms are small compared to the dominant one. The semiclassical Einstein equation (4.9), in particular, is only valid if the right-hand side is treated as a small perturbation. A nonperturbative treatment would in general not lead to sensible results. This argument also applies to the corrected Schrödinger equation (3.11). One would thus not expect that this semiclassical notion of time can be extrapolated to the full theory. On the contrary, the fact that the Schrödinger equation and with it all the concepts of ordinary quantum theory like Hilbert space and unitarity are only recovered approximately (recall (3.14)) leads me to the suspicion that these concepts do *not* play any fundamental role in quantum gravity. As mentioned before, the presence of a Hilbert space structure on the level of the exact theory may be in conflict with results from semiclassical expansions, see Louko (1993). Interpretations of quantum gravity which only work in the semiclassical regime, like the one by Vilenkin (1989), do in my opinion not really solve the problem of time and may even present an obstacle on the way to finding the correct full theory.

There is one attempt to extrapolate the semiclassical concept of time to full quantum gravity which has gained some attention, see Padmanabhan (1990), Greensite (1990), and Squires (1991). The basic idea there is to extract time from the phase of the *full* wave functional. In the semiclassical regime it agrees, of course, with the WKB time in the approximate Schrödinger equation, but it can be drastically different in other cases. This "phase time" parametrises, at least locally, a flow in the configuration space. This flow does not, in general, have anything to do with the background spacetime which was reconstructed from a given solution to the Hamilton-Jacobi equation (2.29). The main advantage of this approach seems to be that an exact Ehrenfest equation can be written down for the expectation value of an operator depending on the variables which are comoving with respect to the "phase time flow." The expectation value is performed with respect to the Schrödinger inner product of the full wave functional and the comoving variables. There are, however, some problems with this approach. Firstly, the procedure does not work for real wave functionals which can therefore not be interpreted in this framework. Secondly, the concept of phase time depends on the given solution for the Wheeler-DeWitt equation and may thus seem inappropriate as a fundamental concept. Thirdly, phase time does not exist globally and may also lead to the "spacetime problem," i. e. the dependence of its definition upon the choice of the original spacetime foliation in the canonical approach (Kuchař, 1992). I emphasise again that this problem did not occur in the semiclassical approximation. Finally, the Ehrenfest condition with respect to the momentum variables does only hold if the phase of the wave functional varies rapidly along the flow lines which it defines. In the special case, however, where the semiclassical expansion scheme converges, it seems that the sum over all orders leads to a modification in the original WKB time which just yields the above phase time. This is spelled out elsewhere (Kiefer, 1993c).

Can anything be said about the concept of time on the fundamental level of (2.1) itself? The answer is yes, since the structure of the DeWitt metric in front of the kinetic term distinguishes a certain variable in configuration space from the remaining ones by its minus sign. At each space point, (2.1) is thus of the hyperbolic type. But only in the case where *one* global minus sign survives after the diffeomorphism group has been factored out could one say that (2.1) itself is hyperbolic. This is, however, not the case in the full configuration space, but only in special domains (Giulini, 1993). One important example is the neighbourhood of Friedmann-type three-geometries for constant lapse. In this case there exists a global intrinsic timelike variable, the scale factor of the Friedmann space, with respect to which a well-defined initial value problem can be posed for the wave functional (Zeh 1988, 1992a). There is, however, no unitary evolution with respect to this intrinsic time. This fact leads to an interesting tension between intrinsic time and WKB time which has drastic consequences in the case of a recollapsing Friedmann Universe (Zeh, 1992; Kiefer, 1993d): The semiclassical approximation can *not* be valid along the whole region of the corresponding classical trajectory and genuine quantum effects of gravity become important far away from the Planck scale.

I want to finally make a brief comparison of the semiclassical approximation to quantum gravity in the canonical approach with the covariant effective action approach ("background field method"). Formally, the level of the approximate functional Schrödinger equation corresponds to the level of the one-loop effective action. Taking into account the correction terms of (3.11) is analogous to the two-loop level of the effective action, although this has not yet been explicitly demonstrated. There is, however, a conceptual difference. The effective action approach starts from a *given* background spacetime which in the canonical framework was only recovered under appropriate circumstances such as the decoherence between different WKB components. The background field method therefore can be applied only under these circumstances.

References

A. Ashtekar (1991): *Lectures on non-perturbative canonical gravity* (World Scientific, Singapore).

A. Ashtekar(1988):*New perspectives in canonical gravity* (Bibliopolis, Napoli).

R. Balbinot, A. Barletta, and G. Venturi (1990): Matter, quantum gravity, and adiabatic phase. Phys. Rev. D **41**, 1848.

T. Banks (1985): TCP, Quantum Gravity, the Cosmological Constant, and all that .... Nucl. Phys. **B249**, 332.

J. B. Barbour (1992): Private communication.

J. B. Barbour (1993): Time and complex numbers in canonical quantum gravity. Phys. Rev. D **47**, 5422.

E. P. Belasco and H. C. Ohanian (1969): Initial Conditions in General Relativity: Lapse and Shift Formulation. Journ. Math. Phys. **10**, 1503.

O. Bertolami (1991): Nonlinear corrections to quantum mechanics from quantum gravity. Phys. Lett. **A154**, 225.

R. Brout and G. Venturi (1989): Time in semiclassical gravity. Phys. Rev. D **39**, 2436.

R. D. Carlitz and R. D. Willey (1987): Lifetime of a black hole. Phys. Rev. D **36**, 2336.

D. Cosandey (1993): Localized quantum mechanics in a semiclassical universe. University of Bern preprint.

T. Damour and J. H. Taylor (1991): On the orbital period change of the binary pulsar PSR 1913+16. Astrophys. Journal **366**, 501.

D. P. Datta (1993): Semiclassical backreaction and Berry's phase. Mod. Phys. Lett. A **8**, 191 (1993).

B. S. DeWitt (1967): Quantum Theory of Gravity I. The Canonical Theory. Phys. Rev. **160**, 1113.

B. S. DeWitt (1970): "Spacetime as a sheaf of geodesics in superspace", in *Relativity*, ed. by M. Carmeli, S. Fickler, and L. Witten (Plenum, New York).

B. S. DeWitt (1975): Quantum field theory in curved spacetime. Phys. Rep. **19**, 295.

K. Freese, C. T. Hill, and M. Mueller (1985): Covariant functional Schrödinger formalism and application to the Hawking effect. Nucl. Phys. **B255**, 693.

U. H. Gerlach (1969): Derivation of the Ten Einstein Field Equations from the Semiclassical Approximation to Quantum Geometrodynamics. Phys. Rev. **177**, 1929.

D. Giulini (1993): "What is the geometry of superspace?", in *from Newton's bucket to quantum gravity*, ed. by J. B. Barbour and H. Pfister (Birkhäuser, Boston).

J. Greensite (1990): Time and Probability in Quantum Cosmology. Nucl. Phys. **B342**, 409.

J. Guven, B. Lieberman, and C. T. Hill (1989): Schrödinger - picture field theory in Robertson-Walker flat spacetimes. Phys. Rev. D **39**, 438.

S. Habib and R. Laflamme (1990): Wigner function and decoherence in quantum cosmology. Phys. Rev. D **42**, 4056.

J. J. Halliwell (1987): Correlations in the wave function of the Universe. Phys. Rev. D **36**, 3626.

J. J. Halliwell and S. W. Hawking (1985): Origin of structure in the Universe. Phys. Rev. D **31**, 1777.

J. B. Hartle (1986): "Prediction in Quantum Cosmology", in *Gravitation in Astrophysics*, ed. by B. Carter and J. B. Hartle (Plenum Press, New York).

S. W. Hawking (1976): Breakdown of predictability in gravitational collapse. Phys. Rev. D **14**, 2460.

W. Heisenberg and H. Euler (1936): Folgerungen aus der Diracschen Theorie des Positrons. Z. Phys. **98**, 714.

C. J. Isham (1992): "Canonical Quantum Gravity and the Problem of Time", in *Integrable Systems, Quantum Groups, and Quantum Field Theories* (Kluwer Academic Publishers, London).

R. Jackiw (1988a): Berry's phase – Topological Ideas from Atomic, Molecular and Optical Physics. Comments At. Mol. Phys. **21**, 71.

R. Jackiw (1988b): "Analysis on infinite-dimensional manifolds – Schrödinger representation for quantized fields", *Field Theory and Particle Physics*, ed. by O. Eboli, M. Gomes, and A. Santano (World Scientific, Singapore).

E. Joos and H. D. Zeh (1985): The Emergence of Classical Properties through Interaction with the Environment. Z. Phys. **B59**, 223.

C. Kiefer (1987): Continuous measurement of mini-superspace variables by higher multipoles. Class. Quantum Grav. **4**, 1369.

C. Kiefer (1992a): Functional Schrödinger equation for scalar QED. Phys. Rev. D **45**, 2044.

C. Kiefer (1992b): Decoherence in quantum electrodynamics and quantum gravity. Phys. Rev. D **46**, 1658.

C. Kiefer (1992c): "Decoherence in quantum cosmology", in *Proceedings of the 10th Seminar on Relativistic Astrophysics and Gravitation*, ed. by S. Gottlöber, J. P. Mücket, and V. Müller (World Scientific, Singapore).

C. Kiefer (1993a): Topology, decoherence, and semiclassical gravity. Phys. Rev. D **47**, 5414.

C. Kiefer (1993b): "Semiclassical gravity with non-minimally coupled fields", in preparation.

C. Kiefer (1993c): "Semiclassical gravity and the problem of time", to appear in the Proceedings of the Cornelius Lanczos International Centenary Conference.

C. Kiefer (1993d): "Quantum cosmology and the emergence of a classical world", in *Mathematics, Philosophy and Modern Physics*, ed. by E. Rudolph and I.-O. Stamatescu (Springer, Berlin).

C. Kiefer and T. P. Singh (1991): Quantum Gravitational Corrections to the Functional Schrödinger Equation. Phys. Rev. D **44**, 1067.

C. Kiefer, T. Padmanabhan, and T. P. Singh (1991): A comparison between semiclassical gravity and semiclassical electrodynamics. Class. Quantum Grav. **8**, L 185.

C. Kiefer, R. Müller, and T. P. Singh (1993): Quantum Gravity and Nonunitarity in Black Hole Evaporation. Preprint gr-qc/9308024.

J. Kowalski-Glikman and J. C. Vink (1990): Gravity-matter mini-superspace: quantum regime, classical regime and in between. Class. Quantum Grav. **7**, 901.

K. V. Kuchař (1970): Ground State Functional of the Linearized Gravitational Field. Journ. Math. Phys. **11**, 3322.

K. V. Kuchař (1992): "Time and Interpretations of Quantum Gravity", In *Proceedings of the 4th Canadian Conference on General Relativity and Relativistic Astrophysics*, ed. by G. Kunstatter, D. Vincent and J. Williams (World Scientific, Singapore, 1992).

C. Kuo and L. H. Ford (1993): Semiclassical gravity theory and quantum fluctuations. Phys. Rev. D **47**, 4510.

C. Lämmerzahl (1993): Private communication.

L. D. Landau and E. M. Lifshitz (1975): *Quantenmechanik* (Akademie-Verlag, Berlin).

V. G. Lapchinsky and V. A. Rubakov (1979): Canonical quantization of gravity and quantum field theory in curved space-time. Acta Physica Polonica B10, 1041.

J. Louko (1993): Holomorphic quantum mechanics with a quadratic Hamiltonian constraint. Phys. Rev. D **48**, 2708.

V. Moncrief and C. Teitelboim (1972): Momentum Constraints as Integrability Conditions for the Hamiltonian Constraint in General Relativity. Phys. Rev. D **6**, 966.

D. N. Page (1993): "Black Hole Information", to appear in the *Proceedings of the 5th Canadian Conference on General Relativity and Relativistic Astrophysics*, Waterloo, Ontario, May 1993.

T. Padmanabhan (1990): A Definition for Time in Quantum Cosmology. Pramana **35**, L199.

T. Padmanabhan and T. P. Singh : On the semiclassical limit of the Wheeler-DeWitt equation. Class. Quantum Grav. **7**, 411 (1990).

J. P. Paz and S. Sinha (1991): Decoherence and back reaction: The origin of the semiclassical Einstein equations. Phys. Rev. D **44**, 1038.

A. Peres (1962): On Cauchy's Problem in General Relativity. Nuovo Cimento **XXVI**, 53.

E. Schrödinger (1926a): Quantisierung als Eigenwertproblem (Erste Mitteilung). Annalen der Physik **79**, 361.

E. Schrödinger (1926b): Quantisierung als Eigenwertproblem (Vierte Mitteilung). Annalen der Physik **81**, 109.

T. P. Singh (1990): Gravity induced corrections to quantum mechanical wave functions. Class. Quantum Grav. **7**, L149.

T. P. Singh (1993): "Semiclassical gravity", in *Advances in Gravitation and Cosmology*, ed. by B. R. Iyer, A. R. Prasanna, R. K. Varma, and C. V. Vishveshwara (Wiley, Eastern, New Delhi).

E. J. Squires (1991): The dynamical role of time in quantum cosmology. Phys. Lett. **A155**, 357.

A. Vilenkin (1989): Interpretation of the wave function of the Universe. Phys. Rev. D **39**, 1116.

J. Wudka (1990): Remarks on the Born-Oppenheimer approximation. Phys. Rev. D **41**, 712 (1990).

H. D. Zeh (1967): Symmetrieverletzende Modellzustände und kollektive Bewegungen. Z. Phys. **202**, 38.

H. D. Zeh (1986): Emergence of Classical Time from a Universal Wave Function. Phys. Lett. **A116**, 9.

H. D. Zeh (1988): Time in Quantum Gravity. Phys. Lett. **A126**, 311.

H. D. Zeh (1992): *The Physical Basis of The Direction of Time* (Springer, Berlin).

H. D. Zeh (1993a): There are no quantum jumps, nor are there particles! Phys. Lett. **A172**, 189.

H. D. Zeh (1993b): "Decoherence and measurements", to appear in the Proceedings of *Stochastic evolution of quantum states in open systems and measurement processes*, Budapest, March 1993.

W. H. Zurek (1991): Decoherence and the Transition from Quantum to Classical. Phys. Today **44**, 36.

W. H. Zurek (1993): Preferred States, Predictability, Classicality and the Environment-Induced Decoherence. Progr. Theor. Phys. **89**, 281.

Loop Representations

Bernd Brügmann

Max Planck Institute for Physics, 80805 München, Germany

1 Introduction

1.1 Generalities

The task of theoretical physics is to find an adequate mathematical description of physical ideas, and there is always a tension between real world experiments and the mathematical structures modelling them. In quantum gravity this tension appears with a twist, since there is no direct experimental evidence to be explained. Guided by the hypothesis that there exists a unified description of nature, we try to find a theory that could describe phenomena that belong at the same time to the domain of quantum theory and general relativity.

It is generally agreed upon that such phenomena exist even though they have not been observed. An example could be matter near the big bang singularity, and predictions like that of Hawking radiation for quantum fields near a black hole horizon are expected to eventually be valid limiting cases of a full theory of quantum gravity. But there are widely varying opinions on how one should go about constructing a theory of quantum gravity.

If we examine the theories that we want to combine closely, they are not flawless to begin with. One can for example argue that the measurement process of quantum mechanics is not sufficiently explained, while classical general relativity does not allow for quantum matter. In fact, we hope to remedy such problems in the more complete theory of quantum gravity.

It is not the case that the lack of experimental data does not allow us to pick the right theory out of a host of possibilities. The most important fact to remember about quantum gravity is that to date there does not exist a single model for a quantum theory of general relativity that is (i) self-consistent and (ii) contains as a special case a reasonable approximation to the observed physical world. There exist many research programs to construct quantum gravity, but all are incomplete even by their own criteria. However, some interesting partial results have been obtained in certain approaches, and here we will focus our attention on one such approach.

We consider the program of canonical quantization of general relativity. Although as incomplete as other approaches, there has been some recent progress initiated by the discovery of a new set of canonical variables for general relativity by Ashtekar [As86]. The reason for our discussion of loop representations is that the loop representation of Rovelli and Smolin [RoSm88] figures prominently in the quantum theory based on the Ashtekar variables.

There are many facets to canonical quantum gravity. In particular, there is much more to the program of canonical quantization starting from the Ashtekar variables then the loop representation, and the Ashtekar variables are in addition very interesting for the classical theory. The goal of this paper is to demonstrate that similarly there is more to the concept of a loop representation than that it is a useful technique in canonical quantum gravity. To this end we will try to paint a coherent picture — drawing on knot theory, gauge theory and canonical quantum gravity — of why loops find a natural place in canonical quantum gravity.

There exist several excellent reviews of canonical quantum gravity, the Ashtekar variables, and the loop representation (which we will point out as we go along). Our emphasis will be on a complete and up to date development of the main ideas behind loop representations rather then on technical details, in the hope that this way the *motivation* for the loop representation becomes apparent and the reader may answer the question, "Why loops?"

1.2 What we mean by canonical quantum gravity

To put our discussion of quantum gravity into perspective, let us first attach a few labels to what we mean and imply by the term 'canonical quantum gravity' in these lectures. A much more thorough background to what quantum gravity could mean can be found in the articles of Isham in this volume.

1. Starting point is standard general relativity, i.e. the theory of a Lorentzian metric $g_{\mu\nu}(x)$ on a four dimensional differentiable manifold $\mathcal{M}$ defined by the Einstein-Hilbert action,
$$S[g] = \int_{\mathcal{M}} d^4x \sqrt{-g}\, R, \qquad (1)$$
where R is the Ricci scalar and g the determinant of $g_{\mu\nu}(x)$. There are all sorts of reasons why one might want to consider discrete spaces instead of $\mathcal{M}$, or strings instead of points, or actions modified by higher order curvature terms, or other alterations. Whenever we refer to general relativity, we mean standard general relativity.
2. Even though there currently is a great deal of interest in theories in dimension unequal four, here we consider general relativity in four spacetime dimensions if we do not state otherwise. In fact, one of the reasons why loop representations are of interest is that they are tailored to the physically observed number of four dimensions.
3. Other approaches to quantum gravity, in particular path integral quantization and quantum cosmology, are based on an Euclideanization. Since Euclidean quantum gravity cannot in general be extended to Lorentzian quan-

tum gravity, it may be an advantage that canonical quantization and the loop representation are Lorentzian (although they are simple to Euclideanize).

4. Since we consider a canonical formulation, we only allow spacetimes that can be split into space and time, $\mathcal{M} = \Sigma \times \mathbb{R}$. The three-manifold Σ is assumed to be compact for simplicity. Such a choice for $\mathcal{M}$ excludes the possibility of topology change, which can be accommodated, for example, in path integral quantization.
5. Our goal is an inherently non-perturbative formulation, for which canonical quantization is well-suited. This has to be contrasted with standard field theoretic methods, which are based on perturbation theory using Feynman diagrams obtained from path integrals. Simply put, perturbation theory for general relativity fails at two loop, and the characteristic properties of general relativity, e.g. diffeomorphism invariance, seem to make a non-perturbative approach necessary. Let us also point out that a non-perturbative, canonical formulation looks and is in fact very different from perturbation theory. Many of the unconventional features that we encounter in the loop representation are not specific to the loop representation but to the canonical approach in general.
6. We consider gravity without matter, only at one point will we introduce a cosmological constant. Matter can be incorporated into the loop representation, but we will only briefly comment on it in section 5.1. The attitude in particle physics is that gravity can be treated by the same methods that are succesful for the other interactions, the only difference being the energy scale. The most elegant scheme along these lines is certainly string theory, although gravity is introduced via gravitons, which is a concept from perturbation theory. The presence of matter is thought to be essential. A common philosophy among relativists is that we can learn something about the deep conceptual issues in quantum gravity already in the absence of matter. Matter is thought to cloud some of these issues. But for questions like the issue of time in quantum gravity, we expect that we do have to include matter.
7. Canonical quantization of the type considered here is a very conservative approach. No new structures are postulated, like strings or supersymmetry. The idea is to use new techniques rather than new concepts as long as they are not forced upon us, i.e. to stay as close as possible to conventional quantum mechanics.
8. A loop is a map from the circle into a manifold. While mathematically the same objects as closed strings in string theory or loops in Feynman diagrams, loops play a completely different role in the loop representation. Also, their algebraic structure is not what is called a loop group, although a group of loops can be defined.

1.3 Outline

Loop representations have their origin in gauge theories, in fact, any Yang-Mills theory admits formally a loop representation. Let us summarize how a connection to general relativity is established. Canonical Yang-Mills theory is characterized by a gauge constraint, G, while general relativity in the usual metric variables is characterized by a diffeomorphism constraint D, which generates three-dimensional diffeomorphisms in Σ and the Hamiltonian constraint H, which essentially specifies how Σ is imbedded in $\mathcal{M}$. In the canonically quantized theory the constraints are imposed as operator equations on the states ψ.

While quantum gravity in terms of the metric variables and Yang-Mills theory in terms of a connection are disjoint in both variables and invariances, quantum gravity in the Ashtekar variables is formulated in terms of a *connection* and *all three* constraints are present. The relation of quantum gravity to Yang-Mills theory is summarized in the following diagram:

$$\left.\begin{array}{l} \hat{G}\psi = 0 \,]\ \text{YMT} \\ \left.\begin{array}{l}\hat{D}\psi = 0 \\ \hat{H}\psi = 0\end{array}\right]\ \text{QG} \end{array}\right]\ \text{QG in Ashtekar variables} \tag{2}$$

The idea behind the loop representation is to choose a representation of the quantum algebra of observables on a state space that consists of functionals $\psi[\gamma]$ of loops γ (as opposed to $\psi[A]$ or $\psi[g]$). The point is that the constraints can be treated more easily in the loop representation than in other representations: (i) The gauge constraint can be solved by gauge invariant variables (the Wilson loops) already on the classical level. (ii) The diffeomorphism constraint can be solved by considering states that are knot invariants, i.e. states that only depend on the diffeomorphism equivalence class of a loop. And (iii), the Hamiltonian constraint, also known as the Wheeler-DeWitt equation, has non-trivial solutions in terms of loops with intersections. This is, in a nutshell, the message we want to clarify. The loop representation of quantum gravity has also very interesting features beyond the constraints, which we will mention only briefly.

In section 2, we give a naturally very brief history of loops in mathematics (knot theory) and physics (gauge theory), and we argue why loop states can be expected to be useful in quantum gravity. In section 3, we discuss Maxwell theory as an example for a theory that possesses a complete formulation in the loop representation. In section 4, we introduce the loop representation for quantum gravity, and discuss various steps of the program of canonical quantization in the loop representations. In section 5, we conclude with a few comments on loop representations in general and on the status of the loop representation for general relativity.

2 History of Loops

For the purpose of motivating the loop representation of quantum gravity, we first introduce loops in the context of knot theory and gauge theory. A starting point for these two topics can be found in the early work by Faraday and Gauss.

Before entering the discussion, let us define loops. First, we define a path μ as a continuous, piecewise smooth map $\mu : [s,t] \to \Sigma$ from an interval into the three-manifold Σ. We usually choose $\Sigma = \mathbb{R}^3$. A loop is a closed path (figure 1),

$\alpha : [0,1] \to \Sigma$ such that $\alpha(0) = \alpha(1)$. Equivalently, α is a map from the circle into the three manifold.

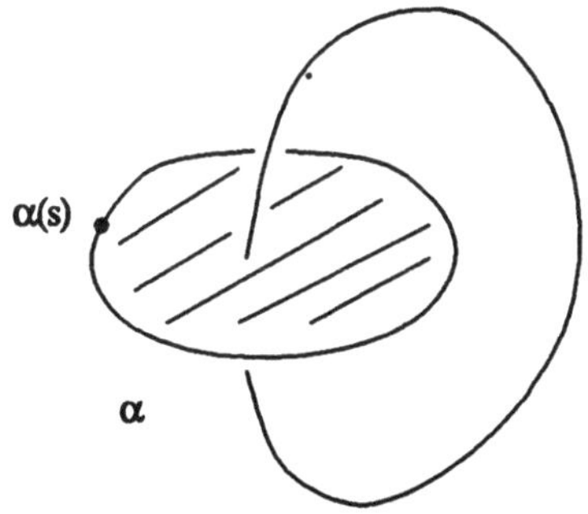

Fig. 1. Two loops α and β that link once.

We denote the path from $\alpha(s)$ to $\alpha(t)$ along α by α_s^t. The parametrization implies an orientation along the loop, e.g. $\alpha(0)$ lies on α_s^t only if $s > t$. The inverse α^{-1} of a loop α is defined by reversing the parametrization,

$$\alpha^{-1}(s) = \alpha(1-s). \tag{3}$$

If two loops α and β intersect at a point, i.e. $\alpha(s) = \beta(t)$ for some $s \neq t$, then we can define a combined loop $\gamma = \alpha_s \circ_t \beta$, as the loop obtained by first going around α from $\alpha(s)$ to $\alpha(s)$ and then around β from $\beta(t)$ to $\beta(t)$. If the point of combination is clear from the context we may just write $\gamma = \alpha\beta$. For example, if $\alpha(1) = \beta(0)$,

$$\gamma(s) = \begin{cases} \alpha(2s) & \text{if } 0 \leq s < 1/2 \\ \beta(2s-1) & \text{if } 1/2 \leq s \leq 1 \end{cases} \tag{4}$$

where we have made the new parametrization explicit. Paths are combined analogously, usually at their endpoints. Finally, we also consider multiloops, which are unordered collections of loops. Given two loops α and β, we denote the multiloop $\eta : S^1 \times S^1 \to \Sigma$ containing α and β once by $\eta = \alpha \cup \beta$ $(= \beta \cup \alpha)$.

2.1 Loops in the Work of Faraday and Gauss

During the years of 1821–32, when M. Faraday was working on electrodynamics, he also developed the concept that the electromagnetic forces are transmitted through a force field, and that this force field has physical reality [Be74]. The competing point of view on electromagnetic forces at that time was that of 'action at a distance', in particular that no additional physical effect comes between charges and their relative forces. The idea to use force fields as intermediaries was not new, but fields were considered to be just a useful mathematical tool. As it often happens when a mathematical construction captures the essentials of a physical phenomenon, it becomes part of our intuition about what is actually physically present. We are justified to think of the electric field as 'real' because we can, for example, store energy in it or compute its propagation in electromagnetic waves. (At a deeper level of 'reality' we will have to deal with QED.)

What is relevant here is that Faraday also noted that in the absence of sources the field lines have to close on themselves, i.e. to form loops, since only then the electric field is divergence free,

$$D_a E^a(x) = 0. \tag{5}$$

Furthermore, we can argue that the elementary excitation of the electric field is based on loops. We cannot construct a divergence free vector field with support on only a point, but it works one dimension up. Of course, we usually consider electric fields with three-dimensional support, but gauge invariance requires only loops, not three-spaces.

This simple fact has applications in modern gauge theories through the so-called Wilson loops, which are gauge invariant variables defined through the parallel transport of spinors around loops. We will discuss the role played by loops in gauge theory in section 2.3.

What may be less known is that electromagnetism also lead to the study of the second type of invariance that we want to consider, namely diffeomorphism invariance. On January 22, 1833, C.F. Gauss found the answer to the following problem [Ga1833], which reads like a recent text book problem, although the implications are quite subtle. What is the work done on a magnetic pole which is moved on a closed curve in the presence of a current loop? Notice that there are two loops in the problem, say α and β, which we assume to be non-intersecting (see figure 1). The answer can be expressed in terms of what is now known as the Gauss linking number,

$$gl(\alpha,\beta) = \frac{1}{4\pi}\int ds \int dt \epsilon_{abc}\dot{\alpha}^a(s)\dot{\beta}^b(t)\frac{\alpha^c(s)-\beta^c(t)}{|\alpha(s)-\beta(t)|^3}. \tag{6}$$

Although not obvious when written this way, $gl(\alpha,\beta)$ is an integer. The Gauss linking number counts (with an appropriate sign) how often two loops are linked, or equivalently, how often one loop winds around the other. If there is no linking, then $gl(\alpha,\beta) = 0$.

The key point is that the Gauss linking number does not change under small, smooth deformations of the loops, that is, the Gauss linking number is invariant under diffeomorphisms. Gauss himself found it quite remarkable that an integral as in (6) has this property. The equivalence classes of loops under diffeomorphism (connected to the identity) is the topic of a branch of mathematics called knot theory.

Historically, the relation between loops and diffeomorphism invariance was developed before loops found their way into gauge theories. We therefore make a few comments on knot theory in section 2.2 and then briefly discuss loops in gauge theories in section 2.3.

2.2 Knot Theory

Knot theory studies the equivalence classes of loops without intersections under diffeomorphisms that are connected to the identity (e.g. [Ka91]). For a single loop the equivalence class is called a knot class, equivalence classes of multiloops are also called link class. If the loops are allowed to intersect, one speaks of generalized knot and link classes. We denote knot and link classes by $\{\eta\}$.

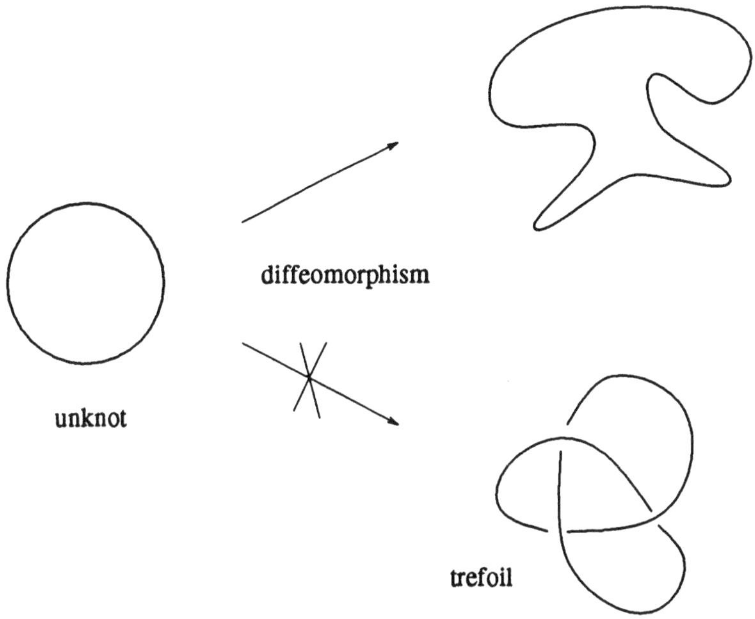

Fig. 2. A diffeomorphism cannot tie a knot into the unknot.

Recall that a diffeomorphism is a C^∞ map between manifolds that is one-to-one, onto, and has a C^∞ inverse. An example for what a diffeomorphism can do with the unknot, which is the trivial knot, is shown in figure 2. A diffeomorphism can deform the unknot quit arbitrarily, but it cannot tie a trefoil knot in the unknot. The simple intuitive reason is that in a continuous transformation from one case to the other, two lines would cross, but a one-to-one and onto map cannot produce intersections or take them apart.

One could say that knot theory was founded in 1877 when P.G. Tait formulated the program of classifying all knots. To this end he introduced knot diagrams. A knot diagram is a non-degenerate projection of a knot that lives in three dimensions onto a two dimensional plane plus a prescription whether an intersection in the plane came from an under crossing or over crossing of two lines in three dimensions (see figure 3). One can show that there always exists a projection that is non-degenerate in that no more than two different points

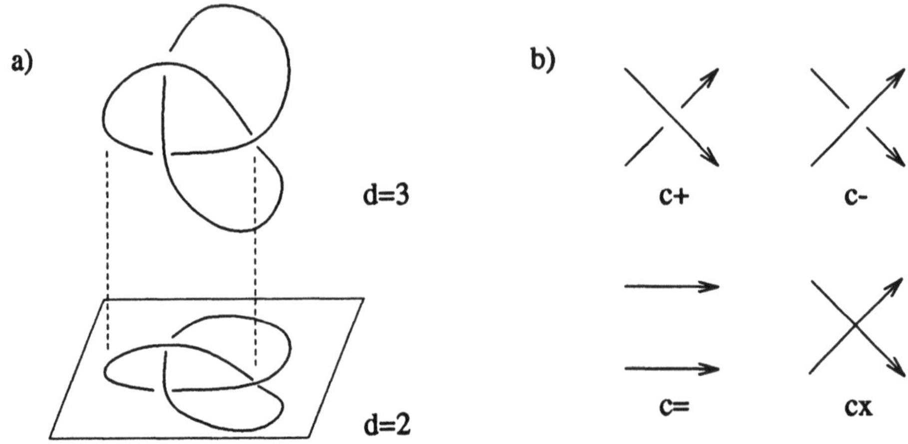

Fig. 3. a) Projecting a knot onto a knot diagram; b) the four crossings in a knot diagram.

are mapped into one. For example, we are not allowed to project the unknot 'sideways' onto a line segment.

The problem of classifying knots in three dimensions can be shown to be equivalent to classifying knot diagrams in two dimensions. An important tool to establish the equivalence of two knot diagrams are the Reidermeister moves. As a first step Tait compiled tables of inequivalent knot diagrams. Figure 4 shows a copy of Tait's original table from 1884 [Ta1877] for the first seven orders of knottiness. Such tables are even today an important source for examples.

As an aside, the first application of knot classes to physics appeared in the work of Lord Kelvin in 1869 [Th1869] (which prompted Tait to start his investigations). He proposed that atoms are 'smoke ring vertices in the ether'. The stability and variety of atoms are to be directly related to the equivalence classes of loops under diffeomorphism. The spectral lines would be created by vibrations of the loop. Today we know about transmutations of elements, and those could have been discussed in terms of line crossings (see below). For about thirty years Kelvin's theory of 'topological matter' was taken seriously, and for example Maxwell concluded that it accommodated more features of atomic physics than other models.

Returning to the classification problem, another important tool are knot invariants. A knot invariant is a functional on the space of loops that assigns to loops in the same knot class the same number, i.e. $\psi[\eta] = \psi[\{\eta\}]$. If ψ is a knot invariant, then

$$\alpha \sim \beta \quad \Rightarrow \quad \psi[\alpha] = \psi[\beta]. \tag{7}$$

Therefore, if $\psi[\alpha] \neq \psi[\beta]$ then the loops are not equivalent. The Gauss linking number is an example for a link invariant.

The hard part is to construct the inverse. Indeed, one of the central, unsolved problems of knot theory is to find a *complete* set of knot invariants, $\{\psi_i\}$, such

Fig. 4. "The first seven orders of knottiness", Tait 1884.

that

$$\psi_i[\alpha] = \psi_i[\beta]\ \forall i \quad \Rightarrow \quad \alpha \sim \beta. \tag{8}$$

However, a complete, indirect classification is possible via the complement of loops in Σ. Also, there exist algorithms to generate all knot classes.

The most important (and most complete) knot invariants arise in the study of knot polymials. Knot polynomials were introduced by Alexander in 1928 [Al28]. A knot polynomial $P_q(\gamma)$ assigns to each knot diagram of a loop γ a Laurent polynomial in a complex variable q such that

(i) $P_q(\gamma)$ is a knot invariant, (9)

(ii) $P_q(\text{unknot}) = 1$, (10)

(iii) The skein relations are satified. (11)

In the skein relations, or crossing change formulas, one considers three knot diagrams that differ only at one crossing. The three different possibilities are an over crossing, c_+, an under crossing, c_-, and no crossing, $c_=$ (see figure 3). If intersections are allowed, one includes the intersections, $c_\times$. Since a reflection at a plane is a diffeomorphism which is not connected to the identity, c_+ and c_- are inequivalent.

For example, the Alexander-Conway polynomial $A_q(\gamma)$ is uniquely determined by the skein relation

$$A_q(c_+) - A_q(c_-) = qA_q(c_=). \tag{12}$$

One can show that such crossing changes are sufficient to reduce any knot diagram to the unknot by recursion. As a simple exercise the reader may show that

$$A_q(\text{trefoil}) = 1 + q^2. \tag{13}$$

While the knot polynomial $A_q(\gamma)$ allows one to distinguish a large number of knots, it is not complete. For example, the mirror images of the trefoil lead to the same polynomial.

2.3 Gauge Theory

The canonical formulation of non-abelian gauge theories and the application of Wilson loops is a very interesting topic, and it is reviewed by Loll in this volume. Here we collect only a minimal set of definitions and facts that fit our discussion of the loop representation.

The canonically conjugate phase space variables of gauge theory are a configuration variable, the connection one-form $A^i_a(x)$, and a momentum variable, the densitized triad $E^{bj}(x)$, on Σ. Here $a, b, ... = 1, 2, 3$ are (co-)tangent space indices of Σ, $i, j, ...$ are the indices of the internal gauge group. For matrices in the Lie algebra of the gauge group we write for example $A_a = A^i_a \tau^i$, where τ^i

are the generators of the group (in the fundamental representation). The Poisson algebra is

$$\{A_a^i(x), A_b^j(y)\} = 0, \qquad \{E^{ai}(x), E^{bj}(y)\} = 0, \tag{14}$$

$$\{A_a^i(x), E^{bj}(y)\} = \delta_a^b \delta^{ij} \delta^3(x, y). \tag{15}$$

Gauge invariance implies the presence of a constraint in the canonical formalism, the Gauss constraint

$$G^i(x) \equiv D_a E^{ai}(x) = 0, \tag{16}$$

where D_a is the covariant derivative constructed from A_a.

There are different ways to deal with the gauge constraint. For example, we can choose to perform a gauge-fixing or not, and we can choose to solve the constraint in the classical theory or in the quantum theory. Given a gauge fixing, one has to check whether the final result depends on the gauge or not, and there may be ambiguities in the quantum theory. In principle, these problems do not appear if one can find a gauge invariant formulation, that is if one is able to solve the constraints classically.

Solving the constraint classically is referred to as reduced phase space quantization, the alternative is to impose the constraints in the quantum theory as in Dirac quantization. In general, the result is *not* the same (e.g. [RoTa89]).

The idea leading to loops is to give a reduced phase space formulation of gauge theories in terms of Wilson loops [Ma62, Wi74, Po79, GaTr80, Mi83] (and Loll in this volume). A Wilson loop, $h[\gamma, A]$, is the trace of the holonomy of γ and A,

$$h[\gamma, A] = \mathrm{tr} U_\gamma(A), \tag{17}$$

$$U_\gamma(A) = P \exp \int_0^1 ds\, \dot\gamma^a(s) A_a(\gamma(s)), \tag{18}$$

where $U_\gamma(A)$ is the matrix for parallel transport around the loop. The P denotes path ordering, i.e. for some one parameter family of matrices $M(s)$,

$$P \exp \int_0^1 ds M(s) = 1 + \int_0^1 ds\, M(s) + \int_0^1 ds \int_0^s dt\, M(t) M(s) + \ldots, \tag{19}$$

such that the products of matrices are always ordered according to the size of the parameter. (There is no factorial factor.)

The Wilson loops are gauge invariant since under a gauge transformation $U \to gUg^{-1}$ and $\mathrm{tr} gUg^{-1} = \mathrm{tr} U$. This is the reason why we consider loops and not paths. What makes the Wilson loops important is that not only are they gauge invariant, but in a sense they span the space of all gauge invariant functionals of the connection. There are reconstruction theorems, for example for $SU(N)$ [Gi81, AsLe93], of the type that if the trace of the holonomy $h[\gamma, A]$ is known for a given A and for all γ, then A is determined up to gauge.

Starting point for the loop representations is that instead of elementary variables A^i_a and E^{bj} we can choose the loop variables

$$T^0[\gamma] = \mathrm{tr} U_\gamma(A), \tag{20}$$
$$T^1[\gamma]^a(s) = \mathrm{tr} U_{\gamma_s}(A) E^a(\gamma(s)), \tag{21}$$

where the T^1 variables are obtained by inserting the matrix E^a at the parameter s into the parallel transport around γ (see also section 4.2). Classically, the choice of variables is equivalent modulo gauge for $SU(N)$ (compare Loll in this volume). The question is whether loop variables offer any advantages over the conventional approaches in the quantum theory.

We could now give the definition of a loop representation for a gauge theory, but since our main objective is the loop representation of quantum gravity, let us first complete the picture of how quantum gravity relates to knot theory and gauge theory.

2.4 Quantum Gravity

Before 1984. Let us summarize the situation in knot theory, gauge theory, and general relativity up to the year 1984 from the perspective of what is important for the loop representation of quantum gravity:

- Knot theory:
  The characteristic invariance is diffeomorphism invariance. The project is to classify all knots. The status is that generating and labeling all knots is well understood, but a deep insight into what a complete set of knot invariants could be is missing.
- Gauge theory:
  The characteristic invariance is gauge invariance. The project is to perform a reduced phase space quantization. The status is that for non-abelian gauge theories it is not known how to treat the equations of motions, and around 1980 most people in the field give up on using loop variables (with the notable exception of the group around Gambini, see references).
- General relativity:
  The characteristic invariance is space-time diffeomorphism invariance, represented in the canonical formalism by the diffeomorphism and Hamiltonian constraints. The project is canonical quantization. The status is that canonical quantization is incomplete, in particular because one does not know how solve the constraints in the metric variables.

Notice that at this time the three topics are completely unrelated. While diffeomorphism invariance plays a central role in both knot theory and general relativity, knots play no role in general relativity. And general relativity is not a gauge theory in the sense explained in section 2.3.

What we are leading up to is, of course, that in the following years a fruitful combination of all these ideas became possible. While the problems mentioned above under status have not been solved, there has been some progress.

After 1984. Let us now summarize some of the recent developments:

- Jones 1985 [Jo85]
  After Alexander, no new knot polynomials had been found until the discovery by Jones of the Jones polynomial, $J_q(\gamma)$. The important point is that his techniques allow a systematic investigation into the space of knot invariants. Roughly speaking, each representation of the braid group that carries a Markov trace defines a knot invariant.
- Ashtekar 1986 [As86, As87, As91]
  Via a canonical transformation from the metric variables, Ashtekar constructs a new set of canonical variables. These are an $SL(2,\mathbb{C})$ connection A^i_a and a conjugate momentum E^{bj}, the same type of variables as in a non-abelian gauge theory. There are as before a spatial diffeomorphism constraint D and a Hamiltonian constraint H, and in addition there is a Gauss constraint G (16) (since $g^{ab} = E^{ai}E^{bj}$, which is invariant under internal gauge transformations).
  One advantage of the Ashtekar variables is that the constraints are considerably simpler as in the metric variables (at least as far as solving the constraints is concerned, see below). The price to be paid is that there is an additional constraint, and that the Ashtekar connection is complex, and suitable reality conditions have to be imposed to recover real general relativity.
  The perhaps most important feature of the new variables is that in the Ashtekar formalism the kinematics of general relativity is imbedded in that of a non-abelian gauge theory for the Ashtekar connection. We have a standard gauge constraint plus the diffeomorphism and Hamiltonian constraint. This imbedding allows us to import ideas from Yang-Mills theory into general relativity, the loop representation being the prime example.
- Rovelli and Smolin 1988 [RoSm88, RoSm90]
  Rovelli and Smolin took the idea to use Wilson loops as gauge invariant variables one step further and constructed a representation of the operator algebra of quantum gravity on wavefunctions that are functionals of loops, $\psi[\eta]$. After all that has been said, it should be obvious that loops have a natural role to play in canonical quantum gravity. Let us consider the constraints:
  1. We can solve the gauge constraint,
  $$G = 0, \tag{22}$$
  on the classical level as in the reduced phase space formulation of gauge theory by using the loop variables T^0 and T^1, (20, 21), for the Ashtekar connection.
  2. We can solve the diffeomorphism constraint of the quantum theory,
  $$\hat{D}\psi[\eta] = 0, \tag{23}$$

by imposing that $\psi[\eta]$ is a knot invariant. Indeed, since a finite diffeomorphism f has a natural representation on the support of the wavefunctions, $(f \cdot \psi)[\eta] = \psi[f^{-1} \circ \eta]$, equation (23) for the generator of diffeomorphisms $\hat{D}$ implies that $\psi[\eta]$ must be a knot invariant, $\psi[\eta] = \psi[\{\eta\}]$. This formal solution of the constraint has to be compared with the metric representation. In that case one can impose that the wavefunctions are functionals of equivalence classes $\{g\}$ of metrics under diffeomorphisms (called three-geometries), $\psi[g] = \psi[\{g\}]$. However, one of the major problems with canonical quantization based on the metric variables is the complicated structure of the space of three geometries. In the case of the loop representation, the situation is simpler since the knot invariants can be easily labeled. A similar simplification occurs for the hydrogen atom, where the energy angular-momentum basis $\psi(n, l, m)$ has many advantages over the position basis $\psi(x)$.

3. A priori it is not clear at all whether the Hamiltonian constraint,

$$\hat{H}\psi[\eta] = 0, \tag{24}$$

can be solved in the loop representation. One reason why the metric representation is incomplete is precisely that we do not know how to factor-order, regularize and solve the Wheeler-DeWitt equation. The encouraging fact in the loop representation is that there exist at least some trivial solutions to the Wheeler-DeWitt equation, while for the metric variables not a single solution to the full equation had been known. As noted by Rovelli and Smolin based on ([JaSm88]), the Hamiltonian constraint trivially annihilates all loop functionals that have support only on non-intersecting loops.

At this point, the loop representation seems to be a promising approach to canonical quantum gravity. The Gauss and diffeomorphism constraints can be solved formally by loop techniques, and the Wheeler-DeWitt equation is at least not completely incompatible with loop functionals.

– Witten 1989 [Wi89]
Witten showed that the Jones polynomial, which is tied to two dimensions through the knot diagrams, can be computed as the vacuum expectation value of the Wilson loops in Chern-Simons field theory, which is a topological field theory in three dimensions:

$$J_q(\gamma) \sim \langle h[\gamma, A_{CS}] \rangle_{CS}, \tag{25}$$

where A_{CS} is the Chern-Simons connection. There are many important aspects to this work, here we want to focus on the new link between invariants in two and three dimensions. While Witten's proof is abstract in the sense that he does not directly evaluate the path integral that defines the expectation value (and mathematicians are still struggling to make his arguments rigorous), one can show that in a perturbation theory based on the path integral, the leading terms satisfy the skein relation that define the Jones polynomial [Sm89, GuMaMi90]. In this expansion one obtains knot invariants in analytic form analogously to the integral (6) for the Gauss linking number, as opposed to the recursion formulas for knot diagrams.

The above results are very interesting and far reaching for many more reasons than just the canonical quantization of gravity. One of the reasons why the author finds the loop representation appealing is that we can tie together all we have said about knot theory and the loop representation, and actually use knot theory (beyond formal solution of the diffeomorphism constraint) to construct non-trivial solutions to the Wheeler-DeWitt equation:

- Brügmann, Gambini and Pullin 1992 [BrGaPu92a, BrGaPu92b, BrGaPu93] One can generalize the Jones polynomial to the case of non-intersecting loops. The action of the Hamiltonian constraint on the analytic knot invariants arising in the perturbation theory of Chern-Simons theory can be computed explicitly. In fact, the second coefficient $a_2(\gamma)$ of the Alexander-Conway polynomial is annihilated by the Hamiltonian constraint. As explained in section 4.6, the non-triviality of this solution lies in the fact that because of the use of intersecting loops it is not simultaneously annihilated by the determinant of the metric.
  Further understanding of the structure of the space of solutions can be gained by considering a transformation between the connection representation and the loop representation. If there exists a suitable transform, then the Jones polynomial (and not just a few coefficients) is a solution to the Hamiltonian constraint with a non-vanishing cosmological constant. This one solution for a cosmological constant leads to several solutions for vanishing cosmological constant.

This concludes the motivational part why the loop representation is natural in quantum gravity. In the remainder of this lecture we discuss some details of the loop representation in the context of an example, Maxwell theory, and for quantum gravity itself.

3 Example for Complete Loop Representation: Maxwell Theory

The loop representation of Maxwell theory is an ideal example for an introduction to loop representations. First of all, in this simple case the program of canonical quantization can be rigorously completed in the loop representation. Furthermore, the standard formulation in terms of Fock space is well understood, and we can compare the two representations and obtain a physical interpretation for the loop operators. In fact, the two representations turn out to be equivalent, since there exists a faithful transform between them.

The earliest work on a loop variable formulation for a $U(1)$ gauge theory is that of Mandelstam [Ma62], and by Gambini and Trias [GaTr80, GaTr83]. We will follow closely Ashtekar and Rovelli [AsRo92], where the loop representation is constructed in a form and with techniques that are directly related to the loop representation of quantum gravity.

3.1 Bargmann Representation of Maxwell Theory

We consider a pure $U(1)$ gauge theory on $\mathbb{R}^3$ with a flat background metric. The canonical variables defining the phase space are a one-form $A_a(x)$ and an electric field $E^a(x)$ such that $\{A_a(x), E^b(y)\} = \delta_a^b \delta^3(x,y)$. There is one first class constraint,

$$D_a E^a(x) = 0. \tag{26}$$

By fixing the gauge, $D^a A_a(x) = 0$, we pass to the reduced phase space which is parametrized by divergence free fields $A_a^T(x)$ and $E_a^T(x)$.

Such transverse fields are conveniently described in the momentum representation. Let us introduce canonical variables $q_j(k)$ and $p_j(k)$, $j = 1,2$, on momentum space with the only non-vanishing Poisson bracket $\{q_m(-k), p_n(k')\} = \delta_{mn}\delta^3(k,k')$. In terms of complex polarization vectors $m_a(k)$, $m_a k^a = m_a m^a = 0$ and $m_a \bar{m}^a = 1$, we can write the transverse variables as

$$A_a^T(x) = \frac{1}{(2\pi)^{3/2}} \int d^3k e^{ik\cdot x}(q_1(k)m_a(k) + q_2(k)\bar{m}_a(k)), \tag{27}$$

$$E_a^T(x) = -\frac{1}{(2\pi)^{3/2}} \int d^3k e^{ik\cdot x}(p_1(k)m_a(k) + p_2(k)\bar{m}_a(k)). \tag{28}$$

In order to make a direct transition from the Fock representation to the loop representation possible, we have to work with Bargmann coordinates, ζ and $\bar{\zeta}$, defined by

$$\zeta_j(k) = \frac{1}{\sqrt{2}}(|k| q_j(k) - i p_j(k)). \tag{29}$$

The non-vanishing Poisson bracket is now

$$\{\zeta_m(k), \bar{\zeta}_n(k')\} = i|k|\delta_{mn}\delta^3(k,k'). \tag{30}$$

The one dimensional analog of these variables are the Bargmann variables z and $\bar{z}$ for the simple harmonic oscillator, $z = \frac{1}{\sqrt{2}}(\omega q - ip)$. The Bargmann variables ζ and $\bar{\zeta}$ are directly related to the positive frequency part of the transverse fields A_a^T and the negative frequency part of E_a^T,

$$^{+}A_a^T(x) = \frac{1}{(2\pi)^{3/2}} \int \frac{d^3k}{|k|} e^{ik\cdot x}(\zeta_1(k)m_a(k) + \zeta_2(k)\bar{m}_a(k)), \tag{31}$$

$$^{-}E_a^T(x) = -\frac{i}{(2\pi)^{3/2}} \int \frac{d^3k}{|k|} e^{ik\cdot x}(\bar{\zeta}_1(-k)m_a(k) + \bar{\zeta}_2(-k)\bar{m}_a(k)). \tag{32}$$

The quantum theory can be formulated in the Bargmann representation, where states are holomorphic functionals $\Psi[\zeta] \equiv \Psi[\zeta_1, \zeta_2]$ of the Bargmann variables. We represent the operators corresponding to the Bargmann variables via

$$\hat{\zeta}_j(k)\Psi[\zeta] = \zeta_j(k)\Psi[\zeta], \tag{33}$$

$$\hat{\bar{\zeta}}_j(k)\Psi[\zeta] = -i\hbar|k|\frac{\delta}{\delta\zeta_j(k)}\Psi[\zeta]. \tag{34}$$

which implies that, as required, the canonical commutation relations are satisfied. Again, compare with the harmonic oscillator where $\hat{z}\Psi(z) = z\Psi(z)$ and $\hat{\hat{z}}\Psi(z) = -i\hbar\omega(d/dz)\Psi(z)$.

To complete the mathematical setup, we have to specify an inner product that turns the space of states into an Hilbert space. This poses in general a non-trivial problem, see below. Here, an inner product exists and is uniquely determined by the reality conditions $\hat{\hat{\zeta}}_j(k) = \hat{\zeta}_j^*(k)$. This inner product is the unique, Poincaré invariant inner product of the Bargmann representation,

$$\langle \Phi[\zeta] | \Psi[\zeta] \rangle = \int d\mu(\zeta, \bar{\zeta}) \overline{\Phi[\zeta]} \Psi[\zeta], \tag{35}$$

$$d\mu(\zeta, \bar{\zeta}) = \prod_j d\!I\zeta_j(k) \wedge d\!I\bar{\zeta}_j(k) \exp\left(-\frac{1}{2\hbar} \int \frac{d^3k}{|k|} |\zeta_j(k)|^2 \right). \tag{36}$$

For the harmonic oscillator, the measure is of the type $dz \wedge d\bar{z} e^{-z\bar{z}}$.

The Bargmann representation has the usual interpretation in terms of Fock states. The vacuum $|0\rangle$ and the one photon state $|k, \epsilon\rangle$ for momentum k and helicity ϵ are given via $\Psi[\zeta] = \langle \zeta | \Psi \rangle$ by

$$|0\rangle : \quad \Psi_0[\zeta] = 1, \tag{37}$$

$$|k, \epsilon\rangle : \quad \Psi_{k,\epsilon}[\zeta] = \zeta_\epsilon(k). \tag{38}$$

The multiplication operator $\hat{\zeta}_j(k)$ acts as creation operator in the Fock basis, the derivative operator $\hat{\hat{\zeta}}_j(k)$ as annihilation operator. For example, the one photon state is created by action with $\hat{\zeta}_j(k)$ on the vacuum, $\Psi_{k,\epsilon}[\zeta] \equiv \hat{\zeta}_j(k)\Psi_0[\zeta]$.

3.2 Loop Representation of Maxwell Theory

Let us consider the same $U(1)$ gauge theory as in the preceding section, but now we choose loop variables instead of ζ and $\bar{\zeta}$. Because the gauge group is abelian, the loop variable T^0 (20) simplifies, and a simpler choice than T^1 (21) is possible. We define

$$h[\alpha] = \exp \oint_\alpha ds^a \, {}^-A_a^T, \tag{39}$$

$$E[f] = \int_{\mathbb{R}^3} d^3x \, f_a \, {}^+E^{Ta}, \tag{40}$$

where α is a loop and f_a a one-form. The loop variable $h[\alpha]$ is the abelian holonomy and takes the place of $T^0[\alpha]$. The analog of $T^a[\alpha](s) = \mathrm{tr} U_\alpha E^a(\alpha(s))$ is $h[\alpha]^+E^a(\alpha(s))$, but since for $U(1)$ gauge invariance no trace is necessary, we can use the loop-independent (smeared) variable $E[f]$. (Note that $\mathrm{tr}E^a = 0$ for $SU(N)$.)

Our choice of loop variables high-lights the fact that there are different possibilities to construct loop variables. In fact, we could as well start with $h[\alpha]$ and $h[\alpha]^+E[f]$ as our elementary variables. Of course, the interpretation of the momentum operators is then different, and not as nicely related to the Fock representation. The choice of the negative frequency connection for $h[\alpha]$ and the positive frequency electric field is necessary, since other frequency splittings are not consistent. Still another choice is to use anti-self-dual connections and self-dual electric fields, as we actually do in gravity.

Notice that the loop variables $h[\alpha]$ and $E[f]$ are in two ways overcomplete. First of all, not each label α and f corresponds to different variables. Two different loops α and β may lead to $h[\alpha] = h[\beta]\ \forall A$, e.g. if $\beta = \alpha\eta\eta^{-1}$ for some path η. And f_a and $f_a + \partial_a g$ give the same $E[f]$. In addition, there is the non-linear identity

$$h[\alpha]h[\beta] = h[\alpha\eta\beta\eta^{-1}] \tag{41}$$

for any path η that connects α and β; we also write $h[\alpha]h[\beta] = h[\alpha\#\beta]$. The latter type of identities play an important role since they differ for different gauge groups (compare (69) for $SL(2,\mathbb{C})$).

The Poisson bracket algebra is (using the same simplectic structure as before)

$$\{h[\alpha], h[\beta]\} = 0, \quad \{E[f], E[g]\} = 0, \tag{42}$$

$$\{h[\alpha], E[f]\} = (\oint_\alpha ds^a f_a)h[\alpha]. \tag{43}$$

Quantization in the loop representation is based on a space of states $\psi[\alpha]$ which are functionals of loops. How can we represent the operators $\hat{h}[\alpha]$ and $\hat{E}[f]$? Since the space of loops is *not* identical with the classical configuration space, which is the space of connections, the representation will not be the 'usual' one in terms of multiplication and derivative operators. Perhaps one should ask the question of how to find a representation with a different emphasis: Is there any chance at all that there exists a sensible representation for our unconventional choice of elementary variables?

Rovelli and Smolin suggested in their work on quantum gravity [RoSm90] that a transform between connection functionals and loop functionals can be used to derive the loop representation from the connection representation. An analog in quantum mechanics is the Fourier transform between the position and the momentum representation, which allows us to transfer both states and operators, e.g.

$$\psi(k) = \int dx e^{ikx}\psi(x), \tag{44}$$

$$ik \leftrightarrow \frac{d}{dx}. \tag{45}$$

The Rovelli-Smolin loop transform is

$$\psi[\gamma] = \int d\mu(A)h[\gamma, A]\psi[A]. \tag{46}$$

Given a functional $\psi[A]$ in the connection representation, the integration over all connections (modulo gauge) leaves only the loop dependence in $h[\gamma, A]$, and the result is a loop functional. The problem in general, and in particular for the $SL(2,\mathbb{C})$ connections of general relativity, is that *the measure* $d\mu(A)$ *is not known*. Without a definition of the measure, the loop transform is completely formal.

For Maxwell theory, however, we do have a measure explicitly available. We therefore can compute the transform from states in the Bargmann representation to the loop representation via

$$\psi[\gamma] = \int d\mu(\zeta,\bar{\zeta})h[\gamma]\Psi[\zeta], \tag{47}$$

where the measure is defined in (36). An explicit calculation leads to the following result for the transform of a Bargmann state, and this result is of central importance for the relation between the two representations. Ashtekar and Rovelli show that

$$\psi[\gamma] = \Psi[F_j(\gamma,k)], \tag{48}$$

where $F_j(\gamma,k)$ is obtained from the Fourier transform of the so-called formfactor $F^a(\gamma,x)$,

$$F^a(\gamma,x) = \int ds\dot{\gamma}^a(s)\delta^3(x,\gamma(s)), \tag{49}$$

$$= \frac{1}{2\hbar(2\pi)^{3/2}}\int d^3k e^{ik\cdot x}(F_1(\gamma,k)m^a(k) + F_2(\gamma,k)\bar{m}^a(k)). \tag{50}$$

In words, given a loop γ we compute the number $\psi[\gamma]$ for a state in the loop representation that was obtained via the transform of a state $\Psi[\zeta]$ in the Bargmann representation by evaluating $\Psi[\zeta]$ for $\zeta = F_j(\gamma,k)$.

The name 'form factor' is appropriate since according to

$$\oint_\gamma ds^a f_a = \int ds\dot{\gamma}^a(s)f_a(\gamma(s)) = \int d^3x F^a(\gamma,x)f_a(x), \tag{51}$$

the factor $F^a(\gamma,x)$ allows us to separate the dependence on the loop from the dependence on the one-form. In the loop transform, the loop is combined with the connection (expressed in terms of the Bargmann coordinates) in precisely the form (51), and after the integration over one-forms in the loop transform, only the loop dependence remains. Therefore it is plausible that, as in (48), the transform of an arbitrary state $\Psi[\zeta]$ depends on the loop only through the form factor.

Since we can explicitly compute the transform of any state, we can define an operator $\hat{O}_L$ in the loop representation by the transform of $\hat{O}_B\Psi[\zeta]$ in the Bargmann representation. For some operators, however, we do not have to be able to perform an explicit calculation. In fact,

$$\hat{h}_L[\alpha]\psi[\gamma] := \int d\mu h[\gamma](\hat{h}_B[\alpha]\Psi)[\zeta] \tag{52}$$

$$= \int d\mu(\hat{h}^*_B[\alpha]h)[\gamma]\Psi[\zeta] \tag{53}$$

$$= \int d\mu h[\alpha]h[\gamma]\Psi[\zeta] \tag{54}$$
$$= \int d\mu h[\alpha\#\gamma]\Psi[\zeta] \tag{55}$$
$$= \psi[\alpha\#\gamma]. \tag{56}$$

All we have to know in this calculation is that $\hat{O}_B$ is self-adjoint with respect to the measure, and we have to know an operator $\hat{O}_L$ such that

$$\hat{O}_L h[\alpha] = \hat{O}_B h[\alpha]. \tag{57}$$

The resulting representation is defined by

$$\hat{h}[\alpha]\psi[\gamma] = \psi[\alpha\#\gamma], \tag{58}$$
$$\hat{E}[f]\psi[\gamma] = \left(i\hbar \oint_\gamma ds^a f_a\right)\psi[\gamma]. \tag{59}$$

The $\hat{h}[\alpha]$ operator acts by adding the loop α to the argument of the loop state, and the $\hat{E}[f]$ acts by multiplying the loop state with the loop integral of f_a.

Several remarks on the construction of the loop representation via the Rovelli-Smolin loop transform are in order.

1. Ashtekar and Rovelli prove that the transform is faithful. Since no information is lost in the transition from the Bargmann representation to the loop representation, we have that *the loop representation is equivalent to the Fock representation.*
2. It is easy to check that we have correctly represented the classical Poisson algebra (42,43), in particular that $[\hat{h}[\alpha], \hat{E}[f]] = i\hbar\widehat{\{h[\alpha], E[f]\}}$. We can therefore define the loop representation *without* ever introducing the Bargmann representation.
3. The loop representation for quantum gravity (but also for other gauge groups) can be constructed by *assuming* that the loop operators are self-adjoint, since for the loop operators we do know transfer relations (57) [RoSm90]. Furthermore, one finds that the operators so obtained have the correct commutator algebra. Consequently, while the loop representation can be motivated by a formal transform, the definition of the loop representation is independent of that transform.

Since for Maxwell theory we know a faithful transform, we can answer the question what the translation of the physical interpretation of the theory in terms of photons is in the loop representation. We find with $\psi[\gamma] = \langle\gamma|\psi\rangle$ that

$$|0\rangle: \quad \psi_0[\gamma] = 1, \tag{60}$$
$$|k,\epsilon\rangle: \quad \psi_{k,\epsilon}[\gamma] = F_\epsilon(\gamma, k). \tag{61}$$

Hence we have discovered a formulation of Maxwell theory in which the elementary quantum excitations of the electric field are based on loops, just as we have argued for Faraday's picture of closed field lines. Is this result helpful for our intuition about physics? Even though the two representations are equivalent, and the loop representation for Maxwell theory is structurally appealing, the photon picture is much closer to how we think about experiments. However, for a non-abelian gauge theory and in the absence of a background metric (so that for example the split into positive and negative frequency parts is not possible), there does not exist a Fock representation. Nevertheless, it is quite possible that a loop representation still exists, and in this case, the loop picture may become a part of our intuition (see section 5.2).

4 Loop Representation of Canonical Quantum Gravity

4.1 Algebraic Quantization

Before we introduce the loop representation for canonical quantum gravity, it is appropriate to recall the program of algebraic quantization and point out the choices that lead to the loop representation. Our discussion of Maxwell theory already serves as an example for algebraic quantization, but the framework described below is designed with the kind of generality appropriate for general relativity.

The program of canonical quantization of constraint systems, in its algebraic form due to Dirac [Di65] and Ashtekar [As91], can be summarized as follows. Given is the classical phase space with the Poisson algebra of classical observables, which are functions from the phase space into the complexes, and constraints C.

→ Choose a set S of elementary variables which is complete in that it coordinatizes the phase space, and which is closed under the Poisson bracket.
- These variables will become the elementary operators of the quantum theory. For a particle on the line we can choose $S = \{1, q, p\}$. On the one hand, S has to be large enough such that the elements of S coordinatize the phase space. On the other hand, S has to be small enough so that we are able to consistently impose the non-commutative structure of the quantum theory. In particular, if S consists of all polynomials in the canonical variables, then there does not exist an irreducible representation of the quantum algebra (van Hove theorem, [Ho51]).
- We do not have to choose the canonical variables, for instance in gauge theory we can choose the loop variables instead.

→ Elevate the elementary variables to operators of the quantum theory, i.e. form the free algebra $\mathcal{A}$ of elements of S, $f \mapsto \hat{f}\ \forall f \in S$, such that

$$[\hat{f}, \hat{g}] = i\hbar\widehat{\{f, g\}} \quad \forall f, g \in S, \tag{62}$$

$$\hat{f}\hat{g} + \hat{g}\hat{f} = 2\widehat{fg} \quad \forall f, g : fg \in S. \tag{63}$$

- The relation (62) is the reason why we call the procedure 'quantization'. The classical Poisson algebra of commuting observables determines a non-commutative algebra of quantum operators.

- Anti-commutation relations like (63) arise if the set S is overcomplete, since if $fg \in S$ we need a unique definition of $fg \mapsto \widehat{fg}$. More general relations are possible. The loop variables are overcomplete in that way, compare (41,69). For a particle on the circle, e.g. $S = \{1, \cos\phi, \sin\phi, p\}$, $\cos^2\phi + \sin^2\phi = 1$.

→ Choose a linear representation of $\mathcal{A}$ on a vector space V.

- The vector space is called the space of states, $V = \{\psi\}$. Again, we have a choice, and for the loop representation we choose $V = \{\psi[\eta]\}$, and we choose a particular representation, e.g. as for Maxwell theory in (58,59). Notice that in the infinite-dimensional case we do not have a Stone-von-Neumann theorem guaranteeing unitary equivalence of representations. Hence, a different choice of elementary variables and representation can in general lead to *inequivalent* quantum theories. In other words, different choices can describe different physical systems, e.g. free electrons and electrons in a superconductor. In the case of the loop representation there are indications that the loop transform may relate the different representations to each other.

→ Represent the constraints,

$$C \mapsto \hat{C}. \tag{64}$$

- While the preceding steps can be performed for a judicious choice of variables, typically it is in the representation of the constraints that serious problems arise. First of all, it may not be trivial to express the constraints in terms of the elementary variables, say in terms of loop variables. Second, we have to define a factor ordering and usually also a regularization.

→ Solve the constraints, i.e. find the space of solutions, $V_{sol} \subset V$, on which

$$\hat{C}\psi = 0. \tag{65}$$

- The space of solutions is sometimes called the space of physical states, but we prefer to reserve the predicate 'physical' to states in the Hilbert space of solutions. Canonical quantization calls for the construction of a Hilbert space of states, and states in V_{sol} may not be normalizable.

→ Define the Hilbert space $\mathcal{H}_{phys}$ of physical states. To this end we have to find an inner product on a subset of V_{sol} such that suitable $*$-relations on $\mathcal{A}$ are satisfied.

- For quantum gravity we do not have a criterion like Poincaré invariance that allows us to pick out an unique inner product. There is a conjecture that if we know a complete set of $*$-relations, for example, if we know which operators are supposed to be self-adjoint (i.e. observables), then the inner product is determined uniquely. Such a criterion suffices in most examples, Maxwell theory included. Its merit is background independence, and therefore it can be applied in quantum gravity. There the main problem is existence rather then uniqueness [Re93].

- One often introduces an inner product at an earlier stage in the quantization program before the constraints have been solved (see Hajicek on pre-quantization in this volume). This may be important for the regularization of the constraint operators and the solution of the constraints. However, such an inner product is in general not related to the inner product of $\mathcal{H}_{phys}$.

→ Define observables of the quantum theory, define the measurement process, give an interpretation, make predictions.

- The previous step has completed the mathematical setup of the theory. Note that for quantum gravity, not even the mathematical part has been completed (the analog is true for other approaches to quantum gravity). This step summarizes all the really important, physical questions, and it is kept so short only because in the case of quantum gravity so little is known about it.

Now that a concrete proposal for a quantization program has been put on the table, let us quickly point out what such a program cannot be. As a matter of principle, there cannot exist a program into which we feed a classical theory and by turning a crank or letting a computer run we produce the corresponding, unique quantum theory. The reason is that quantum theory is the fundamental theory of which the classical world is a special limit. One and the same classical theory may be obtained as the limit of inequivalent quantum theories. Only if we specify information *external* to the classical theory, then we can hope to find a unique quantum theory.

A quantization program is similar to a computer program that requires additional input at various places in order to run at all. Relevant input for the loop representations is the choice of loop variables and the choice of loop states. As already mentioned, different choices may lead to different quantum theories, and only in experiments can we find out whether our theory describes the model that we had in mind.

In this sense the program of algebraic quantization of general relativity is more like a recipe that is justified only if it is successful. If the steps of quantization listed above can be taken, even excluding the 'physical' one, then this would constitute a major success. If algebraic quantization in this form fails, then we have *not*, of course, shown that quantization of general relativity is not possible.

Notice that we have resisted the temptation to number the steps. An obvious possibility is that they may have to be arranged in a different order, or more seriously that they cannot be separated as shown. For example, we may need an inner product to make the discussion of the constraints rigorous. Or we may have to solve the conceptual, physical problems like the meaning of time in quantum gravity (see Isham in this volume) before we can even formulate a quantization procedure.

After these remarks of caution about the nature of the proposed quantization procedure, let us list the steps that we will discuss for quantum gravity in the loop representation in the next sections. We choose loop variables as our elementary variables, and we can form the quantum algebra $\mathcal{A}$ (section 4.2). We define a representation of the loop operators on the space of loop functionals such that the relations (62,63) for the commutators and Poisson brackets are satisfied (section 4.2). We define the constraints in the loop variables (section 4.3) and express the constraint operators as differential operators on loop functionals (section 4.4). We explain how certain analytical knot invariants (section 4.5) can give rise to solutions to the constraints (section 4.6). Finally, in section 5 we give a critical appraisal of the status of the loop representation.

4.2 Loop Variables

The Rovelli-Smolin loop variables for general relativity are ([RoSm90, As91], compare section 2.3)

$$T[\gamma] = \mathrm{tr} U_\gamma, \tag{66}$$

$$T^a[\gamma](s) = \mathrm{tr} U_{\gamma_s^s} E^a(\gamma(s)), \tag{67}$$

$$T^{ab}[\gamma](s,t) = \mathrm{tr} U_{\gamma_t^s} E^a(\gamma(s)) U_{\gamma_s^t} E^b(\gamma(t)), \tag{68}$$

$$\vdots$$

where U_γ is the parallel transport for the $SL(2,\mathbb{C})$ Ashtekar connection and E^a its conjugate momentum. Each T^n variable is characterized by n insertions of the momenta into the trace of the holonomy (see figure 5).

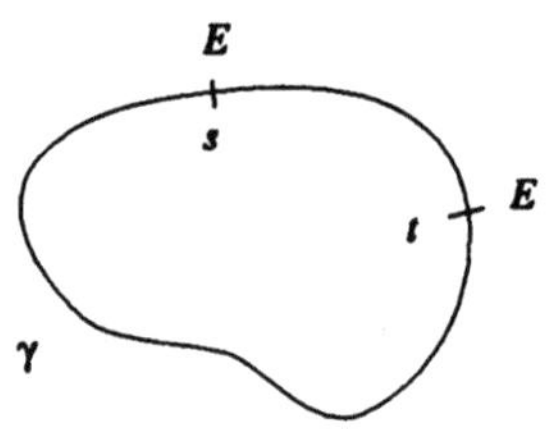

Fig. 5. Insertion of E at parameters s and t along a loop γ.

The loop variables are overcomplete. Since for any $SL(2,\mathbb{C})$ matrices A and B, $\mathrm{tr}A\mathrm{tr}B = \mathrm{tr}AB + \mathrm{tr}AB^{-1}$, we have for example that

$$T[\alpha]T[\beta] = T[\alpha\beta] + T[\alpha\beta^{-1}], \tag{69}$$

if α and β intersect so that $\alpha\beta$ exists. This identity is the Mandelstam identity (also called spinor identity) for $SL(2,\mathbb{C})$, compare (41) for $U(1)$. We will incorporate the spinor identity into the quantum theory via the anti-commutation relations, but in principle one can remove the overcompleteness by a different choice of variables [Lo91].

The Poisson algebra has the structure

$$\{T^m, T^n\} \sim T^{m+n-1}, \tag{70}$$

where '$\sim$' means that the bracket can be expressed in terms of linear combinations and integrals of loop variables of the given order. The algebra of the T^0 and T^1 closes:

$$\{T[\alpha], T[\beta]\} = 0, \tag{71}$$

$$\{T^a[\alpha](s), T[\beta]\} = \int du \dot{\beta}^a(u) \delta^3(\alpha(s), \beta(u)) \\ (T[\alpha_s \circ_u \beta] - T[\alpha_s \circ_{1-u} \beta^{-1}]), \tag{72}$$

$$\{T^a[\alpha](s), T^b[\beta](t)\} = \int du \dot{\alpha}^b(u) \delta^3(\beta(t), \alpha(u)) \\ (T^a[\beta_t \circ_u \alpha] - T^a[\beta_t \circ_{1-u} \alpha^{-1}])(\tilde{u}) \\ - \int du \dot{\beta}^a(u) \delta^3(\alpha(s), \beta(u)) \\ (T^b[\alpha_s \circ_u \beta] - T^b[\alpha_s \circ_{1-u} \beta^{-1}])(\tilde{u}), \tag{73}$$

where $\tilde{u}$ is the parameter of the intersection of the combined loop. Notice that the combination of loops occurs with a sign opposite to (69) and cannot be further simplified. The variables T^0 and T^1 are not expected to be canonical since they are of the type x and xp, respectively.

Initially, we define the space of states to be the space of functionals of multiloops. The definition of the loop operators $\hat{T}^n$ can be motivated via the Rovelli-Smolin transform (46) as explained in section 3.2. We define

$$\hat{T}[\gamma]\psi[\eta] = \psi[\gamma \cup \eta], \tag{74}$$

$$\hat{T}^a[\gamma](s)\psi[\eta] = -\hbar \int du \dot{\eta}^a(u) \delta^3(\gamma(s), \eta(u))(\psi[\gamma_s \circ_u \eta] - \psi[\gamma_s \circ_{1-u} \eta^{-1}]). \tag{75}$$

The $\hat{T}[\gamma]$ operator acts by adding a loop into the multiloop argument of the loop state. The $\hat{T}^a[\gamma](s)$ operator gives a non-zero result only if the loop η in the argument of the state intersects γ at $\gamma(s)$. In this case the result is a linear combination of the loop state evaluated for the two different reroutings of the loop through the intersection. In a graphical short hand,

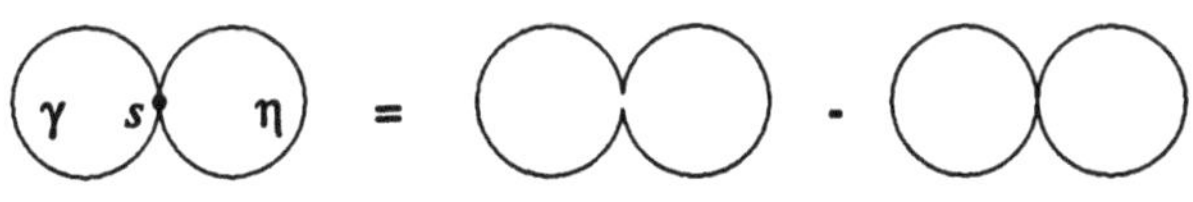

For $n \geq 1$, $\hat{T}^{a\ldots b}[\gamma](s, \ldots, t)\psi[\eta]$ is non-zero only if the loops γ and η intersect in all the distinguished points along γ, and the result is a linear combination of all the possible reroutings of the two loops through the intersection. For example, in the action of $\hat{T}^2$ there appear (with correct signs) the four terms

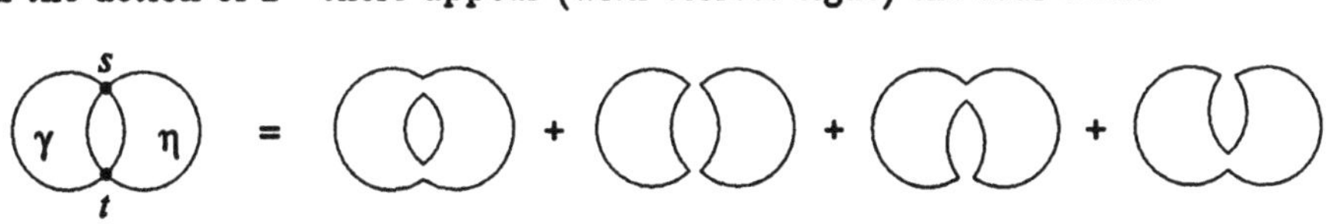

One can check that the relations $[\, , \,] = i\hbar\widehat{\{\, , \,\}}$, (62), are indeed satisfied for the operators $\hat{T}^0$ and $\hat{T}^1$! For the higher order loop operators one finds a deformation of the type

$$[\hat{T}^m, \hat{T}^n] \sim \hbar\hat{T}^{m+n-1} + \hbar^2\hat{T}^{m+n-2} + \dots . \tag{76}$$

Furthermore, we have to require that the action of the $\hat{T}^m$ is consistent with identities for the T^m. This gives rise to several properties of the loop states ψ. First of all, $\psi[\eta]$ has to be invariant under reparametrizations of η. For any loops α and β and any path μ, we require $\psi[\alpha\beta] = \psi[\beta\alpha]$ and $\psi[\alpha\mu\mu^{-1}] = \psi[\alpha]$. In addition there are anti-commutation relations (63) for the non-linear identity (69) where α and β intersect. We have for all η that

$$(\hat{T}[\alpha]\hat{T}[\beta] + \hat{T}[\beta]\hat{T}[\alpha])\psi[\eta] = 2\psi[\alpha \cup \beta \cup \eta], \tag{77}$$

$$2\widehat{T[\alpha]T[\beta]}\psi[\eta] = 2(\hat{T}[\alpha\beta] + \hat{T}[\alpha\beta^{-1}])\psi[\eta] = 2(\psi[\alpha\beta \cup \eta] + \psi[\alpha\beta^{-1} \cup \eta]). \tag{78}$$

Hence we have to impose on the space of states the spinor identity

$$\psi[\alpha \cup \beta] = \psi[\alpha\beta] + \psi[\alpha\beta^{-1}]. \tag{79}$$

As an immediate consequence we obtain that $\psi[\gamma_0 \cup \eta] = 2\psi[\eta]$ for γ_0 the zero loop, and $\psi[\eta^{-1}] = \psi[\eta]$.

There also are identities among T^0 and T^1 variables (and similarly for higher orders), but these imply that the space of states contains only the zero state, $\psi \equiv 0$ [Br93a]. However, in a natural regularization of the $\hat{T}^0$-$\hat{T}^1$ algebra (using strips, see below) such identities are absent, and the spinor identity (79) is the only consequence of the anticommutation relations.

4.3 Constraints from Loop Variables

How can we define the constraints, which are given to us in terms of the Ashtekar variables, in terms of loop variables? The basic observation is that local variables in terms of A and E at a point x can be obtained from the non-local loop variables in the limit that a loop is shrunk to a point. Considering that local and non-local field theories are fundamentally different, the limit of shrinking loops is expected to be non-trivial.

We choose a coordinate system in a neighborhood of x, and we consider a family of loops, γ^δ_{xab}, of size δ at the point x in the a-b-coordinate plane (figure 6) such that

$$\gamma^\delta_{xab}(s) = x + \delta\gamma_{ab}(s), \tag{80}$$

$$\gamma_{ab}(0) = \gamma_{ab}(1) = 0, \tag{81}$$

$$\sigma^{cd}(\gamma_{ab}) = \delta^{[c}_a\delta^{d]}_b, \tag{82}$$

where γ_{ab} is a planar loop at the origin with unit area and area element σ^{cd}, i.e. γ^δ_{xab} has area δ^2. In the limit $\delta \to 0$, γ^δ_{xab} reduces to the zero loop, $\gamma^0_{xab}(s) = x \forall s$.

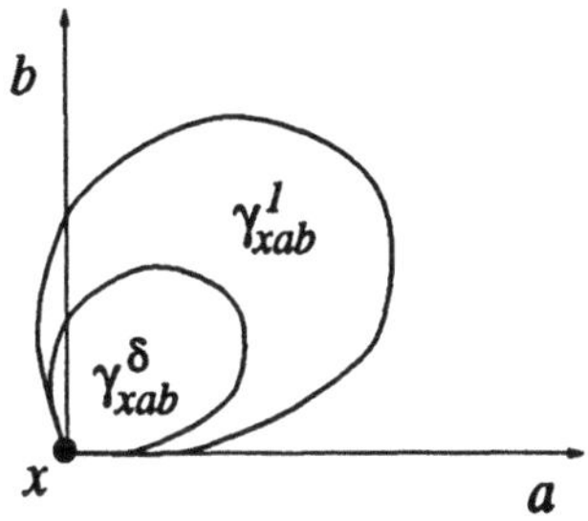

Fig. 6. The family of small planar loops.

The parallel transport around infinitesimal loops γ^{δ}_{xab} gives rise to the curvature F of A at x,

$$U_{\gamma^{\delta}_{xab}} = 1 + \delta^2 F_{ab}(x) + O(\delta^3), \tag{83}$$

where $F^i_{ab} = \partial_a A^i_b - \partial_b A^i_a + \epsilon^{ijk} A^j_a A^k_b$. This is just the equation we need for the transition from from the loop variables T^1 and T^2 to the constraints D and H.

Indeed, in the Ashtekar variables the vector constraint C and the Hamiltonian constraint H are ([As91] or Giulini in this volume)

$$C_a(x) = \mathrm{tr} E^b F_{ab}(x), \tag{84}$$

$$H(x) = \mathrm{tr} E^a E^b F_{ab}(x). \tag{85}$$

The vector constraint generates diffeomorphisms up to gauge. (As an aside, compare the canonical formulation in the metric variables, e.g. Beig in this volume, where H is more complicated, in particular due to the presence of a potential term.)

In order to make the transition from a $T^a = \mathrm{tr} E^a U$ to $C_a = \mathrm{tr} E^b F_{ab}$, we only have to shrink the loop. To be precise,

$$C_a(x) = \lim_{\delta \to 0} \frac{1}{\delta^2} T^b[\gamma^{\delta}_{xab}](0), \tag{86}$$

$$H(x) = \lim_{\delta \to 0} \frac{1}{\delta^2} T^{ab}[\gamma^{\delta}_{xab}](\delta^2, 1), \tag{87}$$

where the summation over indices is for $a \neq b$.

Quantization in the loop representation elevates the loop variables to the loop operators, and since we know how to express the constraints in the loop variables, we obtain immediately the constraints in the loop representation via

$$C_a(T^1) \mapsto \hat{C}_a = C_a(\hat{T}^1), \tag{88}$$

$$H(T^2) \mapsto \hat{H} = H(\hat{T}^2). \tag{89}$$

The question is whether the limit of shrinking loops can actually be computed for the loop operators, and with what result.

The natural differential operator in the context of shrinking loops turns out to be the area derivative $\Delta_{ab}(s)$ of loop functionals. A loop functional is called area differentiable if the following limit exists independently of the choice of small loops γ^{δ}_{xab} and transforms like a two-form under coordinate transformations:

$$\Delta_{ab}(s)\psi[\eta] = \lim_{\delta \to 0} \frac{\psi[\eta_s \circ \gamma^{\delta}_{xab}] - \psi[\eta]}{\delta^2}, \tag{90}$$

where the combination of the loops η and γ^{δ}_{xab} occurs at x. Since we are considering general relativity, one has to show that a background independent definition of the area derivative exists. This can be shown for small parallelograms built from the integral curves of two commuting vector fields [BrPu93]. The beginning of a rigorous calculus on the space of loop functionals can be found in the work of Tavares [Ta93].

The area derivative satisfies

$$\Delta_{ab}(s) = \Delta_{[ab]}(s), \tag{91}$$

$$\frac{\delta}{\delta\eta^a(s)} = \dot{\eta}^b(s)\Delta_{ab}(s), \tag{92}$$

$$\Delta_{ab}(s)\mathrm{tr}U_\gamma = \mathrm{tr}U_{\gamma^s_s} F_{ab}(\gamma(s)). \tag{93}$$

The area derivative is antisymmetric as expected from its definition in terms of an area element (91). The ordinary functional derivative is a special, one-directional case of the area derivative (92). And the area derivative inserts F into the Wilson loop as expected for the limit of shrinking loops (93).

4.4 Constraint Operators

In general terms, the area derivative is defined by a deformation of the loop argument of a loop functional that consists of appending a small loop. And the loop operators insert loops into the argument of loop functionals in the specific way defined in (74,75). The perhaps surprising result is that the terms appearing in the definition of the constraints can be combined into area derivatives. In this calculation we have to make use of the spinor identity for loop states, (79).

For the smeared diffeomorphism constraint $\hat{D}(v) = \int d^3x v^a(x) C_a(x)$ we find that

$$\hat{D}(v)\psi[\eta] \equiv \lim_{\delta\to 0} \frac{1}{\delta^2} \int d^3x v^a(x) T^b[\gamma^{\delta}_{xab}](0) \tag{94}$$

$$= \int ds v^a(\eta(s)) \frac{\delta}{\delta\eta^a(s)} \psi[\eta], \tag{95}$$

which is the natural generator of diffeomorphisms along the vector field v^a, and where we have used (92).

The unregulated result for the smeared Hamiltonian constraint $H(N) = \int d^3N(x)H(x)$ is

$$\hat{H}_{unreg}(N)\psi[\eta] = \int ds \int dt \delta^3(\eta(s), \eta(t)) N(\eta(s)) \dot{\eta}^a(s)\dot{\eta}^b(t) \Delta_{ab}(s, \eta^s_t)\psi[\eta^s_t \cup \eta^t_s]. \tag{96}$$

The second argument of the area derivative indicates on which of the elements of the multiloop the area derivative acts. The result is non-zero only if $\eta(s) = \eta(t)$. This occurs for $s = t$, but also for $s \neq t$ if η intersects itself. If $s = t$, then we expect a zero result since then we have the antisymmetrized product of two identical tangent vectors because the area derivative is antisymmetric in its indices, (91). If there is an intersection, then the action of $\hat{H}$ is to split the loop η at the intersection into a multiloop (figure 7), and to act with the area derivative on one of the component loops.

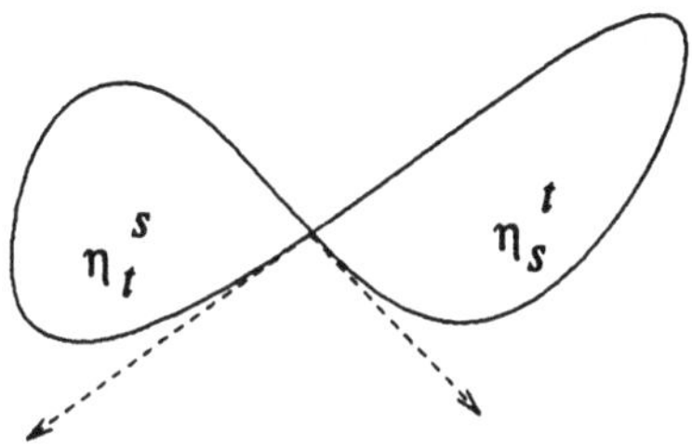

Fig. 7. A loop with one self-intersection.

Notice that the Hamiltonian constraint operator has to be regulated. Since the three-dimensional delta distribution $\delta^3(\eta(s), \eta(t))$ depends on only two parameters, there is a $\delta^1(0)$ divergence. There are different ways known to regulate the constraints, all of which share the same problem that some sort of background dependence is introduced.

Let us comment on the point-splitting regularization most often used in practice [JaSm88, Br93c]. We pick a regulator $f_\epsilon(x, y)$ such that $\lim_{\epsilon \to 0} f_\epsilon(x, y) = \delta^3(x, y)$, and replace the delta distribution by the regulator. The resulting operator, $\hat{H}_\epsilon$, is finite for $\epsilon > 0$. We perform all calculations for $\epsilon > 0$ to leading order in ϵ, which is at order $1/\epsilon$ and corresponds to $\delta^1(0)$. We define the regularized Hamiltonian constraint operator by a multiplicative renormalization as $\hat{H} = \lim_{\epsilon \to 0} \epsilon \hat{H}_\epsilon$.

Background dependence enters since in $f_\epsilon(x, y)$ we measure the separation of x and y by a background metrice, e.g. a common choice is $f_\epsilon(x, y) = \frac{3}{4\pi\epsilon^3} \Theta(\epsilon - |x - y|)$. A background dependence breaks diffeomorphism invariance and is therefore unacceptable in quantum gravity. However, the leading order term in the Hamiltonian constraint at intersections is background independent. The $s = t$ terms vanish because of antisymmetry at order $1/\epsilon^2$, but there is a background dependent contribution at $1/\epsilon$, which however vanishes for diffeomorphism invariant states satisfying a certain regularity conditions. Therefore, it is possible to discuss solutions to the constraints in a background independent way, but there still remain some problems regarding the regularization of the constraint algebra.

Our discussion of the constraints followed closely [BrPu93]. There are two other approaches in the literature, one using a different definition of the constraints in terms of loop variables [Bl90], the other applying the loop transform to the constraints in the connection representation [Ga91]. The resulting constraints in the loop representations are equivalent [Br93c].

4.5 Analytic Knot Invariants

How can we find the simultaneous space of solutions to both the diffeomorphism constraint and the Hamiltonian constraint of quantum gravity? The problem is that very little is known about the Hamiltonian constraint in the loop representation as a differential operator. As a first step we therefore consider a particular class of knot invariants and examine the action of the Hamiltonian constraint on such knot invariants [BrGaPu92a]. At the time of writing, there exists only one class of knot invariants for which we can compute the area derivative, namely the analytic knot invariants that appear as coefficients of knot polynomials in perturbative Chern-Simons theory.

The action of Chern-Simons theory is

$$S_{CS}[A] = \int_{\Sigma} d^3x\, \epsilon^{abc} \mathrm{tr}(A_a \partial_b A_c + \frac{2}{3} A_a A_b A_c), \tag{97}$$

where we choose for simplicity a $SU(2)$-valued one-form $A_a(x)$. At this point there is no relation between the Chern-Simons connection and the Ashtekar connection, the Chern-Simons connection serves only as a calculational tool. The vacuum expectation value of the Wilson loop in Chern-Simons field theory is

$$\langle h[\gamma, A]\rangle = \int DA\, h[\gamma, A] \exp\left(\frac{ik}{4\pi} S_{CS}[A]\right), \tag{98}$$

where $k \in \mathbb{Z}$ is the coupling constant.

We have two ways to evaluate $\langle h[\gamma, A]\rangle$. Witten showed that

$$\langle h[\gamma, A]\rangle = c^{-w(\gamma)} J_q(\gamma) \tag{99}$$

where γ_0 denotes the unknot, $q = \exp \frac{2\pi i}{k+2}$, $c = c(k)$, $w(\gamma)$ is the writhe of γ, and $J_q(\gamma)$ is the Jones polynomial. The right hand side is also known as the Kauffman bracket [Ka91]. The Jones polynomial is defined by the skein relations (compare (12)

$$q J_q(c_+) - q^{-1} J_q(c_-) = (q^{\frac{1}{2}} - q^{-\frac{1}{2}}) J_q(c_=), \tag{100}$$

and $J_q(\text{unknot}) = q^{\frac{1}{2}} + q^{-\frac{1}{2}}$. The Jones polynomial is a knot invariant, or in knot diagram language, the Jones polynomial is invariant under ambient isotopy. The writhe $w(\gamma)$ is defined as the sum over the crossings in a knot diagram counting $+1$ for c_+ and -1 for c_-. The writhe is *not* a knot invariant, but only a regular isotopy invariant (as is the Kauffman bracket).

The reason is that a projection of a knot from three into two dimensions may introduce arbitrary numbers of crossings. For example, depending on the projection we can obtain from an unknotted line segment a line without or with twist, which are equivalent under ambient isotopy, but inequivalent under regular isotopy. The corresponding crossing change formulas for $J_q(\gamma)$ and $\langle h[\gamma, A]\rangle$ are shown in figure 8. The projection dependence of $\langle h[\gamma, A]\rangle$ is known as framing dependence. The framing of a loop can be defined in three dimensions by replacing the loop by a strip, which changes under twists (figure 8).

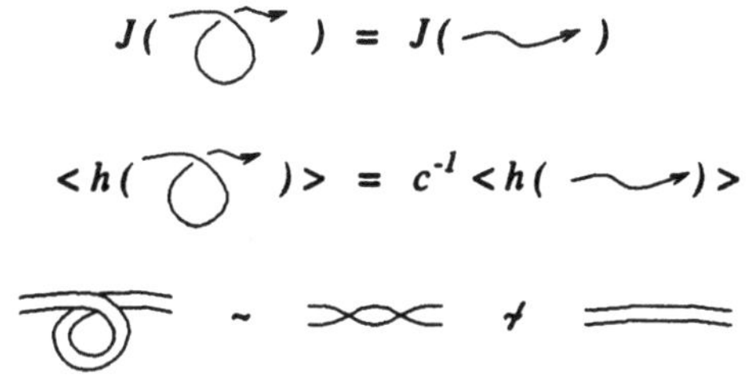

Fig. 8. Framing dependence.

We can also obtain a perturbation expansion for $\langle h[\gamma, A]\rangle$ by inserting for $h[\gamma, A]$ the expansion (19) [GuMaMi90]. The result is

$$\langle h[\gamma, A]\rangle = c_0(\gamma) + c_1(\gamma)\frac{1}{k} + c_2(\gamma)\frac{1}{k^2} + \ldots, \tag{101}$$

where the coefficients $c_i(\gamma)$ are known to be regular isotopy invariants related to the coefficients of the Jones polynomial by Witten's result. The point is that the $c_i(\gamma)$ are expressed as multiple integrals along the loop like the Gauss linking number $gl(\alpha, \beta)$, (6). We have that up to constant factors

$$c_0(\gamma) = 1, \tag{102}$$
$$c_1(\gamma) = gl(\gamma, \gamma), \tag{103}$$
$$c_2(\gamma) = (c_1(\gamma))^2 + \rho(\gamma). \tag{104}$$

$gl(\gamma, \gamma)$ is called the Gauss self linking number. Despite appearance, $gl(\alpha, \beta)$ in (6) is finite for $\alpha = \beta$, but it depends on the coordinates, which is another sign for framing dependence. If we assign to γ a framed loop γ^f, then we can define the framed self linking number for the strip formed by γ and γ^f by $gsl^f(\gamma) = gl(\gamma, \gamma^f)$. The second coefficient, however, contains a framing independent term $\rho(\gamma)$, which is a true knot invariant related to the second coefficient of the Alexander-Conway polynomial $a_2(\gamma)$, $\rho \sim a_2 + \frac{1}{12}$. $\rho(\gamma)$ is the sum of a three-fold and a four-fold integral along γ, but its precise form is not important here, since we are not going to perform an explicit calculation with it. Each coefficient $c_i(\gamma)$ can be shown to contain a framing independent piece, and these constitute the class of knot invariants on which we want to study the action of the Hamiltonian constraint.

4.6 Knot Invariants as Solutions to the Constraints

The history of solutions to the constraints in the loop representation begins with the discovery in [JaSm88] that in the connection representation any $\psi[A] = h[\gamma, A]$ for some loop γ without intersections is a solution to the Wheeler-DeWitt equation. (Such states are clearly not diffeomorphism invariant.) The simple reason is as in (96) that two linearly independent tangent vectors are required at one point for a non-zero result due to the presence of an antisymmetrization.

In fact, the operator corresponding to the determinant of the three-metric, $\widehat{\sqrt{-g}}$, gives zero if there are not three linearly independent tangents at one point, which for smooth loops can only happen at a triple intersection (figure 9). Since the metric itself is not an observable of the theory, one might think that one does not have to worry about its degeneracy on the space of solutions. However, the determinant of the metric typically appears in matter couplings, and it should therefore not be zero. Also, the Hamiltonian for a non-zero cosmological constant Λ is

$$\hat{H}_\Lambda = \hat{H} + \Lambda\widehat{\sqrt{-g}}. \tag{105}$$

States without intersections are therefore solutions to the Wheeler-DeWitt equation for arbitrary cosmological constant. We consider this as an argument that the sector of states for non-intersecting loops is degenerate.

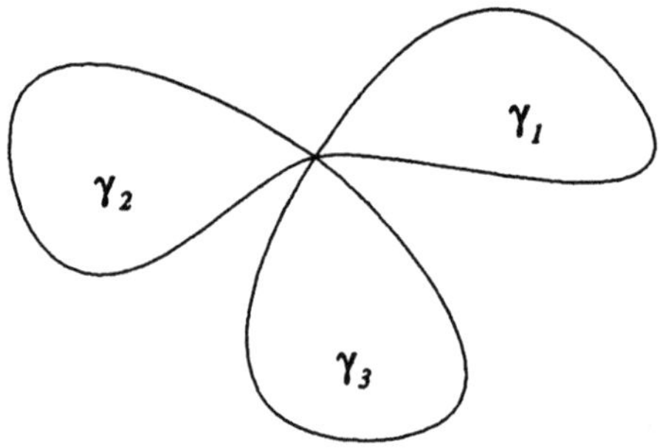

Fig. 9. Generic loop with triple self-intersection and three linearly independent tangents at the intersection (in three dimensions).

Consequently, solutions to the Hamiltonian constraint for intersecting loops were constructed in the connection representation [JaSm88, Hu89, BrPu91]. However, in [BrPu91] it is shown that all such solutions are necessarily annihilated by $\widehat{\sqrt{-g}}$. Since in the connection representation we do not know a general strategy to solve the diffeomorphism constraint anyway, it is natural to look for non-degenrate solutions to the Hamiltonian constraint in the loop representation. One reason why non-degenerate solutions can be found in the loop representation is that the loop representation is based on the opposite factor ordering of the $\hat{A}$ and $\hat{E}$ then the one in which $h[\gamma, A]$ leads to solutions.

Let us summarize the situation in the loop representation for the coefficients $c_i(\gamma)$ [BrGaPu92a]. c_0 is a non-degenerate solution, but not an interesting one. Let us consider loop states with support on loops $\gamma = \gamma_1\gamma_2\gamma_3$, γ_i without self-intersections (figure 9), which have one triple intersection with three linearly independent tangent vectors (the generic case in three dimensions). We find that

$\psi[\gamma]$	$\hat{H}\psi[\gamma]$	$\widehat{\sqrt{-g}}\psi[\gamma]$
$c_1(\gamma)$	nonzero	nonzero
$c_1(\gamma_1 \cup \gamma_2 \cup \gamma_3)$	0	0
$\rho(\gamma)$	0	nonzero!

c_1 (which is not a knot invariant), is non-degenerate but not annihilated by $\hat{H}$. For the more symmetric loop as shown, γ_1 is a degenerate solution. But a non-trivial and non-degenerate solution to both constraints is the second coefficient of the Alexander-Conway polynomial. In [BaGaGrPu93] the same is shown for a part of c_3.

So, using a particular factor ordering and regularization, we can find interesting solutions to both constraints among the coefficients $c_i(\gamma)$. The mere fact that solutions can be found is a success of the loop representation, since as already mentioned, in the traditional variables not a single solution to the full Wheeler-DeWitt equation had been known.

However, it is still completely unclear what kind of structure the *complete* space of solutions might have. The following argument provides at least some insight into what this structure could be. In the connection representation there is exactly one solution known to all constraints [Ko90], and that in the factor ordering that corresponds to the loop representation and for non-vanishing cosmological constant:

$$\psi_\Lambda[A] = \exp(-\frac{6}{\Lambda}S_{CS}[A]), \tag{106}$$

where A is now the Ashtekar connection. This state is gauge invariant for appropriate values of Λ and diffeomorphism invariant since $S_{CS}[A]$ is 'topological'. Since

$$\hat{H}_\Lambda = \epsilon^{ijk}\frac{\delta}{\delta A^i_a}\frac{\delta}{\delta A^j_b}F^k_{ab} + \frac{\Lambda}{6}\epsilon^{ijk}\epsilon_{abc}\frac{\delta}{\delta A^i_a}\frac{\delta}{\delta A^j_b}\frac{\delta}{\delta A^k_c}, \tag{107}$$

$$\frac{\delta}{\delta A^k_c}\exp(-\frac{6}{\Lambda}S_{CS}[A]) = \frac{3}{\Lambda}\epsilon^{cde}F^k_{de}\exp(-\frac{6}{\Lambda}S_{CS}[A]), \tag{108}$$

we immediately have by differentiating only once that

$$\hat{H}_\Lambda\psi_\Lambda[A] = 0. \tag{109}$$

In other words, property (108) of the Chern-Simons action makes it possible to choose the coefficient in the exponent of $\psi_\Lambda[A]$ such that the contributions to $\hat{H}_\Lambda$ from $\hat{H}$ and the cosmological constant term cancel.

The idea in [BrGaPu92b] is to consider the loop transform (46) of $\psi_\Lambda[A]$,

$$\psi_\Lambda[\gamma] = \int d\mu(A) h[\gamma, A] \psi_\Lambda[A]. \tag{110}$$

If the loop transform exists, then $\psi_\Lambda[\gamma]$ is a solution to all the constraints in the loop representation by construction. In general, we cannot compute the transform since we do not know the measure. Let us assume for the moment, that for the transform of $\psi_\Lambda[A]$ we can use the measure of Chern-Simons theory. Then

$$\psi_\Lambda[\gamma] = \langle h[\gamma, A]\rangle_{CS} = c^{-w(\gamma)} J_q(\gamma), \tag{111}$$

for $k = -24/(\pi i \Lambda)$, is a solution to all constraints.

There are several obvious problems with this construction. In gravity, the internal group is $SL(2,\mathbb{C})$ and not $SU(2)$ (in terms of A^i_a, in the former case A^i_a is complex, in the latter it is real). Furthermore, Λ is complex (and takes discrete values). Another problem is that we want to allow loops with intersections, but we can in fact construct an extension of the Jones polynomial to intersecting loops [BrGaPu92b].

While there is no proof that (111) does or does not make sense, it hints at a very interesting relation between quantum gravity and topological field theory. Furthermore, from the single solution (111) for $\Lambda \neq 0$ we can derive a whole tower of solutions for $\Lambda = 0$ by the following simple argument [BrGaPu93]. Consider

$$\hat{H}_\Lambda \psi_\Lambda[\gamma] = (\hat{H} + \Lambda \widehat{\sqrt{-g}})(c_0 + c_1\Lambda + c_2\Lambda^2 + \ldots) = 0, \tag{112}$$

where we have absorbed the factor between Λ and k in the c_i. Since this equation holds for all $\Lambda \neq 0$, we conclude order for order that

$$\hat{H} c_0 = 0, \tag{113}$$

$$\hat{H} c_1 + \widehat{\sqrt{-g}} c_0 = 0, \tag{114}$$

$$\hat{H} c_2 + \widehat{\sqrt{-g}} c_1 = 0, \tag{115}$$

$$\vdots$$

The action of $\widehat{\sqrt{-g}}$ is simpler to compute than that of $\hat{H}$. Since $c_2 = c_1^2 + a_2 + \frac{1}{12}$ and $\hat{H} c_1^2 + \widehat{\sqrt{-g}} c_1 = 0$ we arrive at a simple proof of $\hat{H} a_2 = 0$ for loops as above.

From the structure of the series, we can guess that at each order a part of the coefficients c_i has to be annihilated by $\hat{H}$. Recently it has been argued that indeed $\hat{H} c^{-w(\gamma)} = 0$ [GaPu93], and therefore $\hat{H} J_{q(\Lambda)} = 0$. This explains why each coefficient contains a diffeomorphism invariant part that is a solution to the Hamiltonian constraint (recall that $w(\gamma)$ is not diffeomorphism invariant).

5 Discussion

5.1 Simple loop representations

The main motivation for our discussion of the loop representation comes from quantum gravity. Before we discuss the status of the loop representation for quantum gravity in the next section, let us at least briefly comment on the loop representation of simpler models.

1. Maxwell theory.
   As we have seen in section 3, the loop representation is equivalent to the Fock representation and as complete. Here we can gain some intuition about the physical meaning of loops.
2. Yang-Mills theory (see Loll in this volume).
   As mentioned in section 2.3, loop variables play a natural role in non-abelian gauge theories both on the lattice and in the continuum. The mathematical problems that effectively stopped the continuum approach in the beginning of the 1980's are now being adressed by results stimulated by the loop representation of quantum gravity, see below. On the lattice, we have a complete formulation, and it is well suited for numerical Hamiltonian lattice gauge theory. For example, one can compute the glue ball state and its mass in the high temperature regime of pure 2+1 $SU(2)$ lattice gauge theory, i.e. numerically solve the eigenvalue problem for the Hamiltonian (here not a constraint), $\hat{H}\psi[\gamma] = E\psi[\gamma]$ [Br91] (see also [GaLeTr89]). We can introduce matter into the loop representation by including paths that carry fermions at their ends. Again, on the lattice we can perform numerical computation, see [GaSe93] for $SU(2)$ with fermions in 3+1 dimensions.
3. Linearized gravity.
   Similar to the situation in Maxwell theory, the loop representation is able to reproduce the Fock representation of gravitons [AsRoSm91], and hence the elementary excitations of the gravitational fields are based on loops. Furthermore, one can actually show that the loop representation of quantum gravity, although incomplete, contains the singular limit of gravitons [IwRo93].
4. 2+1 gravity.
   The loop representation can be constructed explicitly [AsHuRoSaSm89]. Many of the problems regarding regularization are absent, notice for example that in (96) there appears only $\delta^2(\gamma(s), \gamma(t))$. Together, the diffeomorphism and the Hamiltonian constraint generate homotopy transformations (that can take intersections apart and transform the unknot to the trefoil, figure 2) [BrVa93]. An important point is that the loop transform in its naive form is degenerate [Ma93], but there exists a non-degenerate generalization [AsLo93].
5. Classical limit in the loop representation — weaves.
   On the kinematic level, i.e. ignoring the Hamiltonian constraint, so-called weave states allow us to define a classical metric [AsRoSm92]. A weave is a multiloop obtained by sprinkling loops randomly into the manifold. Under appropriate conditions, weave states are the eigenstates of a smeared,

diffeomorphism invariant metric operator, and the classical metric arises as its eigenvalue on scales large in comparison to the density of the sprinkling. Remarkably, in this context we can *derive* that physics becomes discrete at the Planck scale. Weaves can acommodate, for example, the classical black hole solution [Ze93].

5.2 Status of the Loop Representations for Quantum Gravity

We have seen in some detail how various steps of the program of algebraic quantization of quantum gravity can be performed in the loop representation. Let us collect the main negative and positive points about this approach, the negative ones first:

$-^{\infty}$ *The program is incomplete.* As long as the program is incomplete, it is totally unclear whether any parts of it will be part of the 'final' theory of quantum gravity. It does not help that all other approaches to quantum gravity are incomplete, too (see for example Isham in this volume). Whenever the claim of progress in full quantum gravity is raised, so far it is only valid with respect to a particular program. More on the positive side, neither has it been shown that the loop representation must necessarily fail.

– The main reason why the loop representation is incomplete is that we do not have an inner product, in particular we do not know how to obtain a complete set of observables that could lead us to an inner product. The absence of an inner product implies, for example, that all we have are examples for $\psi[\gamma] \in V_{sol}$, not for $\psi[\gamma] \in \mathcal{H}_{phys}$.

– The construction of the loop representation ignores the reality conditions. The Ashtekar variables are complex, and certain reality conditions are imposed to obtain real general relativity. We do not know how to impose the reality conditions in the loop representation, although there may be an analog to holomorphicity of $\psi[A]$ in the connection representation. Also, the issue of reality becomes intertwined with that of self-adjointness and the inner product. Therefore, at the level of our discussion we deal with a quantum theory of complex general relativity.

– Regularization is not complete. We know how to regularize the $\hat{T}^0$-$\hat{T}^1$ algebra (using strips for $\hat{T}^1$, e.g. [Ro91]), and the constraints are reasonably well understood in terms of a point-splitting regularization. What is missing is a regularization of the full $\hat{T}^n$ algebra, and of the constraint algebra. Without progress on regularization we cannot hope to decide, for example, whether there are anomalies.

– The loop representation has not lead to a breakthrough regarding the interpretation of quantum gravity, say of observables or the issue of time.

There are, however, also several positive aspects of the loop representation approach:

+ The loop representation is natural for the treatment of the constraints. As argued in depth, the loop representation is well-adapted not only to gauge and diffeomorphism invariance, but also the Wheeler-DeWitt equation seems to become tractable in the loop representation.

+ There exist loop representations for many models that are simpler than full quantum gravity in 3+1 dimensions, that can be called complete to a varying degree (section 5.1). This fact gives us confidence that at least the general idea behind the loop representation is sound. Of course, there are many good ideas that cease to be valid in the case of quantum gravity. Therefore the previous point is important.

+ There has been progress on several mathematical aspects related to loops:
  1. There are two, not in any obvious way related ways to extend the space of loop states to distributions on the space of loops, [AsIs92] and [BaGaGr93]. As usual in quantum field theory, we expect such states rather than just functions on configuration space to be relevant.
  2. A differential calculus for extended loop variables has been developed [Ta93] that allows one to give rigorous meaning to heuristic constructions of operators like the area derivative.
  3. A diffeomorphism invariant measure has been constructed on a completion of the space of connections modulo gauge [AsLe93], see also [Ba93]. This measure may lead to a rigorous definition of the loop representation via the transform.
  4. There are new knot invariants that are more powerful then the knot polynomials, the Vassiliev invariants [Va90]. These can be characterized as the coefficients of knot polynomials (e.g. a_2 corresponds directly to a Vassiliev invariant), which they include as a special case [Bi93]. For the construction of the Vassiliev invariants, intersecting loops are essential. Before the Vassiliev invariants became known, knots with intersections were not studied in knot theory, but given the importance of Vassiliev invariance in knot theory, the application of loops with intersections in quantum gravity has gained additional justification.

+ There are new ideas about quantum gravity physics that have been introduced, or concretized, by the loop representation. Although these ideas refer mostly to limiting situations of full quantum gravity, this is where our intuition originates, and if as often assumed new conceptual ideas are needed for the quantization of gravity, it is good to know that the loop representation produces such ideas. As mentioned in the context of weaves [AsRoSm92], in the classical approximation to full quantum gravity in the loop representation, the discrete structure of space time at the Planck scale can be derived. Furthermore, loops allow one to construct diffeomorphism invariant observables that measure the area of a surface by counting the number of intersections of a loop with this surface (which is diffeomorphism invariant). A prediction of such a framework is that area is quantized.

5.3 Conclusion

As it is often the case with a theory, it is a matter of taste and interests whether one feels that the pro outweighs the contra — especially with a theory as remote from the 'real world' as quantum gravity. Let us draw our conclusion.

We believe that the loop representation is an interesting proposal about how to solve some of the long-standing problems of canonical quantum gravity, in particular solving the constraint equations. While far from complete, the loop representation offers promise for the future.

Here we have focused on three issues. For Maxwell theory we have argued that in the loop representation we can replace the physical picture of photons by that of elementary excitations based on loops, both being equivalent. For full quantum gravity we have shown how to find states that solve all the constraints in the loop representation, something not possible in the traditional approach in terms of metric variables.

The one aspect of the loop representation that arguably is the most important one is the following. The loop representation is *not* a strange idea unrelated to physics found in an appendix to an obscure theory called canonical quantum gravity. Rather the loop representation is a fine example for the surprisingly fruitful interplay between three, initially unrelated theories: knot theory, gauge theory, and quantum gravity.

Acknowledgements

It is a pleasure to thank the organizers of the 117. Heraeus Seminar, and especially Helmut Friedrich, for a well-organized and very interesting meeting. This paper is based on the two lectures that the author presented at the seminar. In my work on the loop representation I have greatly benefitted from stimulating discussions with Abhay Ashtekar, Rodolfo Gambini, Jorge Pullin, and Lee Smolin.

References

[] No attempt was made to cover all the literature related to the loop representation, but let us point out to the non-specialist some of the references that can serve as entry points to the literature. For a guide to the issues of quantum gravity in general see Isham in this volume. For a review of the loop representation in Yang-Mills theory see Loll in this volume. For a list of references on canonical quantum gravity in the Ashtekar variables in general and the loop representation in particular see [Br93b]. Taken together, the latter two sources give a rather complete overview of all the work related to the loop representation.
For a complete and authorative review of quantum gravity in the Ashtekar variables see the book by Ashtekar [As91]. There are also review papers by Rovelli [Ro91] and Smolin [Sm93] that cover the loop representation. For a self-contained introduction to the application of knot theory to quantum gravity along the lines of section 4 see Pullin [Pu93]. For a recent account of knot theory in physics see Kauffman [Ka91].

[Al28] J.W. Alexander. Topological invariants of knots and links. *Trans. Am. Math. Soc.* **30** (1928) 275–306

[As86] A. Ashtekar. New variables for classical and quantum gravity. *Phys. Rev. Lett.* **57** (1986) 2244–7

[As87] A. Ashtekar. New Hamiltonian formulation of general relativity. *Phys. Rev.* **D36** (1987) 1587–1602

[As91] A. Ashtekar. Lectures on non-perturbative canonical gravity. (World Scientific, Singapore 1991)

[AsHuRoSaSm89] A. Ashtekar, V. Husain, C. Rovelli, J. Samuel, and L. Smolin. $2+1$ quantum gravity as a toy model for the $3+1$ theory. *Class. Quan. Grav.* **6** (1989) L185–93

[AsIs92] A. Ashtekar and C. Isham. Representations of the holonomy algebras of gravity and non-abelian gauge theories. *Class. Quan. Grav.* **9** (1992) 1433–85

[AsLe93] A. Ashtekar and J. Lewandowski. Representation theory of analytic holonomy C^* algebras. In "Knots and Quantum Gravity", ed. J. Baez. (Oxford U. Press, in press)

[AsLo93] A. Ashtekar and R. Loll. A new loop transform for 2+1 gravity. In preparation.

[AsRo92] A. Ashtekar and C. Rovelli. A loop representation for the quantum Maxwell field. *Class. Quan. Grav.* **9** (1992) 1121–50

[AsRoSm91] A. Ashtekar, C. Rovelli and L. Smolin. Gravitons and loops. *Phys. Rev.* **D44** (1991) 1740–55

[AsRoSm92] A. Ashtekar, C. Rovelli and L. Smolin. Weaving a classical geometry with quantum threads. *Phys. Rev. Lett.* **69** (1992) 237–40

[Ba93] J.C. Baez. Diffeomorphism-invariant generalized measures on the space of connections modulo gauge transformations. To appear in the proceedings of the "Conference on Quantum Topology", Manhattan, Kansas, March 1993

[BaGaGr93] C. Di Bartolo, R. Gambini, and J. Griego. The extended loop group: an infinite dimensional manifold associated with the loop space. Montevideo preprint IFFI/93.01

[BaGaGrPu93] C. Di Bartolo, R. Gambini, J. Griego, and J. Pullin. In preparation.

[Be74] W. Berkson. Fields of force: the development of a world view from Faraday to Einstein. (Routledge & Kegan Paul, London 1974)

[Bi93] J. Birman. New points of view in knot theory. *Bull. Am. Math. Soc.* **28** (1993) 253–87

[Bl90] M. P. Blencowe. The Hamiltonian constraint in quantum gravity. *Nucl. Phys.* **B341** (1990) 213–51

[Br91] B. Brügmann. The method of loops applied to lattice gauge theory. *Phys. Rev.* **D43** (1991) 566–79

[Br93a] B. Brügmann. Ph.D. thesis. (Syracuse University, 1993)

[Br93b] Bibliography of publications related to classical and quantum gravity in terms of the Ashtekar variables. MPI preprint MPI-Ph/93-68 (Sept. 1993)

[Br93c] B. Brügmann. The Wheeler-DeWitt operator in the loop representation. Unpublished.

[BrGaPu92a] B. Brügmann, R. Gambini and J. Pullin. Knot invariants as nondegenerate quantum geometries. *Phys. Rev. Lett.* **68** (1992) 431–4

[BrGaPu92b] B. Brügmann, R. Gambini and J. Pullin. Jones polynomials for intersecting knots as physical states of quantum gravity. *Nucl. Phys.* **B385** (1992) 587–603

[BrGaPu93] B. Brügmann, R. Gambini and J. Pullin. How the Jones polynomial gives rise to physical states of quantum general relativity. *Gen. Rel. Grav.* **25** (1993) 1–6

[BrPu91] B. Brügmann and J. Pullin. Intersecting N loop solutions of the Hamiltonian constraint of quantum gravity. *Nucl. Phys.* **B363** (1991) 221–44

[BrPu93] B. Brügmann and J. Pullin. On the constraints of quantum gravity in the loop representation. *Nucl. Phys.* **B390** (1993) 399–438

[BrVa93] B. Brügmann and M. Varadarajan. Unpublished notes (1993)

[Di65] P. A. M. Dirac. Lectures on Quantum Mechanics. (Academic Press, NY 1965)

[Ga1833] C. F. Gauss. Note of Jan. 22, 1833. In "Werke", vol. V (Königliche Gesellschaft der Wissenschaften, Göttingen, 1877) 605

[Ga91] R. Gambini. Loop space representation of quantum general relativity and the group of loops. *Phys. Lett.* **B255** (1991) 180–8

[GaLeTr89] R. Gambini, L. Leal, and A. Trias. Loop calculus for lattice gauge theories. *Phys. Rev.* **D39** (1989) 3127–35

[GaPu93] R. Gambini and J. Pullin. The Gauss linking number in quantum gravity. In "Knots and Quantum Gravity", ed. J. Baez. (Oxford U. Press, in press)

[GaSe93] R. Gambini and L. Setaro. SU(2) QCD in the path representation. Montevideo preprint (April 1993)

[GaTr80] R. Gambini and A. Trias. Second quantization of the free electromagnetic field as quantum mechanics in the loop space. *Phys. Rev.* **D22** (1980) 1380–4

[GaTr83] R. Gambini and A. Trias. Chiral formulation of Yang-Mills equations: A geometric approach. *Phys. Rev.* **D27** (1983) 2935–39

[GaTr86] R. Gambini and A. Trias. Gauge dynamics in the C-representation. *Nucl. Phys.* **B278** (1986) 436–48

[Gi81] R. Giles. Reconstruction of gauge potentials from Wilson loops. *Phys. Rev.* **D24** (1981) 2160–8

[GuMaMi90] E. Guadagnini, M. Martellini and M. Mintchev. Wilson lines in Chern-Simons theory and link invariants. *Nucl. Phys.* **B330** (1990) 575–607

[Ho51] L. van Hove. *Acad. Roy. Belg. Bull. Cl. Sci.* **37** (1951) 610

[Hu89] V. Husain. Intersecting loop solutions of the Hamiltonian constraint of quantum general relativity. *Nucl. Phys.* **B313** (1989) 711–24

[IwRo93] J. Iwasaki and C. Rovelli. Gravitons from loops: non-perturbative loop-space quantum gravity contains the graviton-physics approximation. Pittsburgh preprint (April 1993)

[JaSm88] T. Jacobson and L. Smolin. Nonperturbative quantum geometries. *Nucl. Phys.* **B299** (1988) 295–345

[Jo85] V. Jones. A polynomial invariant for links via von Neumann algebras. *Bull. AMS* **12** (1985) 103–10

[Ka91] L. Kauffman. Knots and Physics. (World Scientific, 1991)

[Ko90] H. Kodama. Holomorphic wavefunction of the universe. *Phys. Rev.* **D42** (1990) 2548–65

[Lo91] R. Loll. A new quantum representation for canonical gravity and SU(2) Yang-Mills theory. *Nucl. Phys.* **B350** (1991) 831–60

[Ma62] S. Mandelstam. Quantum electrodynamics without potentials. *Ann. Phys. (N.Y.)* **19** (1962) 1–24

[Ma93] D. Marolf. Loop representations for 2+1 gravity on a torus. Syracuse SU-GP-93/3-1; An illustration of 2+1 gravity loop transform troubles. gr-qc/9305015

[Mi83] A. Migdal. Loop equations and $1/N$ expansion. Phys. Rep. **102** (1983) 199–290

[Po79] A. M. Polyakov. Gauge fields as rings of glue. *Nucl. Phys.* **B164** (1979) 171–88

[Pu93] J. Pullin. Knot theory and quantum gravity: a primer. In "Proceedings of the Vth Mexican School of Particles and Fields", ed. J. Lucio (World Scientific, 1993)

[Re93] A. Rendall. Unique determination of an inner product by adjointness relations in the algebra of quantum observables. To appear in *Class. Quan. Grav.* ; erratum in preparation

[Ro91] C. Rovelli. Ashtekar's formulation of general relativity and loop-space non-perturbative quantum gravity: a report. *Class. Quan. Grav.* **8** (1991) 1613–75

[RoSm88] C. Rovelli and L. Smolin. Knot theory and quantum gravity. *Phys. Rev. Lett.* **61** (1988) 1155–8

[RoSm90] C. Rovelli and L. Smolin. Loop representation of quantum general relativity. *Nucl. Phys.* **B331** (1990) 80–152

[RoTa89] J. Romano and R. Tate. Dirac versus reduced space quantisation of simple constraint systems. *Class. Quan. Grav.* **6** 1487–500

[Sm89] L. Smolin. Invariants of links and critical points of the Chern-Simon path integral. *Mod. Phys. Lett.* **A4** (1989) 1091–112

[Sm93] L. Smolin. What can we learn from the study of non-perturbative quantum general relativity? Syracuse preprint (March 1993)

[Ta1877] P. G. Tait. On knots I, II, III. (Orig. publ. 1877, 1884, 1885, resp.) In "Scientific papers of P. G. Tait" (Cambridge University Press, 1898) 273–347

[Ta93] J. Tavares. Chen integrals, generalized loops and loop calculus. Preprint, U. Porto (April 1993)

[Th1869] W. H. Thomson. On vortex motion. *Trans. R. Soc. Edin.* **25** (1869) 217–60

[Va90] V.A. Vassiliev. Cohomology of knot spaces. In "Theory of singularities and its applications" (V. Arnold, ed.) *Advances in Soviet Math.* **1** (1990) 23–69

[Wi74] K. Wilson. Confinement of quarks. *Phys. Rev.* **D10** (1974) 2445–59

[Wi89] E. Witten. Quantum field theory and the Jones polynomial. *Commun. Math. Phys.* **121** (1989) 351–99

[Ze93] J. Zegwaard. The weaving of curved geometries. *Phys. Lett.* **B300** (1993) 217–22

Gauge Theory and Gravity in the Loop Formulation

Renate Loll

1 Introduction

This article introduces the basic notions and mathematical structures of the so-called loop approach to gauge theory and gravity. The two theories will be treated in parallel as far as this is possible. Emphasis will be put on those aspects of the loop formulation that make it conceptually different from other, local formulations. This contribution is meant to be complementary to other articles in this volume, but some overlap has been unavoidable to ensure a coherent presentation. Many more mathematical details and references for those who want to learn more about loops are contained in a related review (Loll 1993d). Another useful reference focussing on the gravitational application is the review article by Pullin (1993).

The following Section 2 summarizes the classical and quantum description of both gauge theory and gravity as theories on connection space, to set the stage for Sections 3-5 which deal with specific aspects of the loop formulation. In Section 3, I define paths, loops and holonomies, and briefly review past applications in physics. Section 4 is devoted to the identities satisfied by the holonomies and their traces, the latter also known as the Mandelstam constraints. It is shown how a group structure arises on the space of loops. Section 5 discusses the equivalence between the connection and the loop formulation which lies at the heart of the loop approach, and describes the content of the underlying reconstruction theorems. In Section 6, I introduce some ideas from lattice gauge theory which have proved useful in a regularized loop approach and enabled us to do certain calculations explicitly. To round off the discussion, Section 7 describes some features of quantized loop theories, and the final outlook contains a list of mathematical and physical inputs for the loop formulation which still require a more complete understanding.

2 Yang-Mills theory and General Relativity as connection dynamics

This section summarizes the Lagrangian and Hamiltonian formulation of both Yang-Mills gauge theory and gravity, defined on a four-dimensional manifold $M = \Sigma^3 \times \mathbb{R}$ of Lorentzian signature. The gravitational theory will be described in the Ashtekar form, in which it most closely resembles a gauge theory. We will consider only the pure theories, without any matter coupling.

The classical actions defining these theories are (up to overall constants)

$$S_{YM}[^4A] = -\int d^4x\, Tr\, F_{\mu\nu}F^{\mu\nu} = -\int_M Tr(F \wedge *F) \qquad \text{(YM1)}$$

and

$$S_{GR}[^4A, e] = \int d^4x\, e\, e_I^\mu e_J^\nu F_{\mu\nu}{}^{IJ} = \int_M \epsilon_{IJKL}\, e^I \wedge e^J \wedge F^{KL} \qquad \text{(GR1)}$$

for Yang-Mills theory and general relativity respectively. In (YM1), the basic variable A is a $\mathbf{g}$-valued connection one-form on M, where G denotes the compact and semi-simple Lie structure group of the Yang-Mills theory (typically $G = SU(N)$), and $\mathbf{g}$ its Lie algebra. In (GR1), which is a first-order form of the gravitational action, A_μ is a self-dual, $so(3,1)_{\mathbb{C}}$-valued connection one-form, and e^μ a vierbein, defining an isomorphism of vector spaces between the tangent space of M and the fixed internal space with the Minkowski metric η_{IJ}. The e denotes the square root of the determinant of the four-metric $g_{\mu\nu}$. The connection A is selfdual in the internal space, $A_\mu^{MN} = -\frac{i}{2}\epsilon^{MN}{}_{IJ}A_\mu^{IJ}$, with the totally antisymmetric ϵ-tensor. As usual, F denotes in both cases the curvature associated with the four-dimensional connection ^4A,

$$F_{\mu\nu}(x) = \partial_\mu A_\nu(x) - \partial_\nu A_\mu(x) + [A_\mu(x), A_\nu(x)]. \qquad (2.1)$$

Note that the action (GR1) has the unusual feature of being complex. I will not explain here why this still leads to a well-defined theory equivalent to the standard formulation of general relativity. Details may be found in the overview articles by Ashtekar (1991, 1993) and in the contribution by Giulini in this volume.

Our main interest will be the Hamiltonian formulations associated with (YM1) and (GR1). Assuming the underlying principal fibre bundles $P(\Sigma, G)$ is trivial, we can identify the relevant phase spaces with cotangent bundles over the affine spaces $\mathcal{A}$ of $\mathbf{g}$-valued connections on the three-dimensional manifold Σ. (An affine space $\mathcal{A}$ is not itself a linear space, but after fixing a reference point $A_0 \in \mathcal{A}$, the set of all elements of the form $(A - A_0)$ forms a vector space.) We assume the spatial manifold Σ to be orientable and compact, in order not to have to worry about boundary terms. The cotangent bundle $T^*\mathcal{A}$ is coordinatized by canonically conjugate pairs $(A, \tilde{E})$, where A is a $\mathbf{g}$-valued, pseudo-tensorial one-form ("gauge potential") and $\tilde{E}$ a $\mathbf{g}$-valued vector density ("generalized electric field") on Σ. We have $G = SU(N)$, say, for gauge theory, and $G = SO(3)_{\mathbb{C}}$ for gravity. The standard symplectic structure on $T^*\mathcal{A}$ is expressed by the fundamental Poisson bracket relations

$$\{A_a^i(x), \tilde{E}_j^b(y)\} = \delta_j^i \delta_a^b \delta^3(x-y). \tag{2.2}$$

In (2.2), a and b are three-dimensional spatial indices and i and j internal gauge algebra indices.

However, points in $T^*\mathcal{A}$ do not yet correspond to physical configurations, because both theories possess gauge symmetries, which are related to invariances of the original Lagrangians under certain transformations on the fundamental variables. At the Hamiltonian level, this manifests itself in the existence of so-called first-class constraints. They are sets $\mathcal{C}$ of functions C on phase space, which are required to vanish, $C = 0$, for physical configurations and at the same time generate transformation between physically indistinguishable configurations (for more details, see the article by Wipf in this volume). In our case, the first-class constraints are

$$\mathcal{C}_{YM} = \{D_a \tilde{E}^{ai}\} \tag{YM2}$$

and

$$\mathcal{C}_{GR} = \{D_a \tilde{E}^{ai},\ C_a \equiv F_{ab}{}^i \tilde{E}_i^b,\ C \equiv \epsilon^{ijk} F_{ab\,i} \tilde{E}_j^a \tilde{E}_k^b\}. \tag{GR2}$$

The physical interpretation of these constraints is as follows: the so-called Gauss law constraint $D_a\tilde{E}^{ai} = 0$ restricts to configurations whose covariant divergence vanishes, and at the same time generates local gauge transformations in the internal space. The well-known Yang-Mills transformation law corresponding to a finite group element $g \in \mathcal{G}$, the set of G-valued functions on Σ, is

$$(A, \tilde{E}) \mapsto (g^{-1}Ag + g^{-1}dg, g^{-1}\tilde{E}g). \tag{2.3}$$

The second set, $\{C_a\}$, of constraints in (GR2) generates three-dimensional diffeomorphisms of Σ and the last expression, the so-called scalar or Hamiltonian constraint C, generates phase space transformations that are interpreted as corresponding to the time evolution of Σ in M in a space-time picture. Comparing (YM2) and (GR2), we see that in the Hamiltonian formulation, pure gravity may be interpreted as a Yang-Mills theory with gauge group $G = SO(3)_{\mathbb{C}}$, subject to four additional constraints in each point of Σ. However, the dynamics of the two theories is different; we have

$$H_{YM}(A, \tilde{E}) = \int_\Sigma d^3x\; (E^i_a E^a_i + B^a_i B^i_a) \tag{YM3}$$

as the Hamiltonian for Yang-Mills theory (where for convenience we have introduced the generalized magnetic field $B^a_i := -\frac{1}{2}\epsilon^{abc}F_{bc\,i}$). Note the explicit appearance of a Riemannian background metric g on Σ, to contract indices and ensure the integrand in (YM3) is a density. On the other hand, no such additional background structure is necessary to make the gravitational Hamiltonian well-defined. The Hamiltonian H_{GR} for gravity is a sum of a subset of the constraints (GR2),

$$H_{GR}(A, \tilde{E}) = i\int_\Sigma d^3x\; (N^a F_{ab}{}^i \tilde{E}^b_i - \frac{i}{2}\underset{\sim}{N}\epsilon^{ijk}F_{ab\,k}\tilde{E}^a_i\tilde{E}^b_j), \tag{GR3}$$

and therefore vanishes on physical configurations. ($\underset{\sim}{N}$ and N^a in (GR3) are Lagrange multipliers, the so-called lapse and shift functions.) This latter feature is peculiar to generally covariant theories, whose gauge group contains the diffeomorphism group of the underlying manifold. A further difference is the need for a set of reality conditions for the gravitational theory, because we are using the complex coordinate A on phase space.

The space of physical configurations of Yang-Mills theory is the quotient space $\mathcal{A}/\mathcal{G}$ of gauge potentials modulo local gauge transformations. It is non-linear and has non-trivial topology. By excluding the reducible connections from $\mathcal{A}$ (roughly speaking, these are the connections with some additional symmetry properties), and slightly restricting the gauge group $\mathcal{G}$, this quotient can be given the structure of an infinite-dimensional manifold for compact and semi-simple G (Arms et al. 1981). The corresponding physical phase space is then its cotangent bundle $T^*(\mathcal{A}/\mathcal{G})$. Elements of $T^*(\mathcal{A}/\mathcal{G})$ are called the *classical observables* of the Yang-Mills theory.

With Ashtekar's reformulation of general relativity, a remarkable unification has been achieved in the description of the fundamental interactions of physics, which are all described as theories on spaces of connections, subject to a local gauge principle. Moreover, a similar formulation exists for a whole class of so-called topological field theories, examples of which are Chern-Simons theory and 2+1 gravity.

The non-linearities of both the equations of motion and the physical configuration/phase spaces, and the presence of gauge symmetries for these theories lead to numerous difficulties in their classical and quantum description. For example, in the path integral quantization of Yang-Mills theory, a gauge-fixing term has to be introduced in the "sum over all configurations"

$$\int_{\mathcal{A}} [dA]\, e^{iS_{YM}}, \tag{2.4}$$

to ensure that the integration is only taken over one member ${}^4A$ of each gauge equivalence class (see Cheng and Li (1984) for an illustration of this phenomenon). However, on general grounds we know that a unique and attainable gauge choice does not exist (this is the assertion of the so-called Gribov ambiguity (Gribov 1978, Singer 1978)). In any case, since the expression (2.4) can be made meaningful only in a weak-field approximation where one splits $S_{YM} = S_{\text{free}} + S_I$, with S_{free} being just quadratic in A, this approach has not yielded enough information about other sectors of the theory, where the fields cannot assumed to be weak. The situation is even worse in the perturbative treatment of gravity (where the dynamical variables are given by "small" fluctuations around a fixed Minkowskian background structure), which is known to be non-renormalizable.

Similarly, in the canonical quantization, since no good explicit description of $T^*(\mathcal{A}/\mathcal{G})$ is available, one quantizes the theory "à la Dirac". This means that one first "quantizes" on the unphysical phase space $T^*\mathcal{A}$, "as if there were no constraints". That is, one uses a formal operator representation of the canonical variable pairs $(A, \tilde{E})$, satisfying the canonical commutation relations

$$[\hat{A}_a^i(x), \hat{\tilde{E}}_j^b(y)] = i\hbar\delta_j^i\delta_a^b\delta^3(x-y), \tag{2.5}$$

the quantum analogues of the Poisson brackets (2.2). These relations are to be thought of as the infinite-dimensional version of the commutation relations $[\hat{q}^i, \hat{p}_j] = i\hbar\delta_j^i$ of the Schrödinger quantization on $\mathbb{R}^{2n}$ (see Isham (1984) for more details on representation-theoretic considerations of canonical commutation relations). Then a subset of physical wave functions $\mathcal{F}_{\text{phy}} = \{\Psi_{\text{phy}}(A)\}$ is projected out from the space of all wave functions, $\mathcal{F} = \{\Psi(A)\}$, according to

$$\mathcal{F}_{\text{phy}}^{YM} = \{\Psi(A) \in \mathcal{F} \,|\, (\hat{\mathcal{D}E})\Psi(A) = 0\}, \tag{YM4}$$

i.e., $\mathcal{F}_{\text{phy}}$ consists of all functions that lie in the kernel of the quantized Gauss constraint, $\hat{\mathcal{D}E}$. (For a critical appraisal of the use of such "Dirac conditions", see Loll (1992a).) The notation $\mathcal{F}$ and $\mathcal{F}_{\text{phy}}$ indicates that these are just spaces of complex-valued functions on $\mathcal{A}$ and do not yet carry any Hilbert space structure. Whereas this is not required on the unphysical space $\mathcal{F}$, one does need such a structure on $\mathcal{F}_{\text{phy}}$ in order to make physical statements, for instance, about the spectrum of the quantum Hamiltonian $\hat{H}$. Unfortunately, we lack an appropriate scalar product, i.e. an appropriate gauge-invariant measure $[dA]_{\mathcal{G}}$ in

$$< \Psi_{\text{phy}}, \Psi'_{\text{phy}} >= \int_{\mathcal{A}/\mathcal{G}} [dA]_{\mathcal{G}}\ \Psi^*_{\text{phy}}(A), \Psi'_{\text{phy}}(A). \tag{YM5}$$

(More precisely, we expect the integral to range over an extension of the space $\mathcal{A}/\mathcal{G}$ which includes also distributional elements (Ashtekar and Isham 1992)). Given such a measure, the space of physical wave functions would be the Hilbert space $\mathcal{H}^{YM} \subset \mathcal{F}_{\text{phy}}$ of functions square-integrable with respect to inner product (YM5).

Unfortunately, the situation for gravity is not much better. Here the space of physical wave functions is the subspace of $\mathcal{F}$ projected out according to

$$\mathcal{F}^{GR}_{\text{phy}} = \{\Psi(A) \in \mathcal{F} \,|\, (\hat{\mathcal{D}E})\Psi(A) = 0,\ \hat{C}_a\Psi(A) = 0,\ \hat{C}\Psi(A) = 0\}. \quad \text{(GR4)}$$

Again, one now needs an inner product on the space of such states. Assuming that the states Ψ_{phy} of quantum gravity can be described as the set of complex-valued functions on some (still infinite-dimensional) space $\mathcal{M}$, the analogue of (YM5) reads

$$< \Psi_{\text{phy}}, \Psi'_{\text{phy}} >= \int_{\mathcal{M}} [dA]\ \Psi^*_{\text{phy}}(A), \Psi'_{\text{phy}}(A), \quad \text{(GR5)}$$

where the measure $[dA]$ is now both gauge- and diffeomorphism-invariant and projects in an appropriate way to $\mathcal{M}$. If we had such a measure (again, we do not), the Hilbert space $\mathcal{H}^{GR}$ of states for quantum gravity would consist of all those elements of $\mathcal{F}^{GR}_{\text{phy}}$ which are square-integrable with respect to that measure. In the - up to now - absence of explicit measures to make (YM5, GR5) well-defined, our confidence in the procedure outlined here stems from the fact that it can be made meaningful for lower-dimensional model systems (such as Yang-Mills theory in 1+1 and gravity in 2+1 dimensions), where the analogues of the space $\mathcal{M}$ are finite-dimensional.

Note that there is a very important difference in the role played by the quantum Hamiltonians in gauge theory and gravity. In Yang-Mills theory, one searches for solutions in $\mathcal{H}^{YM}$ of the eigenvector equation $\hat{H}_{YM}\Psi_{\text{phy}}(A) = E\Psi_{\text{phy}}(A)$, whereas in gravity the quantum Hamiltonian $\hat{C}$ is one of the constraints and therefore a physical state of quantum gravity must lie in its kernel, $\hat{C}\Psi_{\text{phy}}(A) = 0$.

After all that has been said about the difficulties in finding a canonical quantization for gauge theory and general relativity, the reader may wonder whether one can make any statements at all about their quantum theories. The answer for the Yang-Mills theory is of course in the affirmative. Many qualitative and quantitative statements about quantum gauge theory can be derived in a regularized version of the theory, where the flat background space(-time) is approximated by a hypercubic lattice. Using fairly refined computational methods, one tries to extract (for either weak or strong coupling) properties of the continuum theory, for example to derive an approximate spectrum of the Hamiltonian $\hat{H}_{YM}$ in the limit as the lattice spacing tends to zero. Still there are non-perturbative features, the most important being the confinement property of Yang-Mills theory, which also in this framework are not well understood.

Similar attempts to discretize the (Hamiltonian) theory have so far been not very fruitful in the treatment of gravity, because, unlike in Yang-Mills theory, we cannot choose a fixed background metric (with respect to which we could build a hybercubic lattice), without violating the diffeomorphism invariance of the theory. However, there is a more subtle way in which discrete structures do arise in canonical quantum gravity, which will be mentioned in Section 6 below. One important ingredient in these constructions is a set of non-local loop variables on the space $\mathcal{A}$ of connections, which will be the subject of the next section.

3 Path, loops and holonomies

In order to appreciate the role of the loop representations which have become such an important ingredient in the canonical quantization of theories on connection space, it is important to get a thorough understanding of the classical properties of the loop variables and of how to manipulate them in calculations. This will be done in some detail in this and the following sections. More mathematical properties of loop spaces are discussed in my review article (Loll 1993d).

Given a differentiable, simply connected manifold Σ of dimension d, a *path* in Σ (Fig.1a) is a continuous map w from a closed interval of the real line $\mathbb{R}$ into Σ,

$$\begin{aligned} w :&[s_1, s_2] \rightarrow \Sigma \\ &s \mapsto w^\mu(s). \end{aligned} \tag{3.1}$$

The path w has the properties of a map between two differentiable manifolds, for example, (C^r) differentiability, piecewise differentiability or non-differentiability, and its tangent vector $\frac{dw}{ds}$ may vanish for some or all parameter values s. A *loop* in Σ (Fig.1b) is a closed path, by which we shall mean a continuous map γ of the unit interval into Σ,

$$\gamma : [0,1] \to \Sigma$$
$$s \mapsto \gamma^{\mu}(s), \qquad (3.2)$$
$$\gamma(0) = \gamma(1).$$

We will be using such closed paths in the construction of gauge-invariant quantities in pure gauge theory. Open paths play an important role in gauge theory with fermions, where natural gauge-invariant objects are open flux lines with quarks or Higgs fields "glued to the endpoints" (Kogut and Susskind 1975, Gambini et al. 1988, Gambini and Setaro 1993).

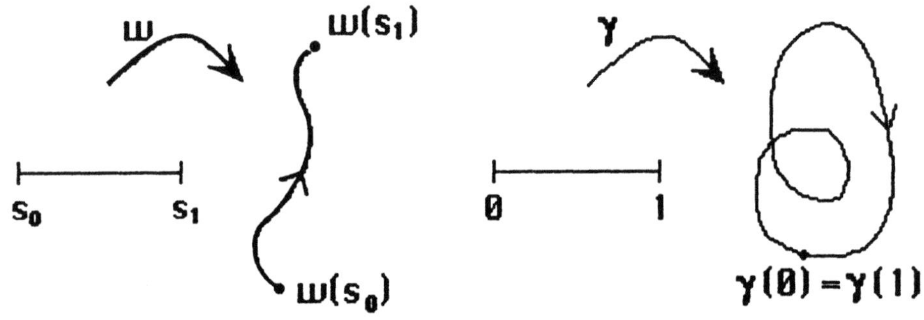

a b

The manifold Σ may be the real vector space $\mathbb{R}^d$, possibly with a Euclidean or Minkowskian metric, but may also be non-linear and topologically non-trivial. In this case the possibility arises of having non-contractible loops, i.e. maps γ that cannot be continuously shrunk to a *point loop*

$$p_x(s) = x \in \Sigma, \quad \forall s \in [0,1]. \qquad (3.3)$$

Note that even if a path is closed it has a distinguished image point, namely, its initial and end point, $\gamma(0) = \gamma(1)$, and that each point in the image of γ is labelled by one or more (if there are selfintersections) parameter values s.

Suppose now that in an open neighbourhood V in Σ we are given a configuration $A \in \mathcal{A}$, the set of all *gauge potentials* $\Sigma \to \Lambda^1 \mathbf{g}$, i.e. a smooth $\mathbf{g}$-valued connection one-form, with $\mathbf{g}$ denoting the Lie algebra of a finite-dimensional Lie group G. We have

$$A(x) = A_\mu(x)\, dx^\mu = A_\mu^a(x) X_a\, dx^\mu, \tag{3.4}$$

where X_a are the algebra generators in the fundamental representation of $\mathbf{g}$ ($a = 1 \ldots \dim G$) and x^μ, $\mu = 1 \ldots d$, a set of local coordinates on V.

The *holonomy* U_w of a path $w^\mu(s)$ with initial point s_0 and endpoint s_1 (whose image is completely contained in V) is the value at s_1 of the solution of the system of differential equations

$$\frac{dU_w(s, s_0)}{ds} = A_\mu(x) \frac{dw^\mu}{ds} U_w(s, s_0), \quad s_0 \leq s \leq s_1, \tag{3.5}$$

with $x = w(s)$, subject to the initial condition

$$U_w(s_0, s_0) = \mathbb{1}, \tag{3.6}$$

where $\mathbb{1}$ denotes the unit element in some matrix representation of G. Note that this definition only makes sense for at least piecewise differentiable paths w. The solution of (3.5) is given by the *path-ordered exponential of A along w*,

$$\begin{aligned} U_w(s_1, s_0) &= \mathrm{P} \exp \int_{s_0}^{s_1} A_\mu(w(t)) \dot{w}^\mu(t) dt := \\ &:= \mathbb{1} + \sum_{n=1}^{\infty} \int_{s_0}^{s_1} dt_1 \int_{t_1}^{s_1} dt_2 \ldots \\ &\qquad \int_{t_{n-1}}^{s_1} dt_n\, A_{\mu_n}(w(t_n)) \ldots A_{\mu_1}(w(t_1)) \dot{w}^{\mu_n}(t_n) \ldots \dot{w}^{\mu_1}(t_1). \end{aligned} \tag{3.7}$$

Note that (3.7) implies the composition law $U_w(s_2, s_0) = U_w(s_2, s_1) U_w(s_1, s_0)$, for $s_0 \leq s_1 \leq s_2$. An alternative definition for U_w that does not need differentiability of the path and employs an approximation of w by n straight line segments $(x_i - x_{i-1})$ is as the limit

$$U_w(s_1,s_0) = \lim_{n\to\infty}(\mathbb{1}+A(x_n)\,(x_n-x_{n-1}))$$
$$(\mathbb{1}+A(x_{n-1})\,(x_{n-1}-x_{n-2}))\dots(\mathbb{1}+A(x_1)\,(x_1-x_0)), \qquad (3.8)$$

with $\sup||x_i - x_{i-1}|| \to 0$ as n increases (Corrigan and Hasslacher 1979), and where $x_0 = w(s_0)$, $x_1,\dots,$ $x_n = w(s_1)$ is a set of $n+1$ points ordered along the path w (Fig.2).

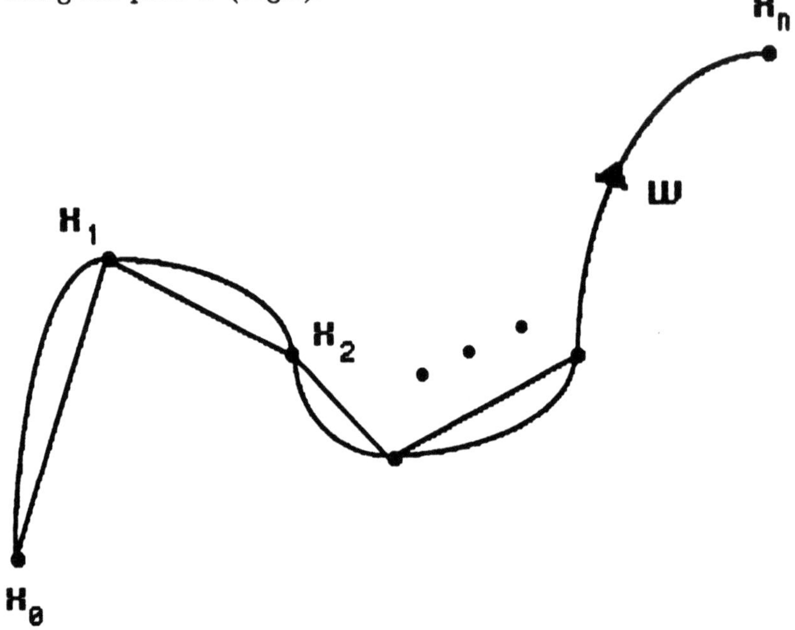

The holonomy U_w takes its values in G and transforms under gauge transformations (smooth functions $g: V \to G$) according to

$$U_w(s_1,s_0)\overset{g}{\longmapsto} g^{-1}(w(s_1))U_w(s_1,s_0)g(w(s_0)). \qquad (3.9)$$

Note this would not be true if we had allowed for discontinuities of the path w. The corresponding change of the gauge potential is straightforwardly computed from equation (3.5),

$$A_\mu(x)\overset{g}{\longmapsto} g^{-1}(x)A_\mu(x)g(x) - \frac{dg^{-1}(x)}{dx^\mu}g(x) \equiv g^{-1}(x)A_\mu(x)g(x) + g^{-1}(x)\frac{dg(x)}{dx^\mu}. \qquad (3.10)$$

Another property of $U_w(s_1, s_0)$ following from (3.5) is its invariance under smooth orientation-preserving reparametrizations f of w, i.e.

$$w(s) = w'(f(s)) \Longrightarrow U_w(s_1, s_0) = U_{w'}(f(s_1), f(s_0)), \tag{3.11}$$

where $\frac{df}{ds} > 0$, $\forall s$. The term "non-integrable" (i.e. path-dependent) "phase factor" for U_w was introduced by Yang (1974), in generalization of its abelian version for $U(1)$-electromagnetism. It is a well-known result in mathematics that the connection A can (up to gauge transformations) be reconstructed from the knowledge of the holonomies of the *closed* curves based at a point $x_0 \in \Sigma$ (see Kobayashi and Nomizu (1969), Lichnerowicz (1976) for details on the concept of holonomy group and related issues). The reader may consult the paper by Barrett (1991) for an account of the mathematical development of these so-called reconstruction theorems, and for more references. The details of such theorems depend on the mathematical setting, for instance whether the underlying fibre bundle is differentiable or just topological. Also generalizations of the concept of connection are possible.

The above description is valid only in a local neighbourhood V in Σ. The appropriate global description is afforded by the mathematical theory of principal fibre bundles $P : G \to P \to \Sigma$ over the base manifold Σ with typical fibre G, with A^P a connection one-form on P. Typically P has no global cross sections (this depends both on the group G and the manifold Σ), and then A^P can be identified with the one-form (3.4) only in a coordinate patch $V \subset \Sigma$, using a local cross section. The global implications of this fact are well known and will not be addressed here. The holonomy variables are global in the sense that because of their gauge-invariance they do not "see" non-trivial transition functions between coordinate patches on the manifold Σ. In the (for the present discussion) important cases of $G = SU(2)$, $SL(2, \mathbb{C})$ and any three-dimensional Σ, global cross sections do exist however and hence there is no need for "patching" on the base manifold. Note that for $F = 0$ (the case of a "flat connection"), where F is the field strength tensor (2.1), U_w depends only on the homotopy of the path w.

Let us now concentrate on the special case when w is a closed path γ in Σ, with associated holonomy,

$$U_\gamma(A) = \mathrm{P} \exp \oint_\gamma A_\mu(x) dx^\mu. \tag{3.12}$$

For a loop γ based at a point $x_0 \in \Sigma$ there is a natural way of constructing a gauge-invariant quantity, namely,

$$h[\gamma, A] := \frac{1}{N} \mathrm{tr}\, U_\gamma = \frac{1}{N} \mathrm{tr}\, \mathrm{P} \exp \oint_\gamma A_\mu(x) dx^\mu, \tag{3.13}$$

the *traced holonomy* or so-called *Wilson loop* (introduced by Wilson (1974) as an indicator of confining behaviour in lattice gauge theory). The trace is taken in a linear representation R of the group G, and the normalization factor in front depends on the dimension N of R. Under a gauge transformation, U_{γ,x_0} transforms homogeneously according to (3.8), but the matrices $g(x_0)$ and $g^{-1}(x_0)$ cancel each other upon taking the trace. Because of the cyclicity of the trace, $h[\gamma, A]$ is independent of the choice of base point $x_0 = \gamma(0) = \gamma(1)$ of γ. Note that $h[\gamma, A]$ is invariant under orientation-preserving reparametrizations of γ and is therefore a function on *unparametrized loops*, i.e. reparametrization equivalence classes $\bar{\gamma}$, where any two members of $\bar{\gamma}$ are related by a smooth orientation-preserving bijection $b : S^1 \mapsto S^1$, $\frac{db}{ds} > 0$. It follows that the Wilson loop is a function on the cartesian product of the quotient spaces, $h[\gamma, A] \in \mathcal{F}(\mathcal{A}/\mathcal{G} \times \Omega\Sigma/Diff^+(S^1))$, where $\Omega\Sigma$ denotes the space of all loops in Σ.

The holonomy of small closed loops measures the *curvature (or field strength) in internal space.* For a small square loop γ of side length ϵ in a coordinate chart around the point $x \in \Sigma$, the base point of γ, the holonomy U_γ can be expanded as

$$U_\gamma = \mathbb{1} + F^a_{\mu\nu}(x) X_a\, \epsilon^2 + O(\epsilon^3), \tag{3.14}$$

where γ is defined by its four corners, $(x, x+\epsilon\vec{e}_\mu, x+\epsilon\vec{e}_\mu+\epsilon\vec{e}_\nu, x+\epsilon\vec{e}_\nu)$, with $\vec{e}_\mu$ denoting the unit vector in μ-direction. Note that in the non-abelian gauge theory, $F_{\mu\nu}(x) = F^a_{\mu\nu}(x) X_a$ is not an observable, since it transforms non-trivially under gauge transformations. A geometric description of the parallel transport associated with the Ashtekar connection in gravity is given in the review paper by Kuchař (1993).

Performing a similar expansion for the traced holonomy of the infinitesimal loop γ, we obtain

$$h[\gamma, A] = \mathrm{tr}\, U_\gamma = 1 + \frac{1}{N} \sum_a F^a_{\mu\nu}(x) \mathrm{tr} X_a \, \epsilon^2$$
$$+ \frac{1}{N} \sum_{a,b} F^a_{\mu\nu}(x) F^b_{\mu\nu}(x) \, \mathrm{tr} X_a X_b \, \epsilon^4 + O(\epsilon^5), \tag{3.15}$$

(no sum over μ, ν). For a semi-simple Lie algebra $\mathbf{g}$ we can always find a basis of generators X_a such that $\mathrm{tr}\, X_a X_b = \delta_{ab}$. Moreover, for $G = SU(N)$ (more generally, for any subgroup of the special linear group $SL(N, \mathbb{C})$, we have $\mathrm{tr}\, X_a = 0$, and the expansion therefore reduces to

$$h[\gamma, A] = 1 + \frac{1}{N} \sum_a F^a_{\mu\nu}(x) F^a_{\mu\nu}(x) \, \epsilon^4 + O(\epsilon^5). \tag{3.16}$$

These expansions illustrate the way in which local gauge-invariant information about the curvature or field strength F is contained in the Wilson loop $h[\gamma, A]$. In concrete calculations involving an unparametrized loop $\bar{\gamma}$ one usually works with a chosen member γ from the equivalence class $\bar{\gamma}$ and then ensures the end result is independent of this choice. There have also been attempts to set up a loop calculus that is intrinsically reparametrization-invariant (Gambini and Trias 1981, 1983, Migdal 1983, Gambini 1991, Tavares 1993).

In gravity, the Wilson loops are not observables, since they are not invariant under the full set of gauge transformations of the theory. However, in this context there is yet another reason to consider variables that depend on closed curves in Σ, since observables in gravity also have to be invariant under diffeomorphisms of Σ. This means going one step further in the construction of such observables, by considering Wilson loops that are constant along the orbits of the $Diff(\Sigma)$-action,

$$h[\varphi \circ \gamma, A] = h[\gamma, A], \; \forall \varphi \in Diff(\Sigma). \tag{3.17}$$

Such a loop variable will just depend on what is called the generalized knot class $K(\gamma)$ of γ ("generalized" since γ may possess self-intersections, which is not usually considered in knot theory). Note that a similar construction for taking care of diffeomorphism invariance is not very contentful if we start from a *local* scalar function, $\phi(x)$, say. The analogue of (3.17) would then tell us that ϕ had to be constant on any connected component of Σ. The set of such variables is not big enough to account for all the degrees of freedom of gravity. In contrast, loops can carry diffeomorphism-invariant information such as their number of self-intersections, number of points of non-differentiability, etc. Although no complete classification of (generalized) knots is known at this point, they are quantities that are easier to visualize and manipulate than their counterparts in the metric formulation of gravity, the equivalence classes $Riem(\Sigma)/Diff_0(\Sigma)$ of Riemannian metrics modulo diffeomorphisms. The relation between quantum gravity and knot theory is the subject of Brügmann's contribution in this volume.

The use of non-local holonomy variables in gauge field theory was pioneered by Mandelstam in his 1962 paper on "Quantum electrodynamics without potentials" (Mandelstam 1962), and later generalized to the non-abelian theory by Bialynicki-Birula (1963) and himself (Mandelstam 1968). Another upsurge of interest in path-dependent formulations of Yang-Mills theory took place at the end of the seventies, following the observation of the close formal resemblance of a particular form of the Yang-Mills equations in terms of holonomy variables and the equations of motion of the two-dimensional non-linear sigma model (Aref'eva 1979, Polyakov 1979). Another set of equations for the Wilson loop were derived by Makeenko and Migdal (1979, 1981). The hope underlying such approaches has always been that one may be able to find a set of basic variables for Yang-Mills theory, which are better suited to its quantization than the local gauge potentials $A(x)$. Due to their simple behaviour under local gauge transformations, the holonomy or the Wilson loop seem like ideal candidates. Hence the aim has been to derive suitable differential equations in loop space for the holonomy U_γ, its trace, $h[\gamma]$, or for their vacuum expectation values, which are to be thought of as the analogues of the usual local Yang-Mills equations of motion. The fact that none of these attempts have been very successful can be attributed to a number of reasons.

(a) To make sense of differential equations on loop space, one first of all has to introduce a suitable topology, and then set up a differential calculus on $\Omega\Sigma$. Even if we give $\Omega\Sigma$ locally the structure of a topological vector space, and are able to define differentiation, this in general will ensure neither the existence of inverse and implicit function theorems nor theorems on the existence and uniqueness of solutions of differential equations. Such issues have only received scant attention in the past.

(b) Since the space of all loops in Σ is so much larger than the set of all points in Σ, one expects that not all loop variables will be independent. This expectation is indeed correct, as will be explained in the next section. Still, the ensuing redundancy is hard to control in the continuum theory, and has therefore obscured many attempts of establishing an equivalence between the connection and the loop formulation of gauge theory. For example, there is no action principle in terms of loop variables, which leaves considerable freedom of how to derive "loop equations of motion". As another consequence, perturbation theory in the loop approach always had to fall back on the perturbative expansion in terms of the connection variable A.

4 Identities satisfied by holonomies and their traces

Central to the loop description of gauge theory and gravity are certain identities satisfied by the holonomies and Wilson loops, which follow from their definitions (3.7) and (3.13). They have to be imposed as constraints if we re-interpret those theories as theories on some loop space, without reference to the space $\mathcal{A}$ of connections. Take G to be the gauge group $GL(N,\mathbb{C})$ or one of its subgroups ($U(N)$, $SU(N)$, $SO(N)$, etc.) in its fundamental representation in terms of $N \times N$ (complex) matrices. For two loops γ_1 and γ_2 based at x, we have

$$U_{\gamma_1 \circ \gamma_2}(A) = U_{\gamma_2}(A) \cdot U_{\gamma_1}(A), \quad \forall A, \tag{4.1}$$

with $\circ$ denoting the loop composition, $\cdot$ matrix multiplication and U_γ defined by (3.12). For the inverse loop γ^{-1} of γ, $\gamma^{-1}(s) := \gamma(1-s)$, we have

$$U_{\gamma^{-1}} = (U_\gamma)^{-1}. \tag{4.2}$$

Here and in the rest of this article, dependence of $U_\gamma(A)$ and $h[\gamma, A]$ on A will not be written explicitly, and we will also use the customary notation $T(\gamma) := h[\gamma, \cdot]$ for the Wilson loops. Since (4.2) holds also for open paths γ, we derive the important *retracing identity*

$$U_\gamma = U_{\gamma'}, \quad \text{for } \gamma' = ((\gamma_1 \circ w) \circ w^{-1}) \circ \gamma_2, \tag{4.3}$$

where $\gamma_1 \circ \gamma_2 = \gamma$, w is an open path "glued to γ" (see Fig.3).

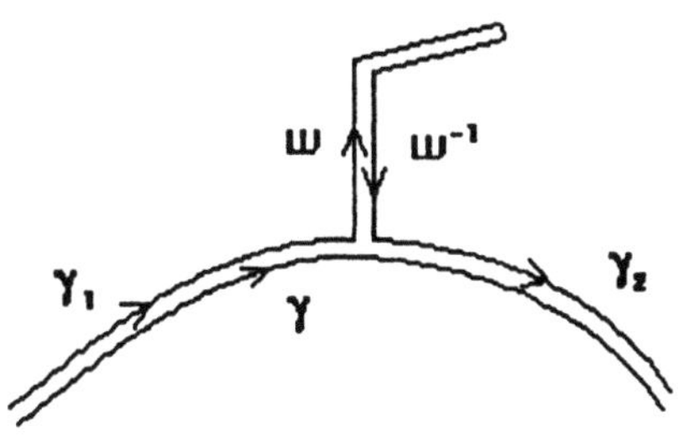

Fig.3

Although the composition of loops is non-associative, it is easily seen that $U_{\gamma'}$ in (4.3) is independent of the order of composition of the loop segments; for example, we could have chosen $\gamma' = \gamma_1 \circ (w \circ (w^{-1} \circ \gamma_2))$ instead. The property (4.3) motivates the introduction of an equivalence relation on paths. A loop γ is called a *thin loop* if it is homotopic to the trivial point loop p by a homotopy of loops whose image lies entirely within the image of γ. Two loops (paths) γ_1 and γ_2 are called thinly equivalent if $\gamma_1 \circ \gamma_2^{-1}$ is a thin loop. Note that two loops that differ by an orientation-preserving reparametrization are always thinly equivalent.

We will denote the equivalence class of a loop γ by $\bar{\gamma}$, and the set of all loops in Σ based at x_0 by $\Omega_{x_0}\Sigma$. Under the above equivalence relation, the set of equivalence classes $\Omega_{x_0}\Sigma/\sim$ acquires a group structure, which is characterized by the relations

$$\begin{aligned} &\bar{\beta} = \bar{\gamma} && \text{if } \beta = \gamma \text{ mod retracing/reparametrization} \\ &\bar{\gamma} \circ \bar{\gamma}^{-1} = \bar{p}_{x_0} && \\ &(\bar{\alpha} \circ \bar{\beta}) \circ \bar{\gamma} = \bar{\alpha} \circ (\bar{\beta} \circ \bar{\gamma}) && \text{(associativity)} \end{aligned} \tag{4.4}$$

with the unit $\bar{p}_{x_0}$, the equivalence class of the constant loop (3.3). This group is usually called the "loop group" or the "group of loops" (not to be confused with the loop groups of maps from S^1 into a Lie group G à la Pressley and Segal (1986)), and will be denoted by $\mathcal{L}\Sigma$. The group $\mathcal{L}\Sigma$ is non-abelian and, as shown by Durhuus and Leinaas (1982), not locally compact, which means it is "very large" even locally. This feature leads to complications when one tries to define a Fourier transform on $\mathcal{L}\Sigma$. The loop group has been discussed by many authors in view of its application to physics (Mensky 1978, Fröhlich 1979, Gambini and Trias 1981, Durhuus and Leinaas 1982, Anandan 1983, Fischer 1986, Barrett 1991, Ashtekar and Isham 1992). Its construction has been around in the mathematics literature for a long time (see the remarks and references in Barrett (1991)). Some toy examples of simpler loop groups in two dimensions are contained in Durhuus and Leinaas (1982). A simplified version for polygonal loops on $\mathbb{R}^4$ is discussed by Gambini and Trias (1981). - Some authors prefer to induce an equivalence relation on loops by defining $\beta \sim \gamma$ if $U_\beta(A) = U_\gamma(A)$, $\forall A$, which explicitly refers to the space of connections (Lewandowski 1993, Ashtekar and Lewandowski 1993b).

Note that by virtue of (4.1), the holonomy mapping U may be viewed as a group homomorphism $U : \mathcal{L}\Sigma \to G$. The Wilson loops (3.13) may therefore be regarded as the characters of a representation of the loop group $\mathcal{L}\Sigma$ (Fröhlich 1979, Durhuus and Leinaas 1982, Anandan 1983). This viewpoint may be profitably exploited in some questions arising in the quantum theory.

There are analogous identities satisfied by the traced holonomies, characterizing them as a particular subset of complex-valued functions on $\Omega\Sigma \times \mathcal{A}/\mathcal{G}$. Independently of the gauge group G, we have the identities

$$T(\gamma_1 \circ \gamma_2) = T(\gamma_2 \circ \gamma_1) \tag{4.5}$$

because of the cyclicity of the trace, and again a retracing identity,

$$T(\gamma) = T(\gamma'), \quad \gamma, \gamma' \text{ related as in (4.3).} \tag{4.6}$$

Another set of identities are the so-called Mandelstam constraints, whose form depends on the dimension N of the group matrices. They can be systematically derived from the identity of N-dimensional δ-functions,

$$\sum_{\pi \in \mathcal{S}_{N+1}} (-1)^{\sigma(\pi)} \delta_{i_1, \pi(j_1)} \ldots \delta_{i_{N+1}, \pi(j_{N+1})} = 0, \quad i_k, j_k = 1 \ldots N, \tag{4.7}$$

with the sum running over all permutations π of the symmetric group of order $N+1$, $\mathcal{S}_{N+1}$, and $\sigma(\pi)$ denoting the parity of the permutation. Contracting $N+1$ holonomy matrices U_γ with (4.7) results in a trace identity for (combinations of) $N+1$ loops. For $N=1$, we have

$$T(\alpha)T(\beta) - T(\alpha \circ \beta) = 0, \tag{4.8}$$

and for $N = 2$,

$$T(\alpha)T(\beta)T(\gamma) - \frac{1}{2}\Big(T(\alpha\beta)T(\gamma) + T(\beta\gamma)T(\alpha) + T(\alpha\gamma)T(\beta)\Big) + \frac{1}{4}\Big(T(\alpha\beta\gamma) + T(\alpha\gamma\beta)\Big) = 0, \tag{4.9}$$

Note that the Mandelstam identities are non-linear algebraic equations on the functions T. If we want to consider traced holonomies of specific subgroups of $GL(N, \mathbb{C})$, there will be more identities satisfied by T, for example, deriving from a condition $\det U_\gamma = 1$ (see Gliozzi and Virasoro (1980), Gambini and Trias (1986) and the next section for some selected cases).

We may use the functions T to induce an equivalence relation on the loop space $\Omega\Sigma$ by defining

$$\beta \sim \gamma \qquad \text{if} \qquad T(\beta) = T(\gamma),\ \forall A. \tag{4.10}$$

The composition law for equivalence classes $\bar{\gamma}$,

$$\bar{\gamma}_1 \circ \bar{\gamma}_2 := \overline{\gamma_1 \circ \gamma_2}, \tag{4.11}$$

induces an abelian group structure on $\Omega\Sigma/\sim$ *if* in addition the relation $T(\alpha) = T(\beta) \Rightarrow T(\alpha \circ \gamma) = T(\beta \circ \gamma)$ is satisfied for all γ (Gambini and Trias 1986, Ashtekar and Isham 1992).

5. Equivalence between gauge potentials and holonomies

The importance of the (traced) holonomies lies in the fact that from them one can reconstruct gauge-invariant information about the gauge potential A. The content of the so-called reconstruction theorems is to specify a set of algebraic and differential conditions on a set of functions on loop space which ensure that from them one can uniquely compute the corresponding equivalence class $[A] \in \mathcal{A}/\mathcal{G}$.

The best-known case is that of holonomies based at a point, for which various versions of the reconstruction theorem are available, depending on the mathematical setting. Although known to mathematicians for a long time, they have been regularly rediscovered by physicists. A very detailed discussion and derivation is contained in the paper by Barrett (1991). I quote here his reconstruction theorem for differentiable principal fibre bundles (related treatments may be found in Fröhlich (1979), Gliozzi and Virasoro (1980), Giles (1981), Anandan (1983)):

Reconstruction Theorem. Suppose Σ is a connected manifold with base-point x_0 and the map $H : \Omega_{x_0}\Sigma \to G$ satisfies the following conditions:

(i) H is a homomorphism of the composition law of loops, $H(\gamma_1 \circ \gamma_2) = H(\gamma_2)H(\gamma_1)$.

(ii) H takes the same values on thinly equivalent loops: $\gamma_1 \sim \gamma_2$ if $\gamma_1 \circ \gamma_2^{-1}$ is thin (cf. Sec.4).

(iii) For any smooth finite-dimensional family of loops $\tilde{\psi} : U \to \Omega_{x_0}\Sigma$, the composite map $H\psi : U \to \Omega_{x_0}\Sigma \to G$ is smooth.

Then there exists a differentiable principal fibre bundle $P(\Sigma, G, \pi)$, a point $p \in \pi^{-1}(x_0)$ and a connection Γ on P such that H is the holonomy mapping of (P, Γ, x_0). (In (iii) above, U is an open subset of $\mathbb{R}^n$ (parametrizing the family), for any n, and $\tilde{\psi}$ is smooth in the sense that the associated map $\psi : U \times I \to \Sigma$ is continuous and piecewise C^∞ with respect to the the loop parameter $t \in I$.)

Note that in this formulation the principal bundle P is not fixed a priori, but only the base space Σ and the fibre G are. Fixing P would amount to fixing the homotopy class of the holonomy mapping.

Let us go back to the discussion of the reconstruction theorem, now for the case of the traced holonomies. Since we already have a reconstruction theorem for the holonomies, it suffices to show one can recover the holonomies from the Wilson loops. The main task is to find algebraic conditions analogous to (i) and (ii) above, characterizing uniquely the traced holonomies as a subset of complex-valued functions on $\Omega\Sigma$. It seems to be much harder to come up with a set of necessary and sufficient conditions, moreover, these conditions now depend on the gauge group G. To my knowledge, the problem of giving a complete set of such conditions for a general gauge group G has not been solved. The most advanced results in this context are those obtained by Giles (1981).

The content of his reconstruction theorem is essentially as follows: Given any complex-valued function $F(\gamma)$ on the loop space $\Omega\Sigma$ satisfying the Mandelstam identities of order N (and possibly some additional identities, characterizing a specific subgroup of $GL(N, \mathbb{C})$), retracing and reparametrization invariance, equation (4.8) and appropriate smoothness conditions, one can construct (modulo a residual gauge freedom) $N \times N$ matrices $U_\gamma \in GL(N, \mathbb{C})$ (or of the subgroup in question) such that the traces of products $U_{\gamma_j} \cdot U_{\gamma_i} \cdot \ldots$ are exactly given by $F(\ldots \circ \gamma_i \circ \gamma_j)$.

Giles gives an explicit way of reconstructing the holonomies from the traced holonomies, which is very useful in practical applications. However, his results are incomplete in at least two aspects. Firstly, in addition to the usual Mandelstam constraints, there are inequalities restricting the range of the Wilson loops, i.e. given a set of n Wilson loops, they can in general not attain arbitrary complex values, even if all the Mandelstam identities are fulfilled (Loll 1993b). Secondly, it has not been proven that by running through all admissible $F(\gamma)$'s one recovers indeed *all* possible holonomy configurations U_γ. In general the U_γ's so obtained will form a subgroup of G. This however can only happen if G is non-compact, a well-studied example being that of $G = SL(2,\mathbb{C})$ (Goldberg et al. 1992).

For the sake of illustration, and because of their importance in many applications, here are the explicit Mandelstam constraints for $G = SL(2,\mathbb{C})$, in the fundamental representation by complex 2×2-matrices. One finds

$$
\begin{array}{ll}
(a) & T(\text{point loop } p) = 1 \\
(b) & T(\gamma_1) = T(\gamma_2), \qquad \text{if } \gamma_1 \text{ and } \gamma_2 \text{ are thinly equivalent} \\
(c) & T(\gamma_1 \circ \gamma_2) = T(\gamma_2 \circ \gamma_1) \\
(d) & T(\gamma) = T(\gamma^{-1}) \\
(e) & T(\gamma_1)T(\gamma_2) = \frac{1}{2}\Big(T(\gamma_1 \circ \lambda \circ \gamma_2 \circ \lambda^{-1}) + T(\gamma_1 \circ \lambda \circ \gamma_2^{-1} \circ \lambda^{-1})\Big).
\end{array}
\tag{5.1}
$$

In (e), λ is a path connecting a point on γ_1 with a point on γ_2, as illustrated by Fig.4. Note that (e) implies both (a) and (d). However, the conditions have been separated in this way since (a)-(c) hold for any group, whereas (d) and (e) are true just for $SL(2,\mathbb{C})$. Presumably the set (9.1) exhaust all the Mandelstam constraints for this particular group and representation, although I am not aware of the existence of a formal proof. In any case, it can be shown that there are holonomies U_γ that cannot be reconstructed from the Wilson loops, namely those that lie in the subgroup of so-called null rotations. However, in a sense this incompleteness is negligible in the physical applications considered so far (Goldberg et al. 1992), and in fact the Wilson loops are "as complete as they could be" (Ashtekar and Lewandowski 1993a).

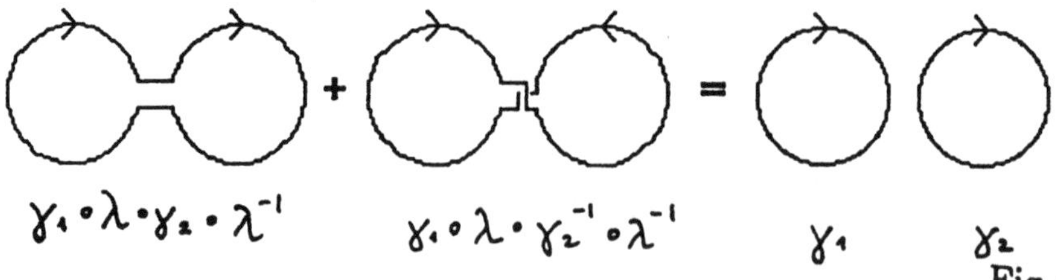

Fig.4.

If one wants to restrict the group G to be described by the Wilson loops to a subgroup of $SL(2,\mathbb{C})$, there are further conditions that have to be imposed on the loop variables, including inequalities. For the subgroup $SU(2) \subset SL(2,\mathbb{C})$, one has

$$
\begin{array}{ll}
(a) & T(\gamma) \text{ real} \\
(b) & -1 \leq T(\gamma) \leq 1 \\
(c) & \left(T(\gamma_1 \circ \gamma_2) - T(\gamma_1 \circ \gamma_2^{-1})\right)^2 \leq 4\left(1 - T(\gamma_1)^2\right)\left(1 - T(\gamma_2)^2\right) \\
 & \vdots
\end{array}
\tag{5.2}
$$

whereas the analogous conditions for the subgroup $SU(1,1) \subset SL(2,\mathbb{C})$ are

$$
\begin{array}{ll}
(a) & T(\gamma) \text{ real} \\
(b) & T(\gamma_1)^2 \leq 1 \text{ and } T(\gamma_2)^2 \leq 1 \Rightarrow \\
 & \left(T(\gamma_1 \circ \gamma_2) - T(\gamma_1 \circ \gamma_2^{-1})\right)^2 \geq 4\left(1 - T(\gamma_1)^2\right)\left(1 - T(\gamma_2)^2\right) \\
 & \vdots
\end{array}
\tag{5.3}
$$

As indicated by the dots, there are other inequalities for more complicated loop configurations, involving more than two loops γ_i. Note that all of the above relations hold independently of A, and hence may be interpreted as *constraints* on loop space functions on $\Omega\Sigma$, rather than as *identities* on the loop-dependent functions on the space of connections, i.e. as functions on $\Omega\Sigma \times \mathcal{A}/\mathcal{G}$. Also at least some of these algebraic conditions can be interpreted as conditions on the underlying loop space. For example, (5.1d) may be interpreted as a condition that T is a function of unoriented loops.

If one wants to reformulate a theory purely in terms of loop variables, without making any reference to the connection variables A, one has to come up with a complete set of conditions of the type discussed above to ensure equivalence with the usual local formulation. This is a non-trivial task, and not always appreciated in physical applications. The difficulty of proving rigorous reconstruction theorems for the Wilson loops has somewhat hampered this kind of "pure loop approach". The most progress in isolating the true degrees of freedom in such an approach has been made on the hypercubic lattice (Loll 1992b,1993a,c). - Once a reconstruction theorem has been proven, the method is potentially very powerful, since any gauge-invariant statement in terms of the connection A is in principle expressible in terms of the traced holonomies $T(\gamma)$.

Alternatively, one may follow a less radical approach which still incorporates the gauge potentials. For example, one may define a (G-dependent) equivalence relation on loops by identifying two loops if their associated Wilson loops agree for all possible configurations $A(x)$ (4.12). Equivalence classes of this kind are sometimes called "hoops". The quotient space of loop space one obtains in this way is of course exactly the one one would like to use as a domain space when constructing a formulation exclusively in terms of loop variables. Another possibility is to construct a "loop transform" in the quantum theory which allows one to translate wave functions and operators from the connection to the loop representation. If it can be shown to exist rigorously for particular spaces of connections, this transform automatically takes care of algebraic constraints among the Wilson loop operators, i.e. quantum analogues of the Mandelstam constraints.

Incidentally, there have also been attempts to reformulate gauge theories in terms of the field strengths F instead of the gauge potentials A, because they have a simpler, homogeneous transformation behaviour. However, it is well known that gauge-inequivalent connection configurations may have the same field strength (although this does not happen for "generic" connections), and hence a rigorous reconstruction theorem does not exist. A discussion of this degeneracy and a guide to the literature is contained in the paper by Mostow and Shnider (1983).

Finally, let me emphasize that the equivalence of the connection and the loop description at the classical level does not necessarily mean that quantization with the Wilson loops as a set of basic variables will be equivalent to a quantization in which the connections A are turned into

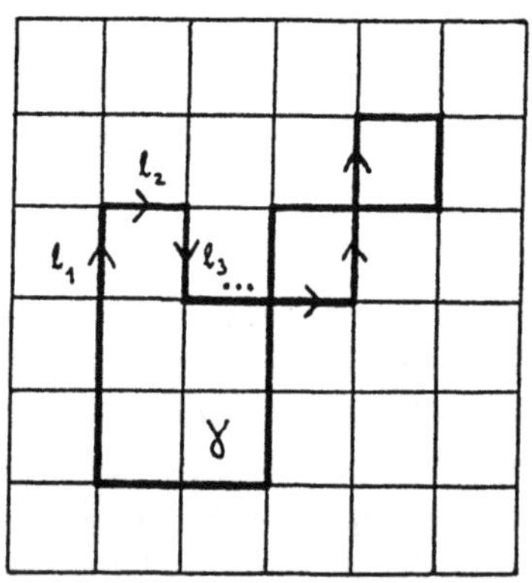

fundamental quantum operators. One may indeed hope that at some level, those two possibilities lead to different results, with the loop representation being better suited for the description of the non-perturbative structure of the theory.

6 Some lattice results on loops

The contents of this section concerns the gauge-theoretic application of taking the Wilson loops as a set of basic variables, although some of the techniques and conclusions may be relevant for gravitational theories too. The lattice discretization is particularly suited to the loop formulation, because the gauge fields are taken to be concentrated on the one-dimensional links l_i of the lattice. The Wilson loops are easily formed by multiplying together the holonomies $U_{l_1} \dots U_{l_n}$ of a set of contiguous, oriented links that form a closed chain on the lattice (see Fig.5),

$$T(\gamma) = Tr_R\, U_{l_1} U_{l_2} \dots U_{l_n}. \tag{6.1}$$

We take the lattice to be the hypercubic lattice N^d, with periodic boundary conditions (N is the number of links in each direction), and the gauge group G to be $SU(2)$, again in the fundamental representation. Although the lattice is finite, there are an infinite number of closed contours one can form on it, and hence an infinite number of Wilson loops. At the same time, the number of Mandelstam constraints (5.1) is infinite. The only constraints that cause problems are those of type (e), because of their non-linear and highly coupled character. Nevertheless, as I showed in Loll (1992b), it is possible to solve this coupled set of equations and obtain an explicit local description of the finite-dimensional physical configuration space,

$$\mathcal{C}_{\text{phy}} = \frac{\mathcal{A}}{\mathcal{G}}|_{\text{lattice}} = \frac{\{T(\gamma),\ \gamma \text{ a lattice loop}\}}{\text{Mandelstam constraints}}. \tag{6.2}$$

in terms of an independent set of Wilson loops. The final solution for the independent degrees of freedom is surprisingly simple: it suffices to use the Wilson loops of small lattice loops, consisting of no more than 6 lattice links. - At this point it may be noted that a similar problem arises in 2+1-gravity (in the connection formulation) on a manifold $\Sigma^g \times \mathbb{R}$, with a compact, two-dimensional Riemann surface Σ^g of genus g. One also has a discrete set of loops, namely, the elements of the homotopy group of Σ^g, and can introduce the corresponding Wilson loops. Again, one is interested in finding a maximally reduced set of such Wilson loops, in terms of which all the others can be expressed, using the Mandelstam constraints (see, for example, Nelson and Regge (1991), Ashtekar and Loll (1994)).

Of course it does not suffice to establish a set of independent loop variables on the lattice; one also has to show that the kinematics can be expressed in a manageable form in terms of these variables. Some important problems are the following.

a) Since the space $\mathcal{C}_{\text{phy}} = \mathcal{A}/\mathcal{G}$ is topologically non-trivial, the set of independent lattice Wilson loops cannot provide a good global chart of $\mathcal{C}_{\text{phy}}$. It would therefore be desirable to find a minimal embedding into a space of loop variables that *is* topologically trivial (together with a finite number of non-linear constraints that define the space $\mathcal{C}_{\text{phy}}$).

b) What is the explicit form of the Jacobian of the transformation from the holonomy link variables to the Wilson loops? ¿From this one could derive an effective action/Hamiltonian for gauge theory, and maybe derive a prescription of how to translate local quantities into their loop space analogues.

c) Is it possible to define a perturbation theory in loop space, once one understands which Wilson loop configurations contribute most in the path integral, say? This would be an interesting alternative to the usual perturbation theory in A and the $\frac{1}{N}$-expansion, and may lead to new and more efficient ways of doing calculations in lattice gauge theory.

d) Lastly, in finding a solution to the Mandelstam constraints it was crucial to have the concept of a smallest loop size on the lattice (the loops of link length 4 going around a single plaquette). What are the consequences of this result for the continuum theory, where there is no obstruction to shrinking loops to points?

Some progress on (a) and (b) has been made for small lattices Loll (1993a), and it is clear that all of the above-posed problems are rather non-trivial, even in the discretized lattice theory where the number of degrees of freedom is effectively finite. One may be able to apply some of the techniques used here to 3+1-dimensional gravity, where also discrete structures appear (the generalized knot classes) after factoring out by the spatial diffeomorphisms.

Attempts to apply lattice gauge theory ideas directly to general relativity have so far failed, basically because the diffeomorphism invariance of the theory is "very badly broken" by the introduction of a lattice in space. This finds expression in the fact that the algebra of constraints does not close on the lattice, and little is known about whether closure can be achieved in the continuum limit as the lattice spacing goes to zero (Renteln 1990). Nevertheless, some lattice-inspired ideas have made their way into gravity. In these cases, the lattice is not to be thought of as a fixed hypercubic lattice imbedded into space, but defined abstractly through its connectivity properties, i.e. a list of data of how pieces of paths in Σ link together to form a generalized lattice or graph. These are properties that are invariant under diffeomorphisms. Using such ideas, one may construct diffeomorphism-invariant measures on the space $\mathcal{A}$ of connections (Ashtekar and Lewandowski 1993b, Baez 1993).

7 Quantization in the loop approach

I will now return to the discussion of the continuum theory of both gravity and gauge theory, and review a few concepts that have been used in their quantization in a loop formulation. The main idea is to base the canonical quantization of a theory with configuration space $\mathcal{A}$ on the non-local, gauge-invariant phase space variables $T(\gamma)$ (and not on some gauge-covariant local field variables). This is a somewhat unusual procedure in the quantization of a field theory, but takes directly into account the non-linearities of the theory.

You may have noted that so far we have only talked about *configuration space* variables, the Wilson loops. For a canonical quantization we obviously need some momentum variables that depend on the generalized electric field $\tilde{E}$. Since $\tilde{E}$ - unlike the gauge potential A - is not a one-form, we cannot introduce holonomy variables that depend only on $\tilde{E}$. This leads to a certain asymmetry in the formulation. One choice of a generalized Wilson loop that also depends on $\tilde{E}$ is

$$T^a_{(A,\tilde{E})}(\gamma, s) := Tr\ U_\gamma(s)\tilde{E}^a(\gamma(s)). \tag{7.1}$$

This variable depends now on both a loop and a marked point, and it is still gauge-invariant (under the transformations (2.3)). Also it is strictly speaking not a function on phase space, but a vector density. This latter fact may be remedied by integrating T^a over a two-dimensional "ribbon" or "strip", i.e. a non-degenerate one-parameter congruence of curves $\gamma_t(s) =: R(s,t)$, $t \in [0,1]$ (Ashtekar and Isham 1992). In any case, the "momentum Wilson loop" depends not just on a loop, but needs an additional geometric ingredient. I will use here the unsmeared version since it does not affect what I am going to say.

The crucial ingredient in the canonical quantization is the fact that the loop variables $(T(\gamma), T^a(\gamma, s))$ form a closing algebra with respect to the canonical symplectic structure on $T^*\mathcal{A}$. This fact was first established by Gambini and Trias (1986) for gauge group $SU(N)$ and later rediscovered by Rovelli and Smolin (1990) in the context of gravity. For the special case of $G = SL(2, \mathbb{C})$ (or any of its subgroups) in the fundamental representation, the Poisson algebra of the loop variables is

$$
\begin{aligned}
&\{T(\gamma_1), T(\gamma_2)\} = 0, \\
&\{T^a(\gamma_1, s), T(\gamma_2)\} = -\Delta^a(\gamma_2, \gamma_1(s))\,(T(\gamma_1 \circ_s \gamma_2) - T(\gamma_1 \circ_s \gamma_2^{-1})), \\
&\{T^a(\gamma_1, s), T^b(\gamma_2, t)\} = \\
&\qquad -\Delta^a(\gamma_2, \gamma_1(s))\,(T^b(\gamma_1 \circ_s \gamma_2, u(t)) + T^b(\gamma_1 \circ_s \gamma_2^{-1}, u(t))\,)+ \\
&\qquad + \Delta^b(\gamma_1, \gamma_2(t))\,(T^a(\gamma_2 \circ_t \gamma_1, v(s)) + T^a(\gamma_2 \circ_t \gamma_1^{-1}, v(s))\,).
\end{aligned} \tag{7.2}
$$

In the derivation of this semi-direct product algebra, the Mandelstam constraints (5.1e) have been used to bring the right-hand sides into a form linear in T. The structure constants Δ are distributional and depend just on the geometry of loops and intersection points,

$$
\Delta^a(\gamma, x) = \oint_\gamma dt\, \delta^3(\gamma(t), x)\dot{\gamma}^a(t). \tag{7.3}
$$

The Poisson bracket of two loop variables is non-vanishing only when an insertion of an electric field $\tilde{E}$ in a T^a-variable coincides in Σ with (the holonomy of) another loop. For gauge group $SU(N)$, the right-hand sides of the Poisson brackets can no longer be written as linear expressions in T.

By a "quantization of the theory in the loop representation" we will mean a representation of the loop algebra (7.2) as the commutator algebra of a set of (self-adjoint) operators $(\hat{T}(\gamma), \hat{T}^a(\gamma, s))$ (or an appropriately smeared version of the momentum operator $\hat{T}^a(\gamma, s)$). Secondly, the Mandelstam constraints must be incorporated in the quantum theory, for example, by demanding that relations like (5.1) hold also for the corresponding quantum operators $\hat{T}(\gamma)$ (for the momentum Wilson loops there are corresponding Mandelstam constraints).

The completion of this quantization program is different for gauge theory and gravity, in line with the remarks made in section 2 above. For the Yang-Mills theory, one has to find a Hilbert space of wave functionals that depend on either the Wilson loops $T(\gamma)$ or directly on the elements γ of some appropriate (quotient of a) loop space. The quantized Wilson loops must be self-adjoint with respect to the inner product on this Hilbert space, since they correspond to genuine observables in the classical theory. Then the Yang-Mills Hamiltonian (YM3) has to be re-expressed as a function of Wilson loops, and finally, one has to solve the eigenvector equation

$$\hat{H}_{YM}(T(\gamma), T^a(\gamma))\Psi(\alpha) = E\Psi(\alpha), \tag{7.4}$$

to obtain the spectrum of the theory. Because of the lack of an appropriate scalar product in the continuum theory, the only progress that has been made along these lines is in the lattice-regularized version of the gauge theory (see, for example, Brügmann (1991), Gambini and Setaro (1993)).

For the case of gravity one does not require the loop variables ($T(\gamma)$, $T^a(\gamma, s)$) to be represented self-adjointly, because they do not constitute observables on the classical phase space. At least one such representation (where the generalized Wilson loops are represented as differential operators on a space of loop functionals $\tilde{\Psi}(\gamma)$) is known, and is described in Brügmann's article in these proceedings. The completion of the loop quantization program for gravity requires an appropriate inclusion of the three-dimensional diffeomorphism constraints C_a and the Hamiltonian constraint C, c.f. (GR2), which again means that they have to be re-expressed in terms of loop variables. Then, solutions to these quantum constraints have to be found, i.e. loop functionals that lie in their kernel,

$$\hat{C}_a(T(\gamma), T^a(\gamma))\Psi(\alpha) = 0 \;\wedge\; \hat{C}(T(\gamma), T^a(\gamma))\Psi(\alpha) = 0. \tag{7.5}$$

Finally, the space of these solutions has to be given a Hilbert space structure to make any physical interpretation possible. No appropriate scalar product is not known at the moment, but there exists a large number of solutions to the quantum constraint equations (7.5), which are related to knot invariants.

There is another interesting mathematical structure in the loop approach which I would like to mention here, which tries to establish a relation between the quantum connection and the quantum loop representation(s). This is the so-called loop transform, first introduced by

Rovelli and Smolin, which is supposed to intertwine the two types of representations according to

$$\tilde{\Psi}(\gamma) = \int_{\mathcal{A}/\mathcal{G}} [dA]_{\mathcal{G}}\, h[\gamma, A]\, \Psi_{\text{phy}}(A) \tag{7.6}$$

and

$$\left(\hat{T}(\alpha)\tilde{\Psi}\right)(\gamma) := \int_{\mathcal{A}/\mathcal{G}} [dA]_{\mathcal{G}}\, h[\gamma, A] \left(\hat{T}(\alpha)\Psi_{\text{phy}}(A)\right), \tag{7.7}$$

and similarly for the action of $\hat{T}^a(\gamma)$. Indeed, it was by the heuristic use of (7.6) and (7.7) that Rovelli and Smolin arrived at the quantum representation of the loop algebra (7.2) mentioned above. The relation (7.6) is to be thought of as a non-linear analogue of the Fourier transform

$$\tilde{\Psi}(p) = \frac{1}{\sqrt{2\pi}} \int dx\, e^{ixp}\Psi(x) \tag{7.8}$$

on $L^2(\mathbb{R}, dx)$, but does in general not possess an inverse. Again, for Yang-Mills theory we do not know how to construct a concrete measure $[dA]_{\mathcal{G}}$ that would allow us to proceed with a loop quantization program, although in this case the loop transform can be given a rigorous mathematical definition.

For the application to gravity, the domain space of the integration will be a smaller space $\mathcal{M}$, as explained in Section 2 above, but for this case we know even less about the well-definedness of expressions like (7.6) and (7.7). There is however a variety of simpler examples, such as 2+1-dimensional gravity (Ashtekar et al. 1989, Ashtekar and Loll 1994) and electrodynamics (Ashtekar and Rovelli 1992), where the loop transform is a useful and mathematically well-defined tool.

8 Outlook

In this article I have reviewed various aspects of the loop formulation of gauge theory and gravity, and explained in some detail the motivation and justification for adopting such an approach in place of a formulation in terms of local field variables. An important aspect of the loop formulation is its constrained character, due to the presence of the Mandelstam constraints. There is as yet no complete proposal for a loop formulation of any non-linear quantum field theory with local degrees of freedom, like the cases of general relativity and Yang-Mills theory we have been considering. The instances where such a program can be completed are typically of topological type, and their reduced phase spaces are finite-dimensional (examples are 2+1 gravity and 1+1 Yang-Mills theory).

It has to be kept in mind that any such proposal involves a number of non-trivial choices along the way, which will consist roughly of the following:

- a choice of an initial space of loops and a set of functions on it corresponding - after imposition of a complete set of constraints - to the physical configurations of the system. These choices are presumably different for the classical and quantum theories.

- A differential calculus on the space of loop functions, specifying the nature of differential operators, and conditions under which differential equations possess solutions and whether those solutions are unique.

- A prescription of how equations of motion are to be derived: whether from some action principle on the space of loop functions or by translating (e.g. by using a "loop transform") expressions from the connection formulation.

- A choice of a (unitary irreducible) representation of an algebra of loop functions in the quantum theory, like the T-algebra of Gambini and Trias or some variation of it. If such representations exist, they are likely to be non-unique, in which case additional physical criteria will be needed to narrow down the choice as much as possible.

- A prescription of how the Mandelstam constraints and inequalities are to imposed on either the classical or the quantum level.

- An appropriate choice of regularization and, if necessary, renormalization of loop space operators; in the case of gravity, this choice should be "as diffeomorphism-invariant as possible".

At this stage we do not have a good description of the set of all such choices, and it is therefore hard to tell if specific choices that have been adopted in ongoing research programs are very special or in fact quite generic and natural. Moreover, for at least some of the above points, the correct choices are likely to be different for gauge theory and general relativity. One hope for the solution of the latter is that the requirement of diffeomorphism invariance may come to our help, because it naturally leads to discrete geometric stuctures which may be easier to handle mathematically. - There still is much to be explored in this most fascinating line of research!

References

Anandan, J. (1983): Holonomy groups in gravity and gauge fields, in Proceedings of the Conference on Differential Geometric Methods in Theoretical Physics, Trieste 1981, eds. G. Denardo and H.D. Doebner, World Scientific

Aref'eva, I. Ya. (1979): The gauge field as chiral field on the path and its integrability, Lett. Math. Phys. 3, 241-247

Arms, J.M., Marsden, J.E. and Moncrief, V. (1981): Symmetry and bifurcations of momentum mappings, Comm. Math. Phys. **78**, 455-478

Ashtekar, A. (1991): Lectures on non-perturbative gravity, World Scientific, Singapore

Ashtekar, A. (1993): Mathematical problems of non-perturbative quantum general relativity, in: Proceedings of the 1992 Les Houches summer school on gravitation and quantization, ed. B. Julia, North-Holland, Amsterdam

Ashtekar, A., Husain, V., Rovelli, C., Samuel, J. and Smolin, L. (1989): 2+1 gravity as a toy model for the 3+1 theory, Class. Quan. Grav. 6, L185-193

Ashtekar, A. and Isham, C.J. (1992): Representations of the holonomy algebras of gravity and non-Abelian gauge theories, Class. Quan. Grav. 9, 1433-1467

Ashtekar, A. and Lewandowski, J. (1993a): Completeness of Wilson loop functionals on the moduli space of $SL(2,\mathbb{C})$ and $SU(1,1)$-connections, Class. Quan. Grav. 10, L69-74

Ashtekar, A. and Lewandowski, J. (1993b): Representation theory of analytic holonomy C^*-algebras, to appear in: Knot theory and quantum gravity, ed. J. Baez, Oxford University Press

Ashtekar, A. and Loll, R. (1994): New loop representations for 2+1 gravity, Penn State University preprint, in preparation

Ashtekar, A. and Rovelli, C. (1992): A loop representation for the quantum Maxwell field, Class. Quan. Grav. 9, 1121-1150

Baez, J. (1993): Diffeomorphism-invariant generalized measures on the space of connections modulo gauge transformations, to appear in the Proceedings of the Conference on Quantum Topology, eds. L. Crane and D. Yetter, hep-th 9305045

Barrett, J.W. (1991): Holonomy and path structures in general relativity and Yang-Mills theory, Int. J. Theor. Phys. 30, 1171-1215

Bialynicki-Birula, I. (1963): Gauge-invariant variables in the Yang-Mills theory, Bull. Acad. Polon. Sci. 11, 135-138

Brügmann, B. (1991): The method of loops applied to lattice gauge theory, Phys. Rev. D43, 566-579

Cheng, T.-P. and Li, L.-F. (1984): Gauge theory of elementary particle physics, Clarendon Press, Oxford

Corrigan, E. and Hasslacher, B. (1979): A functional equation for exponential loop integrals in gauge theories, Phys. Lett. 81B, 181-184

Durhuus, B. and Leinaas, J.M. (1982): On the loop space formulation of gauge theories, Physica Scripta 25, 504-510

Fischer, A.E. (1986): A grand superspace for unified field theories, Gen. Rel. Gravit. 18, 597-608

Fröhlich, J. (1979): Some results and comments on quantized gauge fields, in: Recent developments in gauge theories, Cargèse (ed. 't Hooft et al.)

Gambini, R., Gianvittorio, R. and Trias, A. (1988): Gauge Higgs dynamics in the loop space, Phys. Rev. D38, 702-705

Gambini, R. (1991): Loop space representation of quantum general relativity and the group of loops, Phys. Lett. 255B, 180-188

Gambini, R. and Setaro, L. (1993): SU(2) QCD in the path representation, preprint Montevideo

Gambini, R. and Trias, A. (1981): Geometrical origin of gauge theories, Phys. Rev. D23, 553-555

Gambini, R. and Trias, A. (1983): Chiral formulation of Yang-Mills equations: A geometric approach, Phys. Rev. D27, 2935-2939

Gambini, R. and Trias, A. (1986): Gauge dynamics in the C-representation, Nucl. Phys. B278, 436-448

Giles, R. (1981): Reconstruction of gauge potentials from Wilson loops, Phys. Rev. D24, 2160-2168

Gliozzi, F. and Virasoro, M.A. (1980): The interaction among dual strings as a manifestation of the gauge group, Nucl. Phys. B164, 141-151

Goldberg, J.N., Lewandowski, J. and Stornaiolo, C. (1992): Degeneracy in loop variables, Comm. Math. Phys. 148, 377-402

Gribov, V.N. (1978): Quantization of non-abelian gauge theories, Nucl. Phys. B139, 1-19

Isham, C.J. (1984): Topological and global aspects of quantum theory, in: Relativity, Groups and Topology II, ed. B.S. DeWitt and R. Stora, North-Holland, Amsterdam, 1059-1290

Kobayashi, S. and Nomizu, K. (1969): Foundations of differential geometry, Vol.1, Interscience, New York

Kogut, J. and Susskind, L. (1975): Hamiltonian formulation of Wilson's lattice gauge theories, Phys. Rev. D11, 395-408

Kuchař, K.V. (1993): Canonical quantum gravity, in: Proceedings of the 13th International Conference on General Relativity and Gravitation, ed. C. Kozameh, IOP Publishing, Bristol

Lewandowski, J. (1993): Group of loops, holonomy maps, path bundle and path connection, Class. Quan. Grav. 10, 879-904

Lichnerowicz, A. (1976): Global theory of connections and holonomy groups, Noordhoff International Publishing (French edition published in 1955)

Loll, R. (1992a): Canonical and BRST-quantization of constrained systems, in: Mathematical aspects of classical field theory, eds. M. Gotay, J. Marsden, V. Moncrief, Contemporary Mathematics Series, vol.12, 503-530

Loll, R. (1992b): Independent SU(2)-loop variables and the reduced configuration space of SU(2)-lattice gauge theory, Nucl. Phys. B368, 121-142

Loll, R. (1993a): Yang-Mills theory without Mandelstam constraints, Nucl. Phys. B400, 126-144

Loll, R. (1993b): Loop variable inequalities in gravity and gauge theory, Class. Quan. Grav. 10, 1471-1476

Loll, R. (1993c): Lattice gauge theory in terms of independent Wilson loops, Nucl. Phys. B, Proc. Suppl. 30, 224-227

Loll, R. (1993d): Chromodynamics and gravity as theories on loop space, Penn State University preprint CGPG-93/9-1, hep-th 9309056

Makeenko, Yu.M. and Migdal, A.A. (1979): Exact equation for the loop average in multicolor QCD, Phys. Lett. 88B, 135-137 (E: 89B (1980) 437)

Makeenko, Yu.M. and Migdal, A.A. (1981): Quantum chromodynamics as dynamics of loops, Nucl. Phys. B188, 269-316

Mandelstam, S. (1962): Quantum electrodynamics without potentials, Ann. Phys. (NY) 19, 1-24

Mandelstam, S. (1968): Feynman rules for electromagnetic and Yang-Mills fields from the gauge-independent field-theoretic formalism, Phys. Rev. 175, 1580-1603

Mensky, M.B. (1978): Group of parallel transports and description of particles in curved space-time, Lett. Math. Phys. 2, 175-180

Migdal, A.A. (1983): Loop equations and 1/N expansion, Phys. Rep. 102, 199-290

Mostow, M.A. and Shnider, S. (1983): Does a generic connection depend continuously on the curvature?, Commun. Math. Phys. 90, 417-432

Nelson, J. E. and Regge, T. (1991): 2+1 gravity for genus > 1, Comm. Math. Phys. 141, 211-223

Polyakov, A.M. (1979): String representations and hidden symmetries for gauge fields, Phys. Lett. 82B, 247-250

Pullin, J. (1993): Knot theory and quantum gravity in loop space: a primer, in: Proceedings of the Vth Mexican School of Particles and Fields, ed. J.L.Lucio, World Scientific, Singapore

Pressley, A. and Segal, G. (1986): Loop groups, Clarendon Press, Oxford

Renteln, P. (1990): Some results of SU(2) spinorial lattice gravity, Class. Quan. Grav. 7, 493-502

Rovelli, C. and Smolin, L. (1990): Loop space representation of quantum general relativity, Nucl. Phys. B331, 80-152

Singer, I. (1978): Some remarks on the Gribov ambiguity, Comm. Math. Phys. 60, 7-12

Tavares, J.N. (1993): Chen integrals, generalized loops and loop calculus, Mathematics preprint, University of Porto

Wilson, K. (1974): Confinement of quarks, Phys. Rev. D10, 2445-2459

Yang, C.N. (1974): Integral formalism for gauge fields, Phys. Rev. Lett. 33, 445-447

Reduced Models for Quantum Gravity

T. Thiemann

1 Introduction

As outlined in various other lectures given at this meeting, it seems that a quantum theory of gravity can only be constructed in a non perturbative manner (compare, in particular, Ashtekar's lectures).

Because of that, no calculations, as for example cross sections, decay rates and so on, can be done for the full theory of 3+1 gravity unless one has the full solution space of the quantum constraints and, derived from that, the physical Hilbert space. For a non-gravitational quantum field theory that can be attacked via a perturbative approach one *can* make quantitative predictions and even make estimates of the error due to higher order corrections while for quantum gravity 'one would have to consider *all orders*'. So for the former theories one *does* have a very good idea of how the exact quantum theory should look like and this is important because intuition gives rise to new lines of attack.
Hence unfortunately, for the quantum theory of gravity, we lack this general picture of how the exact theory should look like completely.

The only way that might help to uncover some of the secrets of how to solve the technical and/or conceptual problems of quantum gravity seems to be to study model systems, ideally those that can be solved exactly.

Of course, the lessons that models teach us might be totally misleading and extreme care is due when transferring results from the model to the physical theory of full quantum gravity.

This is the point of view that we adopt in the sequel :
The models that we are going to discuss capture some of the technical and conceptual problems of gravity and we will pin these down. We *will* attempt at drawing some conclusions from the solutions we found but we *stress* the limitations that arise from the various special features of the models we choose and which are not shared by the full theory of quantum gravity.

From the various, completely solvable, models that have been discussed in the literature we choose those that we consider as most suitable for our pedagogical reasons, namely 2+1 gravity and the spherically symmetric model.

The former model arises from a dimensional, the latter from a Killing reduction of full 3+1 gravity. While 2+1 gravity is usually treated in terms of closed topologies without boundary of the initial data hypersurface, the topology for the spherically symmetric system is chosen to be asymptotically flat. Finally, 2+1 gravity is more naturally quantized using the loop representation while spherically symmetric gravity is easier to quantize via the self-dual representation.

Accordingly, both types of reductions, both types of topologies and both types of representations that are mainly employed in the literature in the context of the new variables come into practice.

It is true that both models turn out to have only a finite-dimensional reduced phase space, hence, we are actually dealing only with a quantum mechanical problem and this restricts the usefulness of the results found for the models tremendously when transferring them to quantum gravity which is a theory with an infinite number of physical degrees of freedom. On the other hand, models with a finite number of degrees of freedom make the analysis especially clear and since one of the major motivations of this meeting was that it should serve as an introduction to canonical quantum gravity, we regard it as important to demonstrate the usefulness of the formalism that has been developed in various other lectures of this seminar by means of applying it to models of quantum gravity which are as simple as possible, here formulated in terms of Ashtekar's new variables.

We adopt the abstract index notation and also the notation used by the the other contributors to the seminar.

2 2+1 gravity

2.1 Canonical formulation

The model of 2+1 gravity arises from a dimensional reduction of the Einstein action of full 3+1 gravity. The first order (Palatini) formulation of n+1 gravity is given by

$$S := \frac{1}{2\kappa}\int_M \Omega_{IJ} \wedge \star(e^I \wedge e^J) \tag{2.1}$$

where the notation is as follows : M is the spacetime manifold (dim(M)$= n+1$), κ is the gravitational constant, Ω_{IJ} is the curvature 2-form of the SO(1,n) principal connection ω_{IJ}, $\star$ is the Hodge-duality operator and e^I are the (n+1)-bein fields (that is, they form an orthonormal cobasis field). Internal indices I,J,K,.. run from 0 to n and are raised and lowered with respect to the internal Minkowski metric η_{IJ}.

For the canonical formulation we have to do the n+1 split of the action by choosing a foliation of M into space and time, that is, we assume that M is topologically $\Sigma \times R$ where Σ is an n-dimensional spacelike hypersurface. From now on we will restrict ourselves to the case $n = 2$.

As in [1], we choose coordinates x^0, x^1, x^2 and a foliation such that the label of the various hypersurfaces is given by $t := x^0$. Then we obtain from (2.1)

$$S = \frac{1}{2\kappa}\int_R dt \int_\Sigma [\dot{A}_a^I E_I^a - [-\Lambda^I \mathcal{G}_I + N_I C^I]] \,. \tag{2.2}$$

We now explain the notation : The indices a,b,c,.. are tensor indices with respect to the spatial slice. A_a^I is simply the pull-back to the spatial slice of the 3-dimensional connection ω^I and we have exploited the fact that the defining and adjoint representation of SO(1,2) are isomorphic such that we can work with the quantities $\omega^I := -\frac{1}{2}\epsilon^{IJK}\omega_{JK}$. As is obvious from the action (2.1), in this polarization [1] the SO(1,2) connection A_a^I will play the role of a configuration space variable.
Its conjugate momentum is evidently given by the 'electric fields' $E_I^a := \epsilon^{ab}e_b^I$ where e_b^I is the pull-back of the SO(1,2) triad. Geometrically, the electric fields are so(1,2)-valued vector fields of density weight one due to the 2-dimensional (metric-independent) Levi-Civita density ϵ^{ab}.
The 'Hamiltonian' is a pure constraint. Two kinds of constraints arise. The Gauss constraint

$$\mathcal{G}_I := \mathcal{D}_a E_I^a := \partial_a E_I^a + \epsilon_{IJ}{}^K A_a^J E_K^a \tag{2.3}$$

is manifestly reminiscent of the Gauss constraint for 3+1 gravity while the other 3 constraints

$$C^I := B^I = \frac{1}{2}\epsilon^{ab}F_{ab}^I \tag{2.4}$$

tell us that that the magnetic fields (equivalently the curvature $1/2F_{ab}^I$ of A_a^I) vanish, that is, the constraint surface of the phase space contains only flat connections.
It follows from general arguments that the constraints form a first class subalgebra of the Poisson algebra of functions on the phase space since they are either linear in or independent of the momenta.
Immediately, the question arises what these 'flatness' constraints have to do with the constraints of 3+1 gravity (compare Giulini's lectures). The answer is that, provided that the twice densitized inverted 2-metric

$$E_I^a E_J^b \eta^{IJ} \tag{2.5}$$

is nondegenerate, then we can recast these constraints in an 'Ashtekar-like' form :

$$V_a := \epsilon_{ab} B^I E_I^b : \text{vector constraint and} \tag{2.6}$$
$$C := \epsilon_I{}^{JK} B^I \epsilon_{ab} E_J^a E_K^b : \text{scalar constraint.} \tag{2.7}$$

[1] a polarization is, roughly, a subdivision of a choice of phase space coordinates into momenta and configuration space variables

We can interpret this as follows :
Choose the Lagrange multipliers $\Lambda^I := -\omega_t^I$, $N_I := -e_{tI}$ such that

$$N_I = \frac{1}{2}(\underset{\sim}{N}\epsilon_I{}^{JK}E_J^aE_K^b + N^aE_I^b)\epsilon_{ab} \tag{2.8}$$

where $N := \sqrt{\det(q)}\underset{\sim}{N}$, N^a are lapse and shift functions and the first term in (2.8) is non-vanishing if and only if (2.5) is invertible. Then $N_IC^I = N^aV_a + \underset{\sim}{N}C$.
My personal point of view is that in order to test 3+1 gravity one should actually start from the constraints (2.6) and (2.7) rather than from (2.4) and allow for general, in particular non-flat, connections although the induced 2-metric on the 2-dimensional slice is then in general singular. This is because in 3+1 gravity the first task in the quantization programme is the solution of the constraints and therefore any model should mirror the algebraic form of the constraints as closely as possible. Further, one quantizes 3+1 gravity in the connection representation and thus the state functional will in general have support on non-flat connections.
There has already been done some work in that direction ([2]) : the authors of that paper show that 2 large classes of solutions to the vector and scalar constraint in this 'degenerate' sector allow for a reformulation of (2.4) such that one has the vector constraint and either 1) that the magnetic field B^I is null or 2) that the 2 internal vectors E_I^1, E_I^2 are colinear.
However, since no closed solution to 2+1 gravity for the degenerate sector is known yet, we consider it more appropriate for the present purpose to proceed with the non-degenerate (or Witten-) sector for which the connection is constrained to be flat.
There is still another, quite important, difference between 2+1 gravity and 3+1 gravity : in the 2+1 case the connection is manifestly real whereas in the 3+1 case it is genuinely complex, in general. Since the reality conditions play a major role in the process of selecting an inner product as outlined in earlier lectures (compare those of Giulini, Hajicek and Rendall) we expect that the inner products of the two theories will not be closely related to each other.
To complete the canonical formulation, we have to choose the topology of the initial data hypersurface Σ. In order to avoid technical issues that have to do with the choice of function spaces to which the fields belong (essentially, a choice of fall-off properties at infinity) we will choose a closed topology without boundary as is common in the literature ([3]). The classification of these topologies is well-known. The characterizing parameter is the genus g (number of handles) of Σ.

We will choose later the case of (g=1) (torus) since in this case the quantum theory can be constructed in closed form. We will largely follow ([4]) in the sequel.

2.2 The reduced phase space

We will quantize the present model via the reduced phase space approach, that is, we will determine the gauge invariant information of the phase space before quantizing.

Up to a small degeneracy ([5]) that arises for non-compact gauge groups, the set of traces of the holonomy

$$T^0_\alpha := \mathrm{tr}[h_\alpha[A](t)] := \mathrm{tr}[\mathcal{P}\exp(\oint_\alpha A)] \tag{2.9}$$

is a good coordinate for the reduced configuration space with respect to the gauge constraint. Here, α is any loop (i.e. an embedding of the circle into Σ), $\mathcal{P}$ means path-ordering and the trace is taken with respect to the 2-dimensional fundamental representation of SU(1,1) (thereby exploiting that the Lie algebras su(1,1) and so(1,2) are isomorphic). A suitable basis of su(1,1) is given by

$$\tau_0 := -\frac{i}{2}\sigma_3,\ \tau_1 := \frac{1}{2}\sigma_1,\ \tau_2 := \frac{1}{2}\sigma_2 \tag{2.10}$$

where σ_I are the usual Pauli matrices (the index I is 0,1,2) with respect to which the structure constants are given by $\epsilon_{IJ}{}^K$. The holonomy itself depends on the starting point $\alpha(t)$ of the loop (t ranges from 0 to 1 (to be identified) in our choice of parametrization) however the T^0's are independent of the starting point. The reason why (2.9) is SU(1,1) invariant (even under large, rather than infinitesimal SU(1,1) transformations) is that the holonomy is conjugated by a gauge transformation U which drops out under the trace.

The T^0_α are even invariant under the gauge transformations generated by the constraints C^I because C^I are independent of the momenta so that the Poisson bracket

$\{T^0_\alpha, \int_\Sigma d^2x N_I C^I\}$ vanishes. Accordingly, T^0_α is a Dirac observable by definition.

Next, we need to capture gauge invariant information about the momenta. As is suggested by the construction of the loop variables in the 3+1 case (compare Brügmann's lectures) we consider the 'smeared' version of the so-called T^1 variables :

$$T^1_\alpha := \int_0^1 ds\dot\alpha^b(s)\epsilon_{ab}\mathrm{tr}[E^a(\alpha(s))h_\alpha[A](s)]\,. \tag{2.11}$$

Since $E^a = E^a_I\tau^I$ transforms according to the adjoint representation of SU(1,1), these variables are manifestly gauge invariant. The crucial step is to check its behaviour when taking the Poisson bracket with respect to the flatness constraint. We find

$$\{T^1_\alpha, \int_\Sigma d^2x N_I B^I\} = \int_0^1 ds\dot{\alpha}^a(s)\mathrm{tr}[\tau_I h_\alpha(s)]\mathcal{D}_a N^I(\alpha(s))$$
$$= \lim_{t\to 0}\frac{1}{t}\{T^0_\alpha[A+t\mathcal{D}N] - T^0_\alpha[A]\} = 0 \quad (2.12)$$

because T^0 is gauge invariant and $A^I_a \to A^I_a + t\mathcal{D}_a N^I$ is an infinitesimal gauge transformation. Hence the T^1's are also Dirac observables. Finally, again referring to [5], we learn that up to a small degeneracy the T^0, T^1 capture all the information on the reduced phase space.

On the constraint surface, both Dirac observables have support only on flat connections. It follows that T^1_α only depends on the homotopy class of the loop α, denoted $[\alpha]$, when restricted to the constraint surface. This can be seen as follows : deform the loop α infinitesimally to get a new loop α'. It then follows from the Ambrose-Singer theorem (e.g. [6]) that $T^0_{\alpha'} - T_\alpha$ can be expanded in positive powers of coefficients of the curvature of A^I_a which vanishes on the constraint surface. Therefore, T^0_α is (weakly) nontrivial only if α is not contractible. The structure of the reduced phase space will therefore be largely governed by the choice of genus which determines the dimension of the homotopy group of Σ.

Choose a basepoint * in Σ and consider all loops starting and ending at *. The composition of loops equips this set of loops with the structure of a semigroup (it is not a group since $\alpha \circ \alpha^{-1} \neq *$, the trivial loop, and α^{-1} is the loop α traversed in opposite direction). However, $h_\alpha h_{\alpha^{-1}} = 1$. Thus, when identifying loops α, β according to the rule $\alpha \approx \beta$ iff $h_\alpha(h_\beta)^{-1} = 1$ we get a group homomorphism

$$\alpha \to h_\alpha[A] \text{ such that } h_{\alpha^{-1}} = (h_\alpha)^{-1} \quad (2.13)$$

from the so constructed group of loops modulo $\approx$ (called the hoop group) into SU(1,1) for any connection A.

Now, as derived above, the holonomy actually only depends on the homotopy class of the hoop α, that is $h_\alpha = h_{[\alpha]}$. Therefore, we obtain a homomorphism from the homotopy group of hoops (called the set of equitopic hoops in [4]) into SU(1,1).

We now make use of the following fact ([7], Barrett's theorem) : there is a bijection between (smooth, in Barret's topology) homomorphisms from the group of hoops into the gauge group under consideration and (smooth, in the usual sense of smooth functions) connections of the associated principal fibre bundle up to gauge equivalence (that means that we have only a one to one correspondence between homomorphisms and gauge equivalence classes of connections). Accordingly, we may think from now on of (the gauge equivalence class of) a given connection as given by the set of all possible smooth homomorphisms h_α from the fundamental group $\pi_1(\Sigma)$ of the hypersurface into SU(1,1).

Up to now, the discussion was valid for arbitrary genus. We now specialize to the torus. Exploiting that the homotopy group of the torus is abelian (in terms of generators and relations, for genus g the homotopy group consists of 2g generators and one relation), compare [1], we have that for any *flat* connection A the holonomies of 2 hoops commute :

$$h_\alpha h_\beta = h_{[\alpha]\circ[\beta]} = h_{[\beta]\circ[\alpha]} = h_\beta h_\alpha \tag{2.14}$$

which is only possible if for any flat A all the $h_\alpha[A]$ lie in the same abelian subgroup of SU(1,1). We conclude that the homomorphism therefore must take the form $h_\alpha = \pm\exp(a(\alpha)t^I\tau_I)$ where $a(\alpha)$ is some real number depending on the loop α and t^I is some constant (loop-independent) internal vector. The 2 possible signs of the exponential capture the fact that in SU(1,1) not every group element can be written as the exponential of an element of the Lie algebra. This is due to the fact that the group element $U = -1$ can be written in such a way only if the vector t^I is timelike : $-1 = \exp(\tau_I t^I 2\pi)$.
Let us write down, for illustrative reasons, a connection in a particular gauge such that we get the above homomorphism with the positive sign :
$A_a^I(x) = f_a(x)t^I$. The flatness condition shows that f is a closed one form. It can therefore be labelled by the cohomology class to which it belongs. The 2 generators of the first cohomology group of Σ are just the two anglar coordinates x^i, $i = 1,2$ of the torus such that up to a total differential $f_a = a_i x^i_{,a}$ and a_i are real numbers.
Explicitely we compute

$$h_\alpha[A] = \exp(t^I\tau_I a_i \oint_\alpha dx^i) \tag{2.15}$$

which under a SU(1,1) transformation $U\ :\ \Sigma \to SU(1,1)$ becomes

$$h_\alpha[A] \to U(*)h_\alpha[A]U^{-1}(*) = \exp(t^I U(*)\tau_I U^{-1}(*)a_i \oint_\alpha dx^i)\ ,$$

that is, the internal vector t^I undergoes a constant (since U(*) only depends on the basepoint *) gauge transformation thereby preserving only its timelike, spacelike or null character.
Note that we could choose the function f in $A_a^I = f_{,a}t^I$ in such a way along the loop α that it is smooth everywhere, vanishing at the parameter values 0 and 1/2 and so that t^I is a constant timelike vector on the first half of the loop and a constant spacelike vector on the second half of the loop while $\int_0^{1/2} dt\dot{f}(\alpha(t)) = 2\pi$. Then the holonomy of that connection is still of the general form $\pm\exp(a(\alpha)\tau_I t^I)$ and there are indeed connections that accomodate for both signs.
Returning to the general case, in order to characterize our homomorphism completely it is sufficient to give the image of the two generators $[\alpha_i]$ of the homotopy group $\pi_1(T^2) = Z \times Z$. This is given by either $+\exp(a_i\tau_I t^I)$ or $-\exp(a_i\tau_I t^I)$ where a_i are again two real numbers.

Let us choose standard internal vectors with norm, respectively, $\pm 1, 0$ in the spacelike, timelike or null sector respectively. In the non-spacelike case these can also be taken to be future directed while in the spacelike case this concept is not gauge invariant.
We now want to divide out by the gauge transformations generated by $U(*) \in SU(1,1)$ to obtain the physically relevant range of the 2 real numbers a_i. Given a pair a_i, t^I one can show that by a SU(1,1) gauge transformation one can get another pair b_i, s^I such that in case that t^I is
a) timelike : $b_i = a_i$ and t^I is any normalized future directed timelike vector,
b) spacelike : $b_i = \pm a_i$ and t^I is any normalized spacelike vector and
c) null : $b_i = sa_i$ and t^I is any future directed null vector.
So we see that t^I is pure gauge up to its causal characterization and its normalization and that in the timelike, spacelike and null regime repectively, the space coordinatized by the a_i is topologically $S^1 \times S^1$, $(R^1 \times R^1)/Z_2$ and S^1 respectively (to see this in the timelike case, observe that the holonomy is given then up to a sign in terms of sines and cosines).
Another way to see this is by explicit calculation of the exponential part of the homomorphism

$$\exp(a\tau_I t^I) = \left\{ \begin{array}{rcl} \cosh(a/2)1 + t^I\tau_I\sinh(a/2) & : & \text{spacelike sector} \\ \cos(a/2)1 + t^I\tau_I\sin(a/2) & : & \text{timelike sector} \\ 1 + t^I\tau_I a & : & \text{null sector} \end{array} \right\} \quad (2.16)$$

(where we may think of a as $a = a_i \oint_\alpha dx^i$ in the above mentioned gauge). Taking the trace we infer that the gauge invariant information captured by the exponential part of the holonomy is given by, for the spacelike sector in $a_i \in R^2/Z_2$, for the timelike sector in $a_i \in T^2$ and for the null sector only the angle between a_1, a_2 is gauge invariant since t^I gets not only rotated but also scaled by a positive scale factor so that the reduced configuration space has then the topology of the circle S^1. Note that T^0 does not capture this piece of information in the null regime ([5]). We are not interested in the null regime in the sequel.
Now, let us invoke the piece of information that is captured by the sign of the homomorphism in the non timelike case. Obviously, 4 distinct assignments of signs to the 2 generators are possible, so we get 4 distinct copies of either R^2/Z_2 or S^1. These 4 copies are disconnected since there is no continous way to get, say, a boost of type (+,+) from a boost of type (-,+). Finally, we should observe that the timelike sector is connected to the two other causal sectors at the zero connection. Hence this special point does not correspond to either of these sectors since the causal nature of the zero connection is degenerate. So one should discard this point from either of the three causal sectors to get finally the topology of the reduced configuration space as :

timelike : $T^2 - 0$, spacelike : $(R^2/Z_2 - 0) \times Z_4 \approx (R^2 - 0) \times Z_4$, null : $(S^1 - 0) \times Z_4$ and
zero : 0. Thus, the reduced phase space is just the cotangent bundle over the disjoint union of these sectors of the reduced configuration space.
An interesting question is now which of these sectors corresponds to geometrodynamics, that is, such that the intrinsic 2-metric is nondegenerate and has euclidean signature.
Here it should be stressed that the Witten formulation of 2+1 gravity differs tremendously from the geometrodynamical formulation : For instance, Witten shows ([1]) that we can write the action (2.2) as the action of a ISO(1,2) or ISU(1,1) Chern-Simons field theory whose connection $\omega_I = A^I_a \tau_I + e^I_a T_I$ (T_I are the generators of the translation subgroup of the Poincaré subgroup) is flat on shell. That means locally we can always gauge ω and in particular the triad to zero, corresponding to a vanishing (!) metric. In other words, the Witten constraints generate more general gauge transformations than only spacetime diffeomorphisms and issues like degeneracy and signature of the induced hypersurface become gauge dependent.
At this point the paper by Mess ([8]) should be mentioned. Mess shows that if one starts from the geometrodynamical formulation for the torus, i.e. a nondegenerate metric on a spacelike hypersurface, then the holonomies are either boosts or the identity map, which means that the associated connection is either spacelike or zero. The interested reader is referred to that paper and also to the papers by Louko and Marolf ([9]).

2.3 Quantization

As outlined in reference [9], the quantization for the timelike sector is much easier than for the spacelike and null sector. In fact, it turns out that the kernel of the naively defined loop transform (compare Brügmann's lectures) for the spacelike sector is dense in the Hilbert space that is defined by the loop algebra. Although the authors of that paper suggest a way how to define a loop transform for the spacelike sector such that it becomes an isomorphism between the loop representation and the connection representation[2], the techniques involved would veil the main ideas of the formalism and so we stick to the timelike sector. This, again, does not simulate the 3+1 case because there the T^0's are unbounded.

Let us start with the construction of the loop representation. The idea of the loop representation is to use a *non-canonical* subalgebra of the Poisson-

[2] essentially, these authors show that there is a dense subspace of the connection representation w.r.t. which the naively defined loop transform (compare also (2.21)) is faithful; they then Cauchy complete this image w.r.t. the inner product that coincides with that for the connection representation in the pre-image; the two representations are thus unitarily equivalent by construction, for still another procedure refer to [10]

algebra as the basic set of classical variables that are to be quantized by demanding that the commutation relations among these variables (which in case of the loop representation are just the T^0, T^1's) mirror those of the classical analogues. By doing that, one gets rid of the connections and the electric fields. They should, however, be reconstructed by means of the so-called loop transform (refer, for example, to [11] for more details). Let us now make these ideas concrete.
First, we need to compute the classical Poisson algebra among the T^0, T^1 induced by that for E^a_I, A^I_a. We obtain

$$\begin{aligned}
\{T^0_\alpha, T^0_\beta\} &= 0 \\
\{T^0_\alpha, T^1_\beta\} &= \sum_i \Delta_i(\alpha,\beta)[T^0_{\alpha\circ_i\beta} - T^0_{\alpha\circ_i\beta^{-1}}] \\
\{T^1_\alpha, T^1_\beta\} &= \sum_i \Delta_i(\alpha,\beta)[T^1_{\alpha\circ_i\beta} - T^1_{\alpha\circ_i\beta^{-1}}]
\end{aligned} \tag{2.17}$$

where i labels points of intersection of the two loops involved, $\circ_i$ denotes composition at the intersection point and we have defined the quantity Δ_i which takes values in $\{\pm 1\}$ by

$$\sum_i \Delta_i(\alpha,\beta) := \int_{\alpha\times\beta} dx \wedge dy \delta^{(2)}(x,y) \,. \tag{2.18}$$

Note also that the T variables are classically real.
We now quantize this algebra by demanding that the commutators among the T's produce $i\hbar$ times the right hand side of the Poisson brackets and by imposing the following *-relations with respect to an abstract involution * : $(\hat{T}^0)^* = \hat{T}^0$, $(\hat{T}^1)^* = \hat{T}^1$. We choose the loop representation, that is, we represent the T operators on a complex vector space of functions f of loops. Secretely, one should think of such a function as arising from a function $\tilde{f}$ of connections via a heuristic loop transform

$$f(\{\beta\}) = \int_{\mathcal{A}/\mathcal{G}} d\mu[A] \prod_{\beta\in\{\beta\}} T^0_\beta[A]\tilde{f}(A) \tag{2.19}$$

where $\{\beta\}$ is a set of single loops β and the integration domain is the moduli space $\mathcal{A}/\mathcal{G}$ of (flat) connections modulo gauge transformations.
The analysis is now simplified by the fact that for any subgroup of SL(2,C) that is generated by some real form of sl(2,C) there holds the SL(2,C) Mandelstam identity

$$T^0_{\alpha\circ\beta} + T^0_{\alpha\circ\beta^{-1}} = T^0_\alpha \hat{T}^0_\beta \tag{2.20}$$

which implies that any analytic function of traces of holonomies can be written as a linear combination of traces of holonomies for single loops. We therefore need only to define the action of our loop operators on functions of single loops :

$$f(\alpha) = \int_{\mathcal{A}/\mathcal{G}} d\mu[A] T^0_\alpha[A]\tilde{f}(A) \,. \tag{2.21}$$

It turns out that one can implement the algebra (2.17) then as follows

$$(\hat{T}^0_\alpha f)(\beta) := f(\alpha \circ \beta) + f(\alpha \circ \beta^{-1})$$
$$(\hat{T}^1_\alpha f)(\beta) := i\hbar \sum_i \Delta_i(\alpha,\beta)[f(\alpha \circ_i \beta) - f(\alpha \circ_i \beta^{-1})] \tag{2.22}$$

where the composition $\circ$ is at the basepoint.

In order to make sure that the T observables 'come from a connection', we have to impose further the following identities on our representation space

$$\begin{aligned} i) \quad & \hat{T}^1_\alpha = \hat{T}^1_{\alpha^{-1}},\ \hat{T}^1_{\alpha\circ\beta} = \hat{T}^1_{\beta\circ\alpha} \\ ii) \quad & \hat{T}^0_{\alpha\circ\beta} + \hat{T}^0_{\alpha\circ\beta^{-1}} = \hat{T}^0_\alpha \hat{T}^0_\beta \\ iii) \quad & \hat{T}^0_* = 2,\ \hat{T}^1_* = 0\,. \end{aligned} \tag{2.23}$$

It is easy to check that the analogue of i) for T^0 follows from the SL(2,C) Mandelstam identity ii).

Thus we restrict the representation space further such that the relations i)-iii) hold. One can check that it is enough to demand that (a_i are real quantities)

$$\sum_i a_i f(\alpha_i \circ \beta) = 0 \text{ whenever } \sum_i a_i T^0_{\alpha_i} = 0. \tag{2.24}$$

This, again, follows trivially from the existence of a loop transform (2.19). Next, we restrict the loop algebra and its representation space to depend only on homotopy classes. For the torus this amounts to the fact that T-operators and the functions of our complex vector space depend only on the winding numbers $n_1, n_2 \in Z$ of the 2 generators of the abelian fundamental group of Σ. If we denote by $[\alpha_i]$, $i = 1,2$ the generators of $\pi_1(\Sigma)$ then the action (2.22) becomes

$$(\hat{T}^0_{\alpha_1} f)(n_1,n_2) = f(n_1+1,n_2) + f(n_1-1,n_2)$$
$$(\hat{T}^1_{\alpha_1} f)(n_1,n_2) = i\hbar n_2[f(n_1+1,n_2) - f(n_1-1,n_2)] \tag{2.25}$$

where we have made use of the relation

$$f(n_1,n_2) = f(-n_1,-n_2),\ (n_1,n_2) \sim (n_1',n_2') \Leftrightarrow (n_1',n_2') = (-n_1,-n_2) \tag{2.26}$$

which follows from (2.23). The action of the T operators for the other homotopy class is analogous. The reason for the factor of n_2 in the second line of (2.25) is that any representative of $[\alpha_1]$ will intersect any representative of $[\alpha_1]^{n_1}[\alpha_2]^{n_2}$ precisely $|n_2|$ times with the orientation of this intersection captured by the sign of n_2.

The final task is now to find an inner product such that our basic operators become self-adjoint. The obvious choice is

$$< f,g > := \sum_{Z^2/\sim} \bar{f}(n_1,n_2) g(n_1,n_2) \tag{2.27}$$

and obviously accomplishes our aim.

2.4 Loop transform

In order to explicitely construct the loop transform of the present model, we need also the connection representation.
Physical states depend only on the moduli space of flat connections modulo gauge transformations, labelled by the two parameters $a_1, a_2 \in [-2\pi, 2\pi]$. We realize our basic operators in the connection representation as follows

$$\begin{aligned} (\hat{T}^0_{\alpha_i}\tilde{f})(a_1,a_2) &:= 2\cos(a_i/2)\tilde{f}(a_1,a_2) \\ (\hat{T}^1_{\alpha_i}\tilde{f})(a_1,a_2) &:= -4i\hbar\sin(a_i/2)\frac{\partial\tilde{f}}{\partial a_j}(a_1,a_2)\ ,i\neq j \end{aligned} \tag{2.28}$$

and it is straightforward to check that the commutation relations (2.17) are satisfied.
An inner product that makes these operators self-adjoint can be constructed as follows : we make the ansatz

$$(\tilde{f},\tilde{g}) := \int_{[-2\pi,2\pi]^2} da_1 \wedge da_2 \rho(a_1,a_2)\bar{\tilde{f}}(a_1,a_2)\tilde{g}(a_1,a_2) \tag{2.29}$$

and try to find a weight factor ρ such that the T operators are self-adjoint. We find up to a multiplicative positive constant $\rho = 1$.
Alternatively we may apply the following procedure which is appropriate whenever one is confronted with a non-canonical algebra of basic operators (the interested reader is referred to [13]) :
Choose some Riemannian background metric on the torus and choose physical states as densities of weight 1/2. Define the vector fields

$$v_i := 4\sin(a_i/2)\frac{\partial}{\partial a_j} \tag{2.30}$$

and define the action of the T observables as in (2.28) on half-densities except that $\hat{T}^1_{\alpha_i}\tilde{f} := -i\hbar\mathcal{L}_{v_i}\tilde{f}$ where $\mathcal{L}$ denotes the Lie derivative.
It is then easy to check that the scalar product (2.29) with $\rho = 1$ is well-defined (that is, frame independent) and that all operators are self-adjoint with respect to it.
We are now finally in the position to look at the loop transform. Taking the Lebesgue measure on the torus as the measure μ in (2.21) we obtain

$$f(n_1,n_2) = \int_{[-2\pi,2\pi]^2} d^2a\cos([n_1a_1 + n_2a_2]/2)\tilde{f}(a_1,a_2) \tag{2.31}$$

and one can explicitely check that any of the elementary operators $\hat{O}$ in (2.25) for the loop representation arises from the operators $\hat{\tilde{O}}$ in (2.28) for the connection representation via the loop transform, that is

$$(\hat{O}f)(\alpha) = \int_{A/\mathcal{G}} d\mu(A)T^0_\alpha(A)(\hat{\tilde{O}}\tilde{f})(A)\ . \tag{2.32}$$

Note that the space of functions which are either odd or even under reflection of the a_i is left invariant under the action of the T operators, so the representation space splits into 2 irreducible representations. It follows that the loop transformation (2.31) is a representation isomorphism for the even sector and has the odd sector as its kernel. Since for the even sector (2.31) is just the Fourier transform, it follows that the Hilbert spaces that we have constructed are unitarily equivalent, that is $< f, g >= (\tilde{f}, \tilde{g})$.

2.5 Discussion

The model of 2+1 pure gravity is a dimensional reduction of 3+1 gravity. It can be cast into a form such that the algebraic structure of the constraints is very similar to that of the full theory. The gauge group, SU(1,1), is not compact and therefore simulates the fact that the traced holonomy for the complex Ashtekar connection in 3+1 gravity is an unbounded function (in the topology of complex numbers). Both, the connection representation and the loop representation, together with the loop transform can be constructed in closed form. This gives one some confidence that a loop representation of the 3+1 theory can be constructed as well.
These were the positive remarks. However, sharp criticism is in order :
The 2+1 theory is manifestly real whereas the 3+1 case is genuinely complex. Since it is fair to say that some of the major difficulties of the physical theory rest in the non-trivial reality structure, one should not expect that the constructions that were used in the model are available in the 3+1 case. This has nothing to do with the fact that we only quantized the sector in which the T^0's are bounded because, as mentioned, one *can* take care of the spacelike sector. Rather what we point out here is that a genuinely complex connection gives rise to additional difficulties, for example the T^0, T^1 Poisson algebra is not even a *-algebra in this case.
Next, in the above analysis we restricted ourselves to the Witten sector. Thus, we did not use the form of the constraints that actually equals that of 3+1 gravity. The connection of 3+1 gravity is not flat in general, the theory is not topological and, most seriously, maybe the major difficulty of the full theory is that the scalar constraint is *quadratic* in the momenta. However, we used only constraints that are linear in or independent of the momenta. Accordingly, 2+1 gravity neither accounts for this issue.
Last but not least, the fact that the reduced phase space is finite dimensional is a severe deviation from the 3+1 case because all the divergencies of quantum field theory occur due to the infinite number of degrees of freedom. We were thus unable to deal with this technical problem.

3 Spherically symmetric gravity

3.1 Canonical formulation

The model of spherically symmetric gravity arises from a Killing reduction of 3+1 gravity. We will follow largely reference [14] in the sequel.
We first summarize the Killing reduction first given in [15] [3].
One imposes spherical symmetry on the geometrodynamical phase space by requiring that the induced metric on the Cauchy hypersurface Σ (foliated by SO(3) orbits) and its extrinsic curvature be Lie-annihilated by three Killing vector fields K_i which form an SO(3) subalgebra of the Lie algebra of all vector fields on Σ. In case of the Ashtekar phase space, one has to look for analogous conditions for the triad and the Ashtekar connection. Let us start with the triad. We have

$$\mathcal{L}_{K_I} q_{ab} = 2\delta_{ij} e^i_{(a} \mathcal{L}_{K_I} e^j_{b)} = 0$$

and the most general solution of this equation is given by

$$\mathcal{L}_{K_I} e^i_a = \epsilon^i{}_{Ik} e^k_a \, , \tag{3.1}$$

that is, a rotation of the triad in the tangent space in either of the three Killing directions is compensated by a corresponding rotation in the internal space.
The idea is now as follows : obtain the general solution of (3.1). This is then the general form of a 1-form that transforms according to the defining representation of SO(3). Since the defining and the adjoint representation of SO(3) are isomorphic, the solution of (3.1) is also the general form of an so(3)-valued 1-form. Finally, since the three infinitesimal internal rotations parametrized by $\Lambda^{(I}_i = \delta^I_i$ involved in (3.1) are global ones, the inhomogenous term in the transformation law of the (pull-back to the base manifold of a) connection $\omega^i_a \to -\partial_a \Lambda^i + \epsilon_{ijk}\Lambda^j \omega^k_a$ drops out so that the solution to (3.1) also gives us the general form of a SO(3)-connection. To obtain then the general form of a SO(3) valued vector or vector density, one simply contracts the solution to (3.1) with q^{ab} times an appropriate power of $\sqrt{\det(q)}$ since the three-metric is Lie-derived by the Killing-fields. The form of q_{ab} is already known from the solution to (3.1).
We will shortly sketch the calculations. Fix a local frame (r, θ, ϕ) on Σ where r is a local radial coordinate and θ, ϕ are global angular coordinates. We can then write the rotational Killing fields as follows (compare any textbook in quantum mechanics) :

$$\begin{aligned} -K_1 &= -\sin(\phi)\partial_\theta - \cot(\theta)\cos(\phi)\partial_\phi \\ -K_2 &= \cos(\phi)\partial_\theta - \cot(\theta)\sin(\phi)\partial_\phi \\ -K_3 &= \partial_\phi \end{aligned}$$

[3]In order to make contact with ref. [15], one has to exchange the labels I=2 and I=3 and to replace $A_3 + \sqrt{2}$ there by A_3 in order to get A_3 here

and plugging these expressions into (3.1) we obtain a system of partial differential equations of first order. One solves it for the partial derivatives of e^i_a wrt θ, ϕ and after some algebraic manipulations one sees that one can decouple the equations. The final, unique solution is then

$$(e^i_r, e^i_\theta, e^i_\phi) = (fn^i_r, gn^i_\theta + hn^i_\phi, gn^i_\phi - hn^i_\theta)$$

where n^i_a is the standard orthonormal base in internal space (that is $n^i_r = (\sin(\theta)\cos(\phi), \sin(\theta)\sin(\phi), \cos(\theta))$, $n^i_\theta = \partial_\theta n^i_r$, $\sin(\theta) n^i_\phi = \partial_\phi n^i_r$) and f, g and h are arbitrary real functions of the radial (r) and time (t) variable only. From this we conclude, as expected, the general form of the 3-metric

$$q_{ab} = q_{rr}(r,t) r_{,a} r_{,b} + q_{\theta\theta}(r,t) h_{ab} \tag{3.2}$$

where h_{ab} is the standard line element on the 2-sphere.
Inverting (3.2), computing $E^a_i = \sqrt{\det(q)} q^{ab} e^i_b$ we see that we can parametrize the general form of the conjugate variables of the Ashtekar phase space as follows (the numerical factors are chosen for convenience)

$$(E^r_i, E^\theta_i, E^\phi_i) = (E^1 n^i_r \sin(\theta), \frac{\sin(\theta)}{\sqrt{2}}(E^2 n^i_\theta + E^3 n^i_\phi), \frac{1}{\sqrt{2}}(E^2 n^i_\phi - E^3 n^i_\theta)) \tag{3.3}$$

$$(A^i_r, A^i_\theta, A^i_\phi) = (A_1 n^i_r, \frac{1}{\sqrt{2}}(A_2 n^i_\theta + (A_3 - \sqrt{2}) n^i_\phi), \tag{3.4}$$
$$\frac{\sin(\theta)}{\sqrt{2}}(A_2 n^i_\phi - (A_3 - \sqrt{2}) n^i_\theta)) \, .$$

The arbitrary complex functions $E^I = E^I(t,r)$, $A_I = A_I(t,r); I = 1,2,3$ depend on t and r only. One now simply inserts this into Ashtekar's action of full gravity (compare Giulini's lectures), integrates out the angles (in particular the factor $\sin\theta$ contained in $\underset{\sim}{N}$ drops out) and finally ends up with the following action

$$S = \frac{4\pi}{\kappa}\int_R dt (\int_\Sigma dr[-i\dot{A}_I E^I - [i\Lambda\mathcal{G} - iN^r V + \underset{\sim}{N} C]] + b) \, . \tag{3.5}$$

The explanation of the various quantities involved in (3.5) is as follows :
All quantities only depend on r,t (expanding all quantities in terms of spherical functions, only the angle-independent terms survive after integrating over S^2). The Lagrange multipliers and constraint functions are related to those of the full theory by $\Lambda = \Lambda^i (n_r)^i$, $\sin(\theta)\mathcal{G} = \mathcal{G}_i (n_x)^i$, $N^r = N^a r_{,a}$, $\sin(\theta) V = V_a r_{,a}$, $\underset{\sim}{N} = \sin(\theta) \underset{\sim}{N}_{full}$, $2\sin^2(\theta) C = C_{full}$; $\kappa/8\pi$ is Newtons constant.
The constraint functionals take the following form

$$\mathcal{G} = (E^1)' + A_2 E^3 - A_3 E^2 \text{ : Gauss constraint,} \tag{3.6}$$
$$V = B^2 E^3 - B^3 E^2 \text{ : Vector constraint,} \tag{3.7}$$
$$C = \frac{1}{2}(E^2(2B^2 E^1 + B^1 E^2) + E^3(2B^3 E^1 + B^1 E^3)) \text{ : Scalar constraint} \tag{3.8}$$

Here we have used the reduction to spherical symmetry of the magnetic fields $B_i^a = 1/2\epsilon^{abc}F_{ab}^i$ (where F denotes the field strength of the Ashtekar connection)

$$(B_i^r, B_i^\theta, B_i^\phi) = (B^1 n_r^i \sin(\theta), \frac{\sin(\theta)}{\sqrt{2}}(B^2 n_\theta^i + B^3 n_\phi^i), \frac{1}{\sqrt{2}}(B^2 n_\phi^i - B^3 n_\theta^i)) \tag{3.9}$$

and one can check that

$$(B^1, B^2, B^3) = (\frac{1}{2}((A_2)^2 + (A_3)^2), (A_3)' + A_1 A_2, -(A_2)' + A_1 A_3) \tag{3.10}$$

where a prime denotes differentiation with respect to r.
Since we are interested in asymptotically flat topologies in contrast to the first model, there is also a boundary term b involved in the action which is to make the action functionally differentiable. It reads

$$b = \int_{\partial\Sigma} iN^r(A_2E^2 + A_3E^3) + \underset{\sim}{N}(A_2E^3 - (A_3 - \sqrt{2})E^2)E^1 \tag{3.11}$$

and can be recognized as the sum of ADM momentum and energy. It is also obvious from the action (3.5) that A_I, E^I form a canonical pair. With this boundary term we are able to derive the following equations of motion

$$\begin{aligned}
\frac{d}{dt}A_1 &= i[(-i\Lambda' + \underset{\sim}{N}(B^2E^2 + B^3E^3)], \qquad (3.12)\\
\frac{d}{dt}A_2 &= i[(-i\Lambda A_3 + iN^rB^3 + \underset{\sim}{N}(B^2E^1 + B^1E^2)],\\
\frac{d}{dt}A_3 &= i[(+i\Lambda A_2 - iN^rB^2 + \underset{\sim}{N}(B^3E^1 + B^1E^3)],\\
\frac{d}{dt}E^1 &= -i[-iN^r(A_2E^3 - A_3E^2) + \underset{\sim}{N}(A_2E^2 + A_3E^3)E^1],\\
\frac{d}{dt}E^2 &= -i[+i(\Lambda - N^rA_1)E^3 + i(N^rE^2)' + \underset{\sim}{N}(A_1E^1E^2 + \frac{1}{2}EA_2)\\
&\quad +(\underset{\sim}{N}E^1E^3)'],\\
\frac{d}{dt}E^3 &= -i[-i(\Lambda - N^rA_1)E^2 + i(N^rE^3)' + \underset{\sim}{N}(A_1E^1E^3 + \frac{1}{2}EA_3)\\
&\quad -(\underset{\sim}{N}E^1E^2)'] .
\end{aligned}$$

We can also display the classical canonical constraint algebra (we abbreviate $f \circ g := \int_\Sigma dr f(r)g(r)$, $\xi := V - A_1\mathcal{G})$:

$$\begin{aligned}
\{\Lambda_1 \circ \mathcal{G}, \Lambda_2 \circ \mathcal{G}\} &= 0, \qquad (3.13)\\
\{N \circ \xi, \Lambda \circ \mathcal{G}\} &= -iN\Lambda' \circ \mathcal{G},\\
\{\underset{\sim}{N} \circ H, \Lambda \circ \mathcal{G}\} &= 0,\\
\{M \circ \xi, N \circ \xi\} &= i(MN' - M'N) \circ \xi,\\
\{M \circ \xi, \underset{\sim}{N} \circ H\} &= i(M\underset{\sim}{N}' - M'\underset{\sim}{N}) \circ H,\\
\{\underset{\sim}{M} \circ H, \underset{\sim}{N} \circ H\} &= i(\underset{\sim}{M}\underset{\sim}{N}' - \underset{\sim}{M}'\underset{\sim}{N}) \circ (E^1)^2 H_x
\end{aligned}$$

and it is obvious that the model is still first class (recall Wipf's lectures). The set of equations (3.12) and (3.13) allow for the following interpretation :

The diffeomorphisms have been frozen to the r-direction, the internal rotations to the n_r-direction. A_1 plays the role of an $O(2)$ gauge potential, E^1 is $O(2)$-invariant while the vectors $(E^2, E^3), (A_2, A_3)$ transform according to the defining representation of $O(2)$. A_1, E^2, E^3 are densities of weight one in one dimension, while E^1, A_2, A_3 are scalars.

For the discussion of the reality conditions the reduction to spherical symmetry of the spin connection is needed:

$$
\begin{aligned}
(\Gamma_r^i, \Gamma_\theta^i, \Gamma_\phi^i) &= (\Gamma_1 n_r^i, \sqrt{\frac{1}{2}}(\Gamma_2 n_\theta^i + (\Gamma_3 - \sqrt{2}) n_\phi^i), \\
&\qquad \sqrt{\frac{1}{2}}(\Gamma_2 n_\phi^i - (\Gamma_3 - \sqrt{2}) n_\theta^i) \sin(\theta)), \qquad (3.14)\\
(\Gamma_1, \Gamma_2, \Gamma_3) &= (-\beta', -(E^1)'\frac{E^3}{E}, (E^1)'\frac{E^2}{E}), \quad \beta' = \frac{E^2 E^{3\prime} - E^3 E^{2\prime}}{E} \qquad (3.15)
\end{aligned}
$$

and we have introduced the following quantities :

$$
\alpha := \arctan(\frac{A_3}{A_2}), \quad \beta := \arctan(\frac{E^3}{E^2}), \quad A := (A_2)^2 + (A_3)^2, \quad E := (E^2)^2 + (E^3)^2. \tag{3.16}
$$

The metric is given, in these variables, by

$$
q_{ab} = \frac{E}{2E^1} r_{,a} r_{,b} + E^1 h_{ab} \,. \tag{3.17}
$$

Then the reality conditions become simply

$$
E^I = \text{real and } A_I - \Gamma_I = \text{imaginary.} \tag{3.18}
$$

To complete the Hamiltonian formulation we have to agree on the topology of the 1-dimensional hypersurface Σ as well as on the fall-off properties of the fields at spatial infinity.

First note that the 3-dimensional hypersurface is related to our 1-dimensional one by $\Sigma^{(3)} = \Sigma \times S^2$. Now we can choose Σ either to be closed or open. That is, we choose its topology to be either S^1, the circle, or R^1, the real line. In the first case we are dealing with a compactified wormwhole, in the second with a black whole with two asymptotic regions. One can, at the price of inventing additional assumptions, generalize to more than 2 asymptotic ends ([14]) but we refrain from doing that here for the sake of brevity.

In the asymptotically flat case we also have to deal with boundary conditions on the fields.

First, we choose the hypersurface label $r \in [-\infty, \infty]$ to become asymptotically $r^2 = (x^1)^2 + (x^2)^2 + (x^3)^2$ with respect to an asymptotical cartesian

frame $\{x^a\}$ which means that it is appropriate to describe fall-off properties in powers of r.
Next, following reference [16], we adopt the requirements of that paper to our situation. The guiding principle for choosing fall-off properties are :
1) finiteness of the symplectic structure,
2) finiteness and functional differentiability of the constraint functionals.
Requirement 2) further depends on the set of asymptotic symmetries that one is willing to allow.
In [16] (which is based on the old (ADM) variables) these requirements 1) and 2) including asymtotical Poincare transformations can be satisfied as follows :

$$\begin{aligned} q_{ab} &\rightarrow \delta_{ab} + \frac{f_{ab}(x^c/r,t)}{r} + O(1/r^2) \\ p^{ab} &\rightarrow \frac{k^{ab}(x^c/r,t)}{r^2} + O(1/r^3) \end{aligned} \qquad (3.19)$$

as $r \rightarrow \infty$. Furthermore, it must be required that the functions f_{ab} and k^{ab} respectively are even and odd respectively under reflections of the asymptotically flat frame.
It is clear that for spherical symmetry we are not able to impose the above parity conditions because the reduction to spherical symmetry excludes all modes of the fields (regarded as expanded into spherical harmonics) which have angular momentum different from zero. Hence we have to modify the strategy slightly.
Comparing the spherically symmetric metric (3.17) with the Euclidean metric in spherical coordinates

$$\delta_{ab} = r_{,a}r_{,b} + r^2 h_{ab} \; , \qquad (3.20)$$

we conclude the following fall-off properties :

$$\begin{aligned} &(E^1, E^2, E^3) \rightarrow (r^2[1 + \frac{f^1(t)}{r} + O(1/r^2)], \\ &\sqrt{2}r[\bar{E}^2 + \frac{f^2(t)}{r} + O(1/r^2)], \sqrt{2}r[\bar{E}^3 + \frac{f^2(t)}{r} + O(1/r^2)]) \end{aligned} \qquad (3.21)$$

whereby $(\bar{E}^2)^2 + (\bar{E}^3)^2 = 1$. Inserting this into the formula $A_a^i = \Gamma_a^i + iK_a^i$ (recall Giulini's lecture), using (3.15), (3.19) and (3.3) one concludes that

$$(A_1, A_2, A_3 - \sqrt{2}) \rightarrow (\frac{a_1(t)}{r^2} + O(1/r^3), \frac{a_2(t)}{r} + O(1/x^2), \frac{a_3(t)}{r} + O(1/r^2)) \qquad (3.22)$$

Since, as we noted before, there is no parity freedom left, the requirements 1) and 2) discussed above will not be satisfied yet. Let us explore what further restrictions are there to be imposed.

The symplectic structure on the large phase space can be read off from the action (we can drop the prefactor of the action for the case of pure gravity) :

$$\Omega = \int_\Sigma dr[-idE^I \wedge dA_I]$$
$$= \int_\Sigma dr(\frac{-i}{r}[da_1 \wedge df^1 + \sqrt{2}(da_2 \wedge df^2 + da_3 \wedge df^3)] + O(1/r^2))(3.23)$$

Hence we can satisfy requirement 1) by restricting the variations to be such that

$$da_1 \wedge df^1 + \sqrt{2}(da_2 \wedge df^2 + da_3 \wedge df^3) = 0. \qquad (3.24)$$

As for requirement 2) we first have to agree on the set of allowed symmetries at infinity. We want to incorporate only asymptotic translations. Why do we not consider asymptotic boosts of the 2-dimensional flat structure (rotations do not exist in 1 dimension anyway) ? In the literature, one looks at Schwarzschild-solutions in arbitrarily boosted frames (see ref. [16], for example). However, these boosts are really boosts with respect to the 4-dimensional spacetime which violate spherical symmetry of the initial data. The 'boosts' that we were able to discuss here must be meant with respect to the effective 2-dimensional spacetime coordinatized by the variables r and t in order not to violate spherical symmetry, they are thus not physical anyway. But since we do not have this parity freedom at our disposal our 'boost' generator diverges. So we would have to impose much more restrictive fall-off conditions than above which, in particular, would exclude Schwarzschild configurations and for that reason we refrain from doing so. The same is actually also true for asymptotic spatial translations : only radial translations preserve the spherical symmetry of the fields, that is, translations of the form $x^a \to x^a + cx^a/r$ where c is a constant but these are then position-dependent (on the sphere) and do not correspond to the translation subgroup of the Poincaré group. However, we will keep them for completeness sake.
Obviously, we have then *for symmetry transformations* the following fall-off behaviour of the Lagrange multipliers :

$$(\Lambda, N^x, \underset{\sim}{N}) \to (\frac{\text{const.}}{r^2} + O(1/r^3), \text{const.} + O(1/r), \frac{\text{const.}}{r^2} + O(1/r^3)) \quad (3.25)$$

while *for gauge transformations* we require, for simplicity, that the Lagrange multipliers are of compact support.
We now compute the leading order behaviour of the integrands of the constraint functionals :

$${}^E\mathcal{G} \to 2r(1 - \bar{E}^2) + f^1 + \sqrt{2}(a_2\bar{E}^3 - \sqrt{2}f^2 - \bar{E}^3 a_3) + O(1/x) \qquad (3.26)$$

which becomes a finite and differentiable functional when imposing $\bar{E}^2 = 1$ i.e. $\bar{E}^3 = 0$.

Note that weakly (i.e. on the constraint surface) we have from the Gauss constraint at infinity

$$f^1 - 2f^2 - \sqrt{2}a_3 = 0 \,. \tag{3.27}$$

It is convenient first to compute the asymptotic form of the magnetic fields

$$\begin{aligned} B^1 &\rightarrow -\frac{\sqrt{2}a_3}{r} + O(1/r^2), \\ B^2 &\rightarrow -\frac{a_3}{r^2} + O(1/r^3), \\ B^3 &\rightarrow \frac{a_2 + \sqrt{2}a_1}{r^2} + O(1/r^3) \end{aligned} \tag{3.28}$$

to conclude for the vector constraint

$$V \rightarrow \sqrt{2}\frac{a_2 + \sqrt{2}a_1}{r} + O(1/r^2) \,. \tag{3.29}$$

Hence we have to impose

$$a_2 + \sqrt{2}a_1 = 0 \tag{3.30}$$

in order to make this functional finite and differentiablility can be achieved by adding the ADM-momentum. Finally, it is easy to see that with this restriction the scalar constraint functional is already finite and functionally differentiable when adding the ADM-energy.
Now it is possible to make the restriction that comes from requirement 1) more concrete. We have

$$\begin{aligned} 0 &= da_1 \wedge df^1 + \sqrt{2}(da_2 \wedge df^2 + da_3 \wedge df^3) \\ &= -\frac{1}{\sqrt{2}}da_2 \wedge d(f^1 - 2f^2) + \sqrt{2}da_3 \wedge df^3) \\ &= -\frac{1}{\sqrt{2}}da_2 \wedge d(f^1 - 2f^2 - \sqrt{2}a_3) + da_3 \wedge d(\sqrt{2}f^3 + a_2) \,. \end{aligned} \tag{3.31}$$

Note that the bracket of the 1st wedge product in the last line of (3.31) vanishes weakly according to (3.27). Hence it is consistent with the constraint equations to impose

$$\sqrt{2}f^3 + a_2 = 0 \,. \tag{3.32}$$

This completes the analysis of the boundary conditions.
It is clear that the present model bears a strong resemblance to the full theory : the algebraic structure of the constraints is very similar, the reality conditions are non-trivial, the constraints are quadratic in the momenta.

3.2 Symplectic reduction

Let us first recall some basic facts from the theory of symplectic reduction (for an extensive treatment, see ref. [17] and [18]).

We showed in the previous section that the present model is a field theory with first class constraints. Let $\Gamma, \bar{\Gamma}$ and $\hat{\Gamma}$ denote the full phase space, its constraint surface (where the constraints are identically satisfied) and its reduced phase space (i.e. the constraint surface, but points in it are identified provided they are gauge related). The (local) existence of the latter follows from general theorems that are valid for first class systems. Let

$$\iota \, : \, \bar{\Gamma} \to \Gamma \text{ and } \pi \, : \, \bar{\Gamma} \to \hat{\Gamma} \tag{3.33}$$

denote the (local) imbedding and projection into the large phase space and onto the reduced phase space respectively. Call the symplectic structures on the 2 phase spaces Ω and $\hat{\Omega}$ respectively. Then the presymplectic structure on $\bar{\Gamma}$ is defined by the pull-backs

$$\pi^*\hat{\Omega} := \bar{\Omega} := \iota^*\Omega \, . \tag{3.34}$$

(More precisely, in practise one computes the constraint surface and thus obtains the imbedding. One then defines the presymplectic structure by the pull-back under the imbedding. After that one computes the gauge orbits and obtains the projection. The reduced symplectic structure is then defined by the pull-back under the projection).
On the other hand, if Θ and $\hat{\Theta}$ are the symplectic potentials for the symplectic structures, we obtain

$$d \wedge (\iota^*\Theta - \pi^*\hat{\Theta}) = \iota^*\Omega - \pi^*\hat{\Omega} = 0 \tag{3.35}$$

whence

$$dS := \iota^*\Theta - \pi^*\hat{\Theta} \tag{3.36}$$

is (locally) is exact. S is the Hamilton-Jacobi functional, it is the generator of a singular canonical transformation from the large to the reduced phase space. Substituting the momenta on Γ by the the functional derivatives of S with respect to the coordinates on Γ solves the constraints because by doing this substitution one pulled back the momenta to $\bar{\Gamma}$. Hence, one way of obtaining the reduced phase space is to solve the Hamilton-Jacobi equation for constrained systems.
Another method is suggested by looking at formula (3.36) : it says that, up to a total differential, one obtains the reduced symplectic potential simply by inserting the solution of the constraint equations into the full symplectic potential. For field theories, there might also be boundary terms involved in this reduction process, whose contribution to the reduced symplectic structure does not vanish. They may be neglected at a first stage because they will be recovered when one checks whether the observables of the reduced phase space are finite and functionally differentiable.

3.3 The reduced phase space

It will turn out that for this model the second method is more appropriate. We are thus interested in the solutions of constraint equations.
We first take the following linear combinations of the vector and the scalar constraint functional

$$\begin{aligned} E^1E^2V + E^3C &= E(E^1B^3 + \frac{1}{2}E^3B^1) \\ -E^1E^3V + E^2C &= E(E^1B^2 + \frac{1}{2}E^2B^1), \end{aligned} \tag{3.37}$$

where $E = (E^2)^2 + (E^3)^2$.
Setting these expressions strongly zero we obtain 2 possible solutions :
Case I : $E = 0$ (degenerate case)
Looking at the formula for the metric (3.17) we see that there is no radial distance now. From the reality of the triads we conclude further that $E^2 = E^3 = 0$ whence we conclude $E^1 = E^1(t)$ via setting the Gauss constraint equal to zero. Obviously this solution of the constraint equations is not valid in the asymptotic ends since it violates the asymptotic conditions on the fields. It can therefore only hold inside the hypersurface and we should glue it to a solution of the constraints appropriate for the asymptotic regions. For compact topologies it is a global solution of the constraints. Applying the framework of the previous subsection we obtain for the reduced symplectic potential

$$\hat{\Theta}[\partial_t] = -iE^1\frac{d}{dt}\int_\Sigma drA_1 \,. \tag{3.38}$$

Case II : $E \neq 0$ (nondegenerate case)
We now conclude

$$\begin{aligned} 0 &= E^1B^3 + \frac{1}{2}E^3B^1 \\ 0 &= E^1B^2 + \frac{1}{2}E^2B^1 \end{aligned} \tag{3.39}$$

and can further distinguish between a) $B^1 = 0$ and b) $B^1 \neq 0$.
Subcase a)
Then either $B^2 = B^3 = 0$ or $E^1 = 0$. Consider first the case $B^I = 0$. Then an elementary calculation shows (recall the abbreviations (3.16)) $(B^2)^2 + (B^3)^2 = (A')^2/A + A(A_1 + \alpha')^2 = 2(A_1 + \alpha')^2 = 0$ whence $\gamma := A_1 + \alpha' = 0$. Now, by writing the symplectic potential in terms of 'cylindrical coordinates', that is, by plugging $(A_2, A_3) = \sqrt{A}(\cos(\alpha), \sin(\alpha))$ and $(E^2, E^3) = \sqrt{E}(\cos(\beta), \sin(\beta))$ into $\Theta = -i\int_\Sigma dr\dot{A}_I E^I$ we easily obtain up to a total differential

$$\Theta = \int_\Sigma dr[\dot{\gamma}E^1 + \dot{B}^1\sqrt{\frac{E}{A}}\cos(\alpha - \beta) + \dot{\alpha}\mathcal{G}] \,. \tag{3.40}$$

Accordingly, the symplectic structure pulled back to the Gauss reduced phase space vanishes identically if $B^1 = 0$. We are not interested in this trivial case of a reduced phase space consisting of only one point any longer.
Subcase b)
We can divide by B^1 (everywhere except for isolated points) to solve eqs. (3.39) for the momenta E^2 and E^3

$$E^2 = -\frac{2E^1}{B^1}B^2 \text{ and } E^3 = -\frac{2E^1}{B^1}B^3 \tag{3.41}$$

and insert this into into the Gauss constraint :

$$0 = B^1(E^1)' + 2E^1(B^1)' . \tag{3.42}$$

Eqn. (3.42) can be integrated :

$$E^1 = \frac{m^2}{(B^1)^2} \tag{3.43}$$

where the constant of integration m takes real values in the asymptotically flat case for the following reason : we will show later that B^1 is real on the constraint surface. Moreover, E^1 becomes r^2 at infinity. Thus, m^2 must be a real positive constant.
The last step is then to pull back the symplectic potential. One can check that modulo a total differential

$$\begin{aligned}(\iota^*\Theta)[\partial_t] &= -im^2[\int_\Sigma dr[\dot{A}_1\frac{1}{(B^1)^2} + \dot{A}_2\frac{-2B^2}{(B^1)^3} + \dot{A}_3\frac{-2B^3}{(B^1)^3}] \\ &= -im^2\frac{d}{dt}\int_\Sigma dr\frac{\gamma}{(B^1)^2} .\end{aligned} \tag{3.44}$$

One can check that the integrand in the last line of (3.44) vanishes *on the contraint surface* as $1/r^2$ at infinity, and thus the integral is well defined. However, it is functionally differentiable only if we require $\delta a_2 = \delta a_3 = 0$ in addition to the requirements derived in section 3.1.

3.4 Reality conditions

The reality conditions in the degenerate case are obscure because the spin connection coefficients are ill-defined (they are homogenous functions of degree zero, so their value on the constraint surface depends on the way the limit is taken). Therefore, we focus on the non-degenerate case in the sequel.
First we prove that the magnetic fields are weakly real :
To begin with we have (recall (3.18))

$$\bar{A} = (A_2-2\Gamma_2)^2+(A_3-2\Gamma_3)^2 = A+4[-A_2\Gamma_2-A_3\Gamma_3+(\Gamma_2)^2+(\Gamma_3)^2] = A+4\frac{(E^1)'}{E}\mathcal{G} \tag{3.45}$$

which proves that B^1 is weakly real. We now solve the constraints for the remaining magnetic fields

$$B^2 = -\frac{E^1}{2E^2}B^1 \text{ and } B^3 = -\frac{E^1}{2E^3}B^1$$

to conclude from the reality of the triads that also B^2, B^3 are weakly real. We now exploit this result to show that γ is weakly imaginary. Since $A\gamma = A_2B^2 + A_3B^3$ we have ($\approx$ means $=$ on the constraint surface)

$$\bar{\gamma} \approx -\gamma + 2\frac{\Gamma_2 B^2 + \Gamma_3 B^3}{A} = -\gamma - 2\frac{(E^1)'}{EA}V \,. \tag{3.46}$$

Accordingly, the momentum conjugate to $P := m^2$, $Q := -i\int_\Sigma dr \frac{\gamma}{(B^1)^2}$, is weakly real. The reduced phase space is therefore the cotangent bundle over the positive real line. However, it proves convenient for the interpretation of our Dirac observables to make a canonical transformation and to describe the reduced phase space in terms of m and $T := 2mQ$.

3.5 Interpretation

Let us explore what the geometrical meaning of m is.
Using the fall-off properties of the fields of section 3.1 and in particular eqn. 3.27 we find that $q_{rr} \to 1 - \sqrt{2}a_3/r + O(1/r^2)$. Comparing this with the asymptotical form of a Schwarzschild metric of mass M we find

$$a_3 = -\sqrt{2}M \tag{3.47}$$

whereas from equation (3.43) we have that (recall that $E^1 \to r^2 + O(r)$), $m^2 = 2(a_3)^2$, whence

$$m = \pm 2M \tag{3.48}$$

i.e. m is twice the Schwarzschild mass of the given solution. If we apply the positive censureship conjecture we find that m has range on half the real line only ([19]).
Next we have a look at the reduced action. Plugging the solution of the constraint equations into the action and using the fall-off properties of the fields we find the reduced Hamiltonian in the asymptotically flat context to be equal to

$$H = (N_\infty - N{-\infty})m \tag{3.49}$$

where $N_{\pm\infty}$ is the lapse at $r = \pm\infty$. Accordingly, the solution of the equations of motion for the canonical pair (m, T) turns out to to be

$$m = \text{const. and } T = \text{const.} + \tau_+ - \tau_- \tag{3.50}$$

where $\dot{\tau}_\pm(t) = N_{\pm\infty}(t)$, that is, $\tau_\pm$ is the eigentime of an asymptotic observer at positive or negative spatial infinity (recall that $ds = \sqrt{-g_{tt}}dt$ is the

eigentime interval associated with the time label interval dt and that $g_{tt} = -N^2$ at spatial infinity for asymptotically vanishing shift). Thus, *on shell* T can be identified with the difference τ of these eigentimes. In particular, T is a constant if and only if both clocks run at equal velocities, that is, $N_\infty = N_{-\infty}$ as is the case for the Kruskal solution.
To summarize, spherically symmetric canonical gravity adopts the form of an integrable system where the role of the action and angle variable respectively is played by the mass and the difference of eigentimes at both asymptotic ends respectively. In the case of closed topologies the reduced Hamiltonian vanishes identically and T is a constant of the motion, too.

3.6 Quantization

In the reduced phase space approach one quantizes the classically reduced phase space. In our case, we have that the reduced phase space is a cotangent bundle over either the real or half the real line. In the latter case one would proceed to quantize the canonical pair $(\ln(\pm m), \pm mT)$ which again provides one with a cotangent bundle over the real line. So there is nothing essentially new coming from this case and we therefore concentrate on the first case.
We choose the representation in which T is diagonal and arrive at the following operator equivalents of our basic variables

$$\hat{m} := -i\hbar\frac{\partial}{\partial T}, \ \hat{T} = T \ . \tag{3.51}$$

The physical Hilbert space consists of the usual complex-valued, square integrable functions of T. The solutions of the Schroedinger equation

$$i\hbar\frac{\partial\Psi}{\partial t}(T) = \hat{H}\Psi(T) = -i\hbar\frac{d\tau}{dt}\frac{\partial\Psi}{\partial T}(T) \tag{3.52}$$

are given by

$$\Psi(t,T) = f(T-\tau(t)) \tag{3.53}$$

i.e. it is an arbitrary function of the argument displayed. Of course, only normalizable functions f should be considered. In particular the eigenfunctions of the Hamiltonian, $f = \exp(ik(T-\tau))$ are not normalizable.

In the operator constraint (Dirac) approach, one solves the constraints after quantizing. Let us follow the steps of this quantization procedure.
Step1) : Quantize a complete set of basic operators such that its commutator algebra mirrors the associated classical Poisson algebra. We choose

$$[\hat{A}_I(x), \hat{A}_J(y)] = [\hat{E}^I(x), \hat{E}^J(y)] = 0, \ [\hat{A}_I(x), \hat{E}^J(y)] = -\hbar\delta_I^J\delta(x,y) \ . \tag{3.54}$$

Step2) : Choose a representation of this algebra on a complex vector space V. We choose the self-dual representation, that is, V consists of holomorphic functionals of the connection. Our operators are then represented as follows :

$$(\hat{A}_I(x)\Psi)[A] := A_I(x)\Psi[A],\ (\hat{E}^I(x)\Psi)[A] := \hbar\frac{\delta\Psi[A]}{\delta A_I(x)} \tag{3.55}$$

where $\delta/\delta A_I(x)$ is the functional derivative[4].

Step3) : Try to find a consistent ordering of the constraints, that is, they should form a commutator subalgebra in the sense that the constraint operators appear always ordered to the right after commuting two constraints (this is is a nontrivial requirement because the structure functions for constraints bilinear in the momenta turn out to be operator valued, compare Wipf's lectures). As we analyzed in section 3.3, for our model it is actually possible to cast the constraints into a form in which they are linear in the momenta. Therefore, we do not have any problems with this step : just order the scalar constraint in such a way that the operators linear in the momenta which are to vanish appear to the right handside and order the momenta to the right in the remaining constraints. Thus, for sector I we would write

$$\hat{C} = \hat{E}^1(\hat{B}^2\hat{E}^2 + \hat{B}^3\hat{E}^3) + \frac{1}{2}\hat{B}^1(\hat{E}^2\hat{E}^2 + \hat{E}^3\hat{E}^3) \tag{3.56}$$

whereas for sector II we would order as

$$\hat{C} = \hat{E}^2(\hat{B}^2\hat{E}^1 + \frac{1}{2}\hat{B}^1\hat{E}^2) + \hat{E}^3(\hat{B}^3\hat{E}^1 + \frac{1}{2}\hat{B}^1\hat{E}^3) \tag{3.57}$$

and one can explicitely check that the commutator algebra closes in the sense that the operators linear in the momenta which are to annihilate the physical states always appear to the right.

Note that we should actually regulate the scalar constraint since it is bilinear in the momenta. However, since we are effectively working with a rewritten version which is linear in the momenta, we can circumvent this step.

Step4) : Solve the constraints, that is, find the physical subspace V_{phys} of the vector space V.

For sector I this amounts to imposing

$$\frac{\delta\Psi}{\delta A_2(x)} = \frac{\delta\Psi}{\delta A_3(x)} = (\frac{\delta\Psi}{\delta A_1(x)})' = 0 \tag{3.58}$$

the unique solution of which is given by

$$\Psi[A] = f(\int_\Sigma dr A_1(r)) . \tag{3.59}$$

[4] If it exists, the functional derivative is defined by $\int_\Sigma dr\xi(r)\delta f[\phi]/\delta\phi(r) := \lim_{s\to 0}(f[\phi+s\xi]-f[\phi])/s$ for any test function ξ of compact support

For sector II the kernel of the constraint operators consists of the functions satisfying

$$\begin{aligned} 0 &= \hat{B}^2\frac{\delta\Psi}{\delta A_1}+\frac{1}{2}\hat{B}^1\frac{\delta\Psi}{\delta A_2} \\ 0 &= \hat{B}^3\frac{\delta\Psi}{\delta A_1}+\frac{1}{2}\hat{B}^1\frac{\delta\Psi}{\delta A_3} \\ 0 &= \hat{B}^1(\frac{\delta\Psi}{\delta A_1})'+2(\hat{B}^1)'\frac{\delta\Psi}{\delta A_1} \end{aligned} \tag{3.60}$$

the unique solution of which is given by

$$\Psi[A]=f(\int_\Sigma dr\frac{A_1+(\arctan(\frac{A_3}{A_2}))'}{(B^1)^2})\,. \tag{3.61}$$

Thus, for both cases the solution consists of arbitrary functions of the functionals displayed.
Step5) : Find a complete algebra of basic quantum observables.
By definition, observables leave the physical subspace invariant. Accordingly, we choose them to be the multiplication and differentiation operators with respect to the argument of the functions of the physical subspace. For sector I we thus have

$$\hat{Q}:=-i\int_\Sigma drA_1,\ \hat{P}:=\hat{E}^1(x) \tag{3.62}$$

whereas for sector II we obtain

$$\hat{Q}:=-i\int_\Sigma dr\frac{A_1+(\arctan(\frac{A_3}{A_2}))'}{(B^1)^2},\ \hat{P}:=(\hat{B}^1)^2(x)\hat{E}^1(x)\,. \tag{3.63}$$

In both cases the argument x of the operator $\hat{P}$ is irrelevant since it is a spatial constant on the physical subspace.
Step6) : Equip V_{phys} with a Hilbert space structure by demanding that the reality conditions *induced* on the quantum observables become adjointness conditions with respect to that inner product.
It follows from the analysis in section 3.6 that for sector II the classical analogues Q and P of the observables found in step 5 are real. Since the observables $\hat{Q},\hat{P}$ found in step5 are canonically conjugate, $[\hat{Q},\hat{P}]=i\hbar$, we will also postulate for sector I that Q is classically real (its imaginary part arises then classically from a canonical transformation).
The classical range of P is positive in case of sector II. Accordingly, we either choose a representation in which $\hat{P}$ is diagonal and proceed along the lines of [12] or we allow for classically not allowed regions of the quantum theory and can stay within the representation such that $\hat{Q}$ is diagonal. Let us choose the latter option.
Then in both sectors, the unique inner product that accomplishes our aim is just $L_2(R,dQ)$.

One could finally form the direct sum of both Hilbert spaces, thus producing sectors in the technical sense of the word because states of different sectors cannot be superposed (they are not annihilated by the same constraint operators).

Altogether, in the present model both quantization procedures give equivalent answers.

3.7 Loop representation

In one dimension, except for the case of a closed topology, there are no loops. However, in order to test the 3-dimensional theory, one should rather look at loop variables for the 3+1 case restricted to spherical symmetry. It turns out that one can find a subset of loops which form a closed loop subgroup such that one can express all O(2)-invariant quantities in terms of them. When expressing the Dirac observables found for the present model in terms of them, the expressions become rather horrible for the non-degenerate sector but become very simple for the degenerate sector. This is in accordance to the fact that all solutions to the constraints in the 3+1 case that have been found so far belong to the degenerate sector but that no solutions are known for the non-degenerate sector.
More details are given in [14]. We refrain from giving them here because issues like a loop transform and an algebra of loop operators have not been worked out yet.

3.8 Discussion

The model of spherically symmetric gravity has been successfully quantized, both, via the reduced phase space and the operator contraint approach. The model captures various technical problems of the full 3+1 case :
Its reality structure is non-trivial, the constraints mirror those of the full 3+1 theory, in particular they are bilinear in the momenta. What comes out as a surprise is that the Dirac observables (3.63) have such a simple reality structure. That raises the hope that also in the full theory the reality structure of the reduced phase space turns out to be rather simple.
Also, all the results given here in terms of Ashtekar's variables can also be written in terms of geometrodynamical (ADM) variables. In particular, it is possible to write down a one parameter family of exact solutions to the Hamilton-Jacobi equation associated with the Wheeler-DeWitt equation ([14]).
These are the positive remarks. Again sharp criticism is in order :
We actually exploited the fact that the scalar constraint could be cast in such a form that it is linear in the momenta. This technical help will not be

available in the full theory and we cannot expect to solve the quantization programme without regularizing and renormalizing the constraint operators.
The other issue is of course that again we are effectively dealing with quantum mechanics rather than quantum field theory.

Acknowledgements

This work was supported by the Graduierten-Programm der Deutschen Forschungsgemeinschaft (DFG).

References

[1] E. Witten, Nucl. Phys. **B311**(1988)46

[2] F. Barbero, M. Varadarajan, Syracuse preprint (1993), Syracuse/NY, USA

[3] S. Carlip, Phys. Rev. **D42**(1990)2647
V. Moncrief, Journ. Math. Phys. **30**(1989)2907
P. Peldan, Class. Quantum Grav. **9**(1992)2097
N. Manojlovic, Nucl. Phys. **B385**(1992)571

[4] A. Ashtekar et al, Class. Quantum Grav. **6**(89)L185

[5] J. N. Goldberg, J. Lewandowski, C. Stornailo, Comm. Math. Phys. **148**(1992)377

[6] Y. Choquet-Bruhat, C. DeWitt-Morette, **Analysis, Manifolds and Physics**; North Holland, New York(1989)

[7] J. W. Barrett, Int. Journ. Theor. Phys. **30**(1991) 1171

[8] G. Mess, Preprint IHES/M/90/28, Inst. des Hautes Etudes Scientifiques (1990)

[9] D. M. Marolf, Syracuse preprint(1993), Syracuse, NY 13244, USA
J. Louko, D. M. Marolf, Syracuse preprint, Syracuse, NY 13244, USA

[10] A. Ashtekar, R. Loll, CGPG preprint, University Park, PA 16802, USA

[11] C. Rovelli, L. Smolin, Nucl. Phys. B **331**(1990)80

[12] C. J. Isham in **Relativity, groups and toplogy II**, Les Houches Lectures (1983), ed. B. S. DeWitt, R. Stora, North Holland, Amsterdam(1994)

[13] A. Ashtekar, Comm. Math. Phys. **71**(1980)59

[14] T. Thiemann, diplom thesis, April 1992, RWTH Aachen, Germany
T. Thiemann, H. A. Kastrup, Nucl. Phys. **B399** (1993) 211
T. Thiemann, PITHA-preprint 93-32, August 1993, RWTH Aachen, Germany
T. Thiemann, PhD thesis, December 1993, RWTH Aachen, Germany
T. Thiemann, H. A. Kastrup, PITHA preprint 93-35, October 1993, RWTH Aachen, Germany

[15] I. Bengtsson, Class. Quantum Grav. **7**(1990)27

[16] R. Beig, N. ó Murchadha, Ann. Phys. **174**(1987)463

[17] N. Woodhouse, **Geometric Quantization**, Clarendon Press, Oxford(1980)

[18] E. T. Newman, C. Rovelli, Phys. Rev. Lett. **69**(1992)1300

[19] E. Witten, Comm. Math. Phys. **80**(1981)381

Adjointness Relations as a Criterion for Choosing an Inner Product

Alan D. Rendall

1. Sufficient Conditions for Uniqueness

In the quantisation of constrained systems it can happen that one obtains a representation of an algebra of quantum operators on a vector space without a preferred inner product. Since an inner product is necessary for the probabilistic interpretation of quantum theory, some way needs to be found of introducing an appropriate inner product on this vector space. If the classical system being quantised possesses some background structure then it may be possible to use this to fix an inner product. In the case of gravity, where no background structure is present, this is not an option. The algebra of quantum observables usually admits a preferred *-operation, often related to complex conjugation of functions on the classical phase space. It has been suggested by Ashtekar that a preferred inner product could be fixed by the requirement that this *-operation is mapped by the representation into the operation of taking the adjoint of an operator with respect to the inner product in question. Discussions of this proposal can be found in [1-3]. The purpose of the following is to discuss the circumstances under which this idea suffices to determine the inner product uniquely.

Let A be an associative algebra with identity over the complex numbers. This is to be interpreted as the algebra of quantum observables. Suppose that a representation ρ of A on a complex vector space V is given. Suppose further that a *-operation $a \mapsto a^*$ is given on A. The defining properties of a *-operation are that it is conjugate linear ($(\lambda a + \mu b)^* = \bar{\lambda} a^* + \bar{\mu} b^*$), that $(ab)^* = b^* a^*$ and that $(a^*)^* = a$. In this paper the origin of these various objects will not be discussed; information on that can be found in [1] and [2]. Instead we take this collection of objects as starting point. The condition which is supposed to characterise the inner product is that

$$\langle \rho(a)x, y\rangle = \langle x, \rho(a^*)y\rangle, \tag{1}$$

for all $a \in A$ and $x, y \in V$. Note that there is a trivial non-uniqueness due to the possibility of multiplying the inner product by a non-zero real constant but this is not physically significant since it leaves expectation values unchanged. In order to ensure the uniqueness of the inner product it is necessary to require that the representation ρ be irreducible in some appropriate sense. In [3] it was argued that the correct concept of irreducibility to use is that of topological irreducibility and this leads to the following definition:

Definition 1 Let A be a complex *-algebra with identity and ρ a representation of A on a complex vector space V. An inner product $\langle\ ,\ \rangle$ on V is called *strongly admissible* if

(i) ρ is a *-representation with respect to this inner product i.e. equation (1) is satisfied

(ii) for each $a \in A$ the operator $\rho(a)$ is bounded with respect to the norm associated to the given inner product so that ρ extends uniquely by continuity to a representation $\hat{\rho}$ on the Hilbert space completion $\hat{V}$ of V with respect to this norm

(iii) $\hat{\rho}$ is topologically irreducible i.e. it leaves no non-trivial closed subspaces of $\hat{V}$ invariant

This definition is tailored to the case of representations by bounded operators; the unbounded case will be discussed later. In [3] it was claimed that if $\langle\ ,\ \rangle_1$ and $\langle\ ,\ \rangle_2$ are two inner products which are strongly admissible for a given representation then there exists a positive real number c such that $\langle\ ,\ \rangle_2 = c\langle\ ,\ \rangle_1$. In fact this is incorrect. An explicit example where the claim fails is given in section 3 below. An additional condition which ensures uniqueness will now be presented but first some terminology is required. Let $\langle\ ,\ \rangle_1$ and $\langle\ ,\ \rangle_2$ be two admissible inner products. The inner product $\langle\ ,\ \rangle_2$ will be said to be *compatible* with $\langle\ ,\ \rangle_1$ if any sequence $\{x_n\}$ in V which satisfies $\langle x_n, x_n\rangle_1 \to 0$ as $n \to \infty$ and which is a Cauchy sequence with respect to $\langle\ ,\ \rangle_2$ also satisfies $\langle x_n, x_n\rangle_2 \to 0$. (This is not a standard definition. It is only introduced for convenience in this paper.) The modified claim is now:

Theorem 1 Let $\langle\ ,\ \rangle_1$ and $\langle\ ,\ \rangle_2$ be inner products on a complex vector space V which are strongly admissible with respect to a representation ρ of a complex *-algebra A. Suppose that $\langle\ ,\ \rangle_2$ is compatible with $\langle\ ,\ \rangle_1$. Then $\langle\ ,\ \rangle_2 = c\langle\ ,\ \rangle_1$ for some positive real number c.

This theorem will be proved in section 2. A question which comes up immediately is: what is the interpretation of the property of compatibility and is it a reasonable condition from the point of view of the original motivation, namely quantisation of certain systems? To make contact with known mathematics, let $\hat{V}_1$ be the completion of V with respect to $\langle\ ,\ \rangle_1$ and consider $\langle\ ,\ \rangle_2$ as an unbounded sesquilinear form on the Hilbert space $\hat{V}_1$ with domain V. In general, if S is an unbounded positive sesquilinear form on a Hilbert space H with domain D then a new inner product can be defined on D by $\langle x, y\rangle_S = \langle x, y\rangle_H + S(x, y)$. The sesquilinear form S is called closed if D is complete with respect to $\langle\ ,\ \rangle_S$. More generally S is called closable if it has an extension to a domain $\bar{D}$ such that the extension is closed. It turns out that $\langle\ ,\ \rangle_2$ is compatible with $\langle\ ,\ \rangle_1$ if and only if $\langle\ ,\ \rangle_2$ is closable when considered as an unbounded sesquilinear form on $\hat{V}_1$ (see chapter 6 of [4]). It is useful to have this information since it puts at our disposal known results on closable sesquilinear forms.

Concerning the relation of the compatibility condition with the original motivation, note first that the kind of situation which is likely to occur in practice is that one inner product is already known and we would like to know if it is characterised by the condition that the *-relations go over to adjointness relations. When the requirement of compatibility is added to the other hypotheses of the theorem the result is not to restrict the class of representations covered but rather to narrow the class of alternative inner products within which uniqueness is shown to hold. In examples it is usually the case that the vector space V is a space of functions on some set X and that the inner product which is known is the L^2 inner product corresponding to some measure $d\mu$ on that set. It is desired to have uniqueness within the class of inner products corresponding to the L^2 norms defined by measures of the form $f d\mu$, where f is a non-negative real-valued function on X. There are general theorems which make it reasonable to expect that these inner products will all be compatible with the original one for conventional choices of the space of functions V. Thus the compatibility condition does not restrict the applicability of Theorem 1 too drastically. To illustrate this some examples will now be discussed.

First let $(X, d\mu)$ be a general measure space and f a non-negative measurable real-valued function on X. Let

$$D_f = \left\{ g \in L^2(X, d\mu) : \int f|g|^2 d\mu < \infty \right\}. \tag{2}$$

Then D_f is dense in $L^2(X, d\mu)$ and the formula

$$S_f(g, h) = \int f g \bar{h} d\mu \tag{3}$$

defines a closed sesquilinear form with domain D_f[4]. It is compatible with the restriction of the L^2 norm to D_f and so this provides a wide class of examples where compatibility is satisfied. It is not necessary to choose $V = D_f$ in this example. Any linear subspace of D_f which is dense in L^2 would also suffice. For instance in the case that $X = \mathbf{R}^n$ and $d\mu$ is Lebesgue measure V could be chosen to be the space of smooth functions with compact support. More generally, it is possible to consider two non-negative real-valued locally integrable functions f_1 and f_2 on $\mathbf{R}^n$. Let $\langle\ ,\ \rangle_1$ and $\langle\ ,\ \rangle_2$ be the L^2 inner products defined by the measures $f_1 d\mu$ and $f_2 d\mu$, restricted to $V = C_c^\infty(\mathbf{R}^n)$. A sufficient condition that these sesquilinear forms be inner products is that the zero sets of f_1 and f_2 have zero measure. Assume that this is the case. V is dense in $L^2(\mathbf{R}^n, f_1 d\mu)$ and it follows from the above that $\langle\ ,\ \rangle_2$ is compatible with $\langle\ ,\ \rangle_1$.

If one inner product is continuous with respect to the other, $\| \|_2 \leq C\| \|_1$ for some constant C, then it is obviously compatible with it. However it is in any case rather easy to prove a uniqueness result for inner products in that case. If V is finite dimensional then the continuity is automatic. It is natural to ask what happens when V has a countable basis. (The word basis is used here in the algebraic and not in the Hilbert space sense.) Suppose that an inner product $\langle\ ,\ \rangle_1$ is given on V. Using the Gram-Schmidt process it is possible to go over to an orthonormal basis. Hence there is an isomorphism of V with the space of complex sequences with finitely many non-zero entries such that the inner product takes the form $\langle\{a_n\},\{b_m\}\rangle_1 = \sum a_n\bar{b}_n$. The completion of V can then be identified with the space l^2 of square summable sequences and this will be done from now on. Any other inner product on V takes the form $\langle\{a_n\},\{b_m\}\rangle_2 = \sum a_n k_{mn}\bar{b}_m$ for some k_{mn}. Deciding whether a given set of coefficients k_{mn} defines an inner product compatible with the original one is a concrete problem on the convergence of sequences. Nevertheless it does not seem easy to give a general solution. It will be shown by example in section 3 that the compatibility condition does not always hold.

An analogue of Theorem 1 for unbounded operators will now be presented. In [3] a procedure was given for reducing the problem of uniqueness of the inner product in the case of a representation by unbounded operators to the corresponding problem for bounded operators under certain circumstances. This reduction process combined with Theorem 1 gives a theorem in the case of unbounded operators which will now be stated.

Definition 2 Let A, ρ and V be as in Definition 1. Let S be a set of elements of A which satisfy $a^* = a$ and which generate A. An inner product $\langle\ ,\ \rangle$ on V is said to be *admissible* if:
(i) ρ is a *-representation with respect to this inner product
(ii) for each $a \in S$ the operator $\hat{\rho}(a)$ is essentially self-adjoint
(iii) $\hat{\rho}$ is irreducible
(iv) $\hat{\rho}$ is closed

Here $\hat{\rho}$ is obtained from ρ by considering the linear maps $\rho(a)$ on V as unbounded operators on $\hat{V}$ with domain V. The meanings of the words 'closed' and 'irreducible' in this context are explained in [3]; suffice it to say that this definition reduces to the definition of 'strongly admissible' in the case that all $\rho(a)$ are bounded.

Theorem 2 Let $\langle\ ,\ \rangle_1$ and $\langle\ ,\ \rangle_2$ be inner products on a complex vector space V which are admissible with respect to a representation ρ of a complex *-algebra A. Suppose that $\langle\ ,\ \rangle_2$ is compatible with $\langle\ ,\ \rangle_1$. Then $\langle\ ,\ \rangle_2 = c\langle\ ,\ \rangle_1$ for some positive real number c.

2. Proof of the Uniqueness Theorem

Let $\langle\ ,\ \rangle_1$ and $\langle\ ,\ \rangle_2$ be inner products satisfying the assumptions of

Theorem 1. Let $\langle\ ,\ \rangle = \langle\ ,\ \rangle_1 + \langle\ ,\ \rangle_2$. Define $\hat{V}$ and $\hat{V}_1$ to be the completions of V with respect to $\langle\ ,\ \rangle$ and $\langle\ ,\ \rangle_1$ respectively. The representation ρ extends uniquely by continuity to representations $\hat{\rho}$ and $\hat{\rho}_1$ on $\hat{V}$ and $\hat{V}_1$ respectively. Now some facts proved in [3] will be recalled. It was shown there under the hypotheses that the inner products $\langle\ ,\ \rangle_1$ and $\langle\ ,\ \rangle_2$ are admissible that there exists a bounded self-adjoint operator L_1 on $\hat{V}$ such that $\langle x, y\rangle_1 = \langle x, L_1 y\rangle$ for all x, y in V. It was also shown that unless the two inner products are proportional the operator L_1 has a non-trivial kernel. Now the sesquilinear form $\langle\ ,\ \rangle_1$ on V is bounded with respect to $\langle\ ,\ \rangle$. Hence it extends uniquely by continuity to a sesquilinear form S on $\hat{V}$. Using continuity again shows that $S(x,y) = \langle x, L_1 y\rangle$ for all x and y in $\hat{V}$. It follows that any vector x in the kernel of L_1 satisfies $S(x,x) = 0$ and hence that S is degenerate. In [3] it was claimed that this is incompatible with the fact that $\langle\ ,\ \rangle_1$ is an inner product. However this is not true, as can be seen explicitly in the example given in section 3 below. The facts that $\langle\ ,\ \rangle_1$ is non-degenerate and that S is an extension by continuity of $\langle\ ,\ \rangle_1$ do not together imply that S is non-degenerate.

Suppose then that $x \in \hat{V}$ satisfies the condition that $S(x,x) = 0$. Since V is dense in $\hat{V}$ there exists a sequence x_n of vectors in V with $\|x - x_n\| \to 0$ as $n \to \infty$. On the other hand, the continuity of S implies that $S(x_n, x_n) \to 0$ and, due to the fact that all x_n belong to V, this is equivalent to the condition that $\|x_n\|_1 \to 0$. The sequence x_n is a Cauchy sequence with respect to $\langle\ ,\ \rangle$ and hence with respect to $\langle\ ,\ \rangle_2$. Now the hypothesis that $\langle\ ,\ \rangle_2$ is compatible with $\langle\ ,\ \rangle_1$ implies that $x_n \to 0$ with respect to $\langle\ ,\ \rangle_2$. We already know that it tends to zero with respect to $\langle\ ,\ \rangle_1$. It follows that $x = 0$. This means that the kernel of L_1 is trivial and, in conjunction with what was said above, completes the proof of the theorem.

3. A Cautionary Example

Let H be the Hilbert space $L^2([-1,1])$ and denote its inner product by $\langle\ ,\ \rangle$. Let V be the space of functions on $[-1,1]$ which extend analytically to a neighbourhood of that interval. If $f \in V$ let M_f be the multiplication operator on H defined by $g \mapsto fg$. This is a bounded operator. If ϕ is an orientation preserving diffeomorphism of $[-1,1]$ with $\phi(0) = 0$ which extends to an analytic mapping on a neighbourhood of $[-1,1]$ define an operator T_ϕ on H by $T_\phi f = f \circ \phi$. This operator is also bounded. Define A to be the algebra of bounded operators on H generated by all the M_f and T_ϕ. If ϕ is a diffeomorphism as above define a function $\tilde{\phi}$ by

$$\tilde{\phi}(x) = d/dx(\phi^{-1}(x)) \tag{4}$$

Then the adjoints of the operators of interest are given by $M_f^* = M_{\bar{f}}$ and $T_\phi^* = M_{\tilde{\phi}} T_{\phi^{-1}}$. It follows that the operation of taking the adjoint defines a

*-operation on A. Each operator belonging to A maps V into itself and so restricting the elements of A to V defines a representation of A on V. It will be shown that there exist two inner products on V which are strongly admissible with respect to ρ and which are not proportional. These are defined as follows.

$$\langle f,g\rangle_+ = \int_0^1 f(x)\bar{g}(x)dx,$$

$$\langle f,g\rangle_- = \int_{-1}^0 f(x)\bar{g}(x)dx.$$

These expressions obviously define sesquilinear forms but it needs to be checked that they are non-degenerate on V. If $\langle f,f\rangle_+ = 0$ then f vanishes almost everywhere on $[0,1]$. But by analyticity this implies that f vanishes identically. The proof for $\langle\ ,\ \rangle_-$ is similar.

It remains to show that both inner products are admissible. Because of the symmetry of the situation it suffices to do this for $\langle\ ,\ \rangle_+$. First note that the completion of V with respect to $\langle\ ,\ \rangle_+$ can be identified with $L^2([0,1])$. The computations which show that ρ is a *-representation with respect to $\langle\ ,\ \rangle_+$ and that the operators M_f and T_ϕ are bounded with respect to the corresponding norm $\|\ \|_+$ are essentially the same as those which are needed to show that they are bounded on H and to compute their adjoints there. Next the irreducibility of $\hat{\rho}_+$ will be examined. Let Π be a projection in the Hilbert space $\hat{V}_+$ which commutes with all operators in the image of $\hat{\rho}_+$. The aim is to show that Π must be zero or the identity. Let $p = \Pi(1)$. This is an L^2 function on $[0,1]$. If f belongs to V then

$$pf = M_f p = M_f\Pi(1) = \Pi M_f(1) = \Pi f \tag{5}$$

Thus on V the operator Π is given by multiplication by p. It can easily be seen by approximating an arbitrary continuous function on $[0,1]$ uniformly by elements of V that in fact $\Pi f = pf$ for any continuous function f. It will now be shown that the function p must be essentially bounded. Let E_n be the set where $|p| \geq n$ and let ϵ_n be the measure of E_n. By Lusin's theorem [5] there exists a continuous function f_n with $|f_n| \leq 1$ such that the measure of the set where f_n is not equal to the characteristic function of E_n is less than $\epsilon_n/2$. Hence $\|pf_n\|_2^2 \geq n^2\epsilon_n/2$. On the other hand $\|pf_n\|_2^2 = \|\Pi f_n\|_2^2 \leq \|f_n\|_2^2 \leq 3\epsilon_n/2$. Hence $\epsilon_n = 0$ for $n \geq 2$ and p is essentially bounded. It follows that multiplication by p defines a bounded operator M_p on $L^2([0,1])$. Since this operator agrees with Π on a dense subspace it follows that $\Pi = M_p$. Now $\Pi^2 = \Pi$ implies that $M_p = M_p^2 = M_{p^2}$. Hence $(p^2-p)f = 0$ for all $f \in L^2([0,1])$. It follows that $p = 0$ or 1 almost everywhere and that p is equal to the characteristic function χ_E of some measurable subset E of $[0,1]$. The condition $\Pi T_\phi = T_\phi\Pi$ will now be used. When worked out explicitly it gives

$$[p(x) - p(\phi(x))]f(\phi(x)) = 0 \tag{6}$$

for any L^2 function f. It follows that $p(\phi(x)) = p(x)$ i.e. that $\phi(E) = E$ up to set of measure zero. It will now be shown that if E has non-zero measure then it must differ from $[0,1]$ by a set of zero measure. If E has non-zero measure there must exist a point $x \in E \cap (0,1)$ which is a point of density. A point of density is roughly speaking a point of E which is almost entirely surrounded by other points of E; the exact definition can be found in [6] where it is proved that any set of non-zero measure contains such a point. This notion is invariant under diffeomorphisms and is insensitive to altering the set E by a set of measure zero. We can therefore conclude that the set of points of density of $E \cap (0,1)$ is invariant under all diffeomorphisms of the type under consideration here. Consider now the vector field on $\mathbf{R}$ given by $X = (1 - \cos(2\pi x))\partial/\partial x$. Exponentiating it gives a one-parameter group of analytic diffeomorphisms ϕ_t. The restriction of each ϕ_t to the interval $[-1,1]$ belongs to the class of diffeomorphisms used in defining the operators T_ϕ. Moreover, if x_1 and x_2 are any two points of $(0,1)$ there is some t for which $\phi_t(x_1) = x_2$. It follows that all points of $(0,1)$ are points of density of E. The definition of points then implies that no point of $(0,1)$ is a point of density of the complement of E. Hence the complement of E has measure zero. This completes the proof of the irreducibility of $\hat{\rho}_+$ and hence of the strong admissibility of $\langle\ ,\ \rangle_+$.

A small modification of this example can be used to establish another point, namely the existence of inner products on a vector space with a countable basis which are not compatible. Let V' be the vector space of polynomials on $[-1,1]$. Define two inner products $\langle\ ,\ \rangle_+$ and $\langle\ ,\ \rangle_-$ on V' as the restrictions of the corresponding inner products on V defined above. Then $\langle\ ,\ \rangle_-$ is not compatible with $\langle\ ,\ \rangle_+$. To see this, let f be a continuous function on $[-1,1]$ which vanishes identically on $[0,1]$ but not on $[-1,0]$. By Weierstrass' theorem there exists a sequence of polynomials converging uniformly to f on the interval $[-1,1]$. This converges to zero with respect to the norm $\|\ \|_+$ and is Cauchy with respect to $\|\ \|_-$. However it does not converge to zero with respect to $\|\ \|_-$. The operator algebra defined above does not act on V' and this example is only meant to illustrate the notion of compatibility of inner products. It is not known to the author whether Theorem 1 remains true for a vector space V with countable basis if the compatibility hypothesis is dropped.

Acknowledgements

I thank Horst Beyer for helpful remarks.

References

1.Ashtekar, A. (1991) Lectures on non-perturbative canonical gravity. World Scientific, Singapore.

2.Ashtekar, A., Tate, R.S. An extension of the Dirac program for the quan-

tisation of constrained systems. Preprint.
3.Rendall, A. D. (1993) Unique determination of an inner product by adjointness relations in the algebra of quantum observables. Class. Quantum Grav. **10**, 2261-2269.
4.Kato, T. (1966) Perturbation Theory for Linear Operators. Springer, Berlin.
5.Rudin, W. (1987) Real and complex analysis. 3rd Edition. McGraw-Hill, New York.
6.Stein, E. M. (1970) Singular integrals and differentiability properties of functions. Princeton University Press, Princeton.

Overview and Outlook

Abhay Ashtekar

When you do some mountaineering...you sometimes... want to climb to some peak but there is fog everywhere...you have your map or some other indication where you probably have to go and still you are completely lost in the fog. Then...all of a sudden you see quite vaguely in the fog, just a few minute things from which you say, "Oh, this is the rock I want." In the very moment that you have seen that, then the whole picture changes completely, because although you still don't know whether you will make the rock, nevertheless for a moment you say, "... Now I know where I am; I have to go closer to that and then I will certainly find the way to go ..."

(Werner Heisenberg, 1963)

1. Introduction

The problem of quantum gravity is an old one and over the course of time several distinct lines of thought have evolved. However, for several decades, there was very little communication between the two main communities in this area: particle physicists and gravitation theorists. Indeed, there was a lack of agreement on even what the key problems are. By and large, particle physics approaches focused on perturbative techniques. The space-time metric was split into two parts: $g_{\mu\nu} = \eta_{\mu\nu} + Gh_{\mu\nu}$, $\eta_{\mu\nu}$ being regarded as a flat kinematic piece, $h_{\mu\nu}$ being assigned the role of the dynamical variable and Newton's constant G playing the role of the coupling constant. The field $h_{\mu\nu}$ was then quantized on the $\eta_{\mu\nu}$-background and perturbative techniques that had been so successful in quantum electrodynamics were applied to the Einstein-Hilbert action. The key problems then were those of handling the infinities. The gravity community, on the other hand, felt that a central lesson of general relativity is that the space-time metric plays a dual role: it is important that one and the same mathematical object determine geometry *and* encode the physical gravitational field. From this perspective, an ad-hoc split of the metric goes against the very spirit of the theory and must be avoided. If one does not carry out the split, however, a theory of quantum gravity would be simultaneously a theory of quantum geometry and the notion of quantum geometry raises a variety of conceptual difficulties. If there is no background space-time geometry –but only a probability amplitude for various possibilities–

how does one do physics? What does causality mean? What is time? What does dynamics mean? Gravity theorists focused on such conceptual issues. To simplify mathematics, they often truncated the theory by imposing various symmetry conditions and thus avoided the field theoretic difficulties. Technically, the emphasis was on geometry rather than functional analysis. It is not that each community was completely unaware of the work of the other (although, by and large, neither had fully absorbed what the other side was saying). Rather, each side had its list of central problems and believed that once these issues were resolved, the remaining ones could be handled without much difficulty. To high energy theorists, the conceptual problems of relativists were perhaps analogous to the issues in foundations of quantum mechanics which they considered to be "unimportant for real physical predictions." To relativists, the field theoretic difficulties of high energy physicists were technicalities which could be sorted out after the conceptual issues had been resolved. Even on the rare occasions when they got together, the two sides seemed to "talk past each other."

Over the last decade, however, there has been a certain rapprochement of ideas on quantum gravity. Each side has become increasingly aware of the difficulties that were emphasized by the other. There are some agreements and increasing clarity. In the poetic imagery of Heisenberg, we have seen faint outlines of some of the rocks we want to climb.

Perhaps the most important among these developments are the following:

1. Recognition that non-perturbative methods are essential. There is a growing sentiment that, at a fundamental level, the theory should not involve fluctuations around a classical geometry. Indeed, there should be no background geometry or any background fields for that matter. The theory should be diffeomorphism invariant.
2. Acceptance that conceptual issues such as the problem of time will have to be handled satisfactorily. The unease with the standard measurement theory of quantum mechanics is brought to the forefront by the absence of a background, classical space-time. The interface of classical and quantum domains and the decoherence processes that make the world seem classical are receiving greater attention.
3. Agreement that the field theoretic divergences have to be faced squarely. There is growing awareness that although the mini-superspaces that feature in quantum cosmology are obviously interesting, from the perspective of full quantum gravity, they are essentially toy models. One has to learn to deal with the infinite number of degrees of freedom "honestly." Time has come to give proper mathematical meaning to the formal equations such as the Wheeler-DeWitt equa-

tion. New ideas and mathematical techniques are needed since the issue of regulating quantum operators in absence of a background geometry has rarely been faced.

4. Reconciliation with the possibility that the "quantum geometry" governing the Planck scale may not at all resemble the Riemannian picture. It is likely that differential geometry, in the standard sense of the term, will itself be inadequate to capture physics. Some discrete structures are likely to emerge and new mathematical tools will be needed to handle them.

Thus, the goals of the two communities have moved closer.

These recognitions do not imply, however, that there is a general consensus on *how* all these problems are to be resolved. Thus, there are again many approaches. But I strongly support belief that this diversity in the lines of attack is very healthy. In a problem like quantum gravity, where directly relevant experimental data is scarce, it would be a grave error if everyone followed the same path. Indeed, the most promising way of enhancing our chances at success is to increase the amount of variety. What is striking is that, in spite of the diversity of methods, some of the results are often qualitatively similar. For example, in the last few years, the fourth point I mentioned above has come up again and again in quite different contexts. The idea that the continuum picture itself is likely to break down is not new. For the first time, however, it is arising as a *concrete result* of calculations within well-defined, coherent schemes which were not explicitly constructed to obtain such a result. It arose from certain computer simulations of 4-dimensional Euclidean gravity, (see e.g. Agishtein & Migdal (1992) and Brügmann & Marinari (1993)), from string theory (see e.g. Gross & Mende (1988), Amati et al (1990), Aspinwall (1993)), and, from canonical quantization of 4-dimensional general relativity (Ashtekar et al (1992)). The detailed pictures of the micro-structure of space-time that arise in these approaches are quite different at least at first sight. Nonetheless, there *are* certain similarities in the results; most of them are obtained by using genuinely non-perturbative techniques. The overall situation makes me believe that we have, so to say, entered a new era in the field of quantum gravity. Through the dense fog, we have caught a fleeting glimpse of the rocks that, we think, will bring us to the distant peak and, in the spirit of Heisenberg, various groups are charting their paths even though they still do not know if they will make it.

In this article, I will attempt to provide a broad overview of the canonical approach based on connections and loops and suggest some directions for future work. I should emphasize that this is not a systematic survey. I will not even attempt to cover all areas in which significant developments have occurred. Rather, I will concentrate on issues which are related to the topics covered by other speakers but which were not discussed in their talks in any detail. I should also emphasize that most of what I will say are personal perspectives which are not necessarily sha by others in the field. (For comprehensive reviews of the canonical approach based on connections and loops see other articles in this volume and, e.g., Ashtekar (1991,1992) and Pullin (1993).)

I begin in section 2 with an assessment of the canonical approach and, using 2+1-dimensional general relativity, illustrate the vision that underlies the program. In section 3, I will summarize some recent mathematical developments which were mentioned by Bernd Brügmann (1994) and Renate Loll (1994) in their talks. These results have made it possible to define the loop transform rigorously for non-Abelian connections with an infinite number of degrees of freedom and they provide a basis for regulating and solving the quantum constraints. Section 4 illustrates how low energy (i.e. laboratory scale) physics can arise from non-perturbative quantum gravity and section 5 presents some promising directions for future work.

2. The canonical approach

This workshop as a whole was devoted to canonical methods in general relativity, with a particular emphasis on quantization. Therefore, in the first part of this section, I will present an assessment of the strengths and weaknesses of these methods in general terms, without restricting myself to the choice of a specific type of dynamical variables or to the details of how the canonical program is carried out. In the second part, I will present my own view of the general framework one can hope to arrive at in quantum gravity through the use of canonical methods.

2.1 An assessment

At the mathematical as well as conceptual level, the quantum theory we best understand is the non-relativistic quantum mechanics of point particles. In this case, the canonical approach provides the "royal road" to quantization. One can even formulate the fundamental kinematic problem as that of obtaining the appropriate representation of the basic canonical commutation relations. The problem of dynamics is then that of defining the Hamiltonian operator and of fully understanding its action on states. Note that the path integral method is *not* a substitute for canonical quantization. It is rather an alternative -and, in the case of scattering problems, a powerful– method of tackling the issue of quantum dynamics. What it provides is the transition amplitudes $\langle \vec{x}', t' \,|\, \vec{x}, t \rangle$ for the particle to go from the position $\vec{x}$ at time t to the position $\vec{x}'$ at time t'. The full quantum mechanical results, however, involve calculating the amplitude for a transition from one *physically realizable* quantum state to another

—say, from a state $\Psi_0(\vec{x})$ with angular momentum 0 to a state $\Psi_m(\vec{x})$ with angular momentum m. And, to obtain these, we need to know the Hilbert space of states, the action of physically interesting operators and the eigenfunctions and eigenvalues of these operators. These have to be supplied externally —typically through canonical quantization. It is for this reason that most courses on non-relativistic quantum mechanics focus on the canonical method.

Coming to field theory, the canonical approach is again the oldest coherent method to tackle the problem of quantization. It is deeply rooted in the Heisenberg uncertainty relation and hence in the fundamental *physical* principles of quantum mechanics. Why is it then that no modern course in field theory develops the subject within the canonical method? The main reasons, I believe, are the following. First, the approach lacks *manifest* covariance. Second, it is not well-suited to the use of Feynman diagrams and other perturbative techniques. The first drawback arises from obvious reasons: The basic canonically conjugate fields satisfy equal time commutation relations and are therefore defined on a Cauchy slice, whence the 4-dimensional covariance is broken at the outset. The origin of the second drawback lies in the fact that the quantum states in the canonical approach arise as suitable functionals of fields and this representation is quite inconvenient if one wants to work with particle states. In perturbative field theory, on the other hand, the emphasis is on particles –real and virtual– and the powerful machinery of Feynman diagrams is tuned to the particle picture. Furthermore, each diagram represents a *space-time* process. The exchange of virtual particles, particles moving forward in time and the anti-particles moving backwards, real particles scattering off each other – each of these processes involves the passage of time and is represented by a 4-dimensional, space-time integral. It is in principle possible but in practice cumbersome to capture all this in the canonical framework. It is hard to prove renormalizability. It is *much* easier to forego the descriptions in terms of functionals of 3-dimensional fields and consider instead the Fock spaces of particle states. Indeed, this is hardly surprising. It is always the case that calculations simplify when one tailors the framework to the dynamics in question. In the perturbative treatment of scattering theory, then, simplifications should occur when we use eigenstates of the asymptotic, free Hamiltonian. And these are precisely the particle states in the Fock space.

One might adopt the viewpoint that from the standpoint of what is true "in principle," both the drawbacks stem from aesthetic considerations. Indeed, although the procedure is not manifestly covariant, it is nonetheless true that, in the final picture, the quantum field theory one obtains, say in the case of a free Maxwell field, is completely equivalent to the more familiar one featuring the Fock space of photons. In the canonical approach, the states are represented by (gauge invariant) functionals $\Psi(A)$ of the vector potential $A_a(\vec{x})$ on a 3-dimensional plane (or, more generally, a Cauchy surface) in Minkowski space. This description breaks the manifest covariance. However, the Poincaré group *is* unitarily implemented. The vacuum state *is* Poincaré invariant. The spectrum of the (free) 4-momentum operator *is* causal and future-directed, and so on. Similarly, as we noted already, using the equivalence with the Fock representation, one can translate, step by step, any perturbative calculation in, say, quantum electrodynamics to the canonical language. This is all true. But the price one would pay would be substantial. It would be analogous to –and *much* more complicated than– insisting on using Cartesian coordinates in a problem where the Hamiltonian has manifest spherical symmetry. Not only would it make calculations enormously more difficult but one would also lose all sorts of *physical* insights. This is not to say that canonical quantization has no place in field theory. It does. Most text books begin the discussion with canonical quantization perhaps because the underlying basic structure is clearest in this framework. Similarly, at the odd place when one is not sure of the measure in the path integral, one returns to the canonical picture and "derives the correct measure" starting with the Liouville form on the phase space. However, for the actual calculations of scattering cross sections or probing general properties such as renormalizability, the language of particles and the 4-dimensional pictures are obviously better-suited.

What about general relativity? The canonical approach has the great advantage that it does not require us to introduce a background metric. It is well suited for a non-perturbative treatment. The final Hamiltonian framework does have a number of features not encountered in Minkowskian field theories. However, in view of the profound conceptual differences between these theories and general relativity, the emergence of such features is but to be expected. In particular, much of the dynamical information of the theory is now contained in constraints. Indeed, as discussed by Robert Beig (1994), in the spatially compact case, the Hamiltonian vanishes identically on the constraint surface, i.e., on the physical states of the classical theory. In quantum theory then, to begin with, there is no Hamiltonian, no time and no evolution. One only has physical states –the solutions to quantum constraints. Yet, it may be possible to "extract" dynamics from these solutions by identifying a suitable physical variable as an internal clock. Indeed, this has already been done in 2+1-dimensional gravity (see e.g. Carlip (1993)). Thus, at least in principle, the approach provides us with a well-defined, precise strategy that is sufficiently sophisticated to extract physical information at the quantum level in spite of the absence of a background geometry. This is its greatest strength.

How does the approach fare with respect to the two key difficulties it faces in Minkowskian quantum field theories? In quantum gravity, the scattering cross sections are not our *prime* concerns. Since there is no background space-time metric, notions such as particle states –and gravitons in particular– which are tied to the Poincaré group would presumably be only approximate concepts. The fact that quantum general relativity (and various modifications thereof) fails to provide us with a consistent, local quantum field theory perturbatively also indicates that the fundamental theory should not be formulated in terms of these particle states. The whole imagery of processes mediated by particles moving forward and backward in time is probably inappropriate. So, the second reason for abandoning the canonical method now loses its force. Indeed, the key question now is if the quantum dynamics of the gravitational field can be made simple in an appropriate representation within the canonical scheme. And, as we have seen in this workshop, there are indeed strong indications that the loop representation is well-suited for this purpose (see e.g. Brügmann, this volume).

The lack of manifest covariance of the canonical scheme, however, is still with us. And indeed it is now a *much more* serious issue. In Minkowskian field theories the issue was an aesthetic one. While the canonical procedure violates *manifest* covariance, it *is* covariant; the Poincaré group *is* unitarily implemented. We can, if we wish, describe dynamics in the space-time picture. In quantum gravity this is no longer the case. In classical general relativity, it is the space-time geometry that is the dynamical variable. A space-time represents a possible history – analogous to a trajectory in particle mechanics. Just as in non-relativistic quantum mechanics the particle trajectories have no basic role to play in the final quantum description, one would expect that in full canonical quantum gravity, space-times would have no distinguished place. There will presumably exist some special ("semi-classical") states which can be approximated by 4-dimensional space-times. Given any such state, one would be able to speak of approximate 4-dimensional covariance in an effective theory which ignores large quantum flucutations away from that state. At a fundamental level, however, there would not be a 4-dimensional geometric entity to replace the classical space-time. And if there is no such entity, how can one even speak of 4-dimensional covariance? It does seem that there is a branching of ways here. Furthermore, it is the qualitative features of the canonical approach that force this branching; it is not tied to the use of specific variables.

To preserve the space-time covariance, it is tempting to choose the path integral approach. Even though in the finished picture there is no preferred space-time, one does have an underlying 4-manifold and the basic objects that one calculates with are 4-metrics. And the diffeomorphism group does act on the space of these metrics. Since the Einstein-Hilbert action is diffeomorphism invariant, one would expect to get diffeomorphism invariant answers for (the correctly formulated) physical questions. Thus, the strategy seems attractive. There are, however, three problems.

First, since path integrals in full quantum gravity are only formal expressions, everything one does *from the very beginning* involves formal manipulations; the level of mathematical precision leaves much to be desired. In practice, one generally proceeds by making some "approximations" which mimic the successful strategies in other field theories. However, it is not at all obvious that these strategies can be taken over to gravity in a meaningful fashion. Indeed, in Minkowskian field theories such as quantum electrodynamics, the only way we know how to make sense of path integrals is through perturbation theory. In fact, the power of the method lies in the fact that it captures the enormous information in the perturbation expansions very succinctly. In a theory that does not exist perturbatively, we are at a loss. It would be surprising indeed if all of the approximation methods that are useful in the perturbative context will continue to be useful in quantum gravity. A much more sophisticated approach is needed. One might imagine using the techniques developed in rigorous quantum field theory. However, in constructive quantum field theories, path integrals are defined in the Euclidean regime and the physical, Lorentzian Green's functions are obtained through a Wick rotation. We all know that, in gravity, this simple route is not available.

The second problem is that while the method is manifestly covariant in the sense of classical general relativity, it lacks *quantum covariance* even in the case of particle dynamics. Let me explain this point in some detail since many of the readers may be unfamiliar with the issue. In the sum over histories approach, one fixes, at the outset, a preferred configuration space —all histories are to be trajectories in this space. In quantum mechanics, this corresponds to fixing a representation (such as the position or the momentum) of the observable algebra. Now, one of the technically powerful and, I feel, conceptually deep features of quantum physics is Dirac's transformation theory[†] which establishes the covariance of the *quantum* theory under the change of representation. This covariance is manifest in the canonical approach. In the path integral approach, not only it is not manifest but it is often cumbersome to incorporate. Con-

---

[†] about which Dirac said in 1977: "I think that is the piece of work which has most pleased me of all the works that I have done in my life...The transformation theory became my darling." (Pais (1987))

sequently, just as it is often hard to phrase "space-time questions" in the canonical approach, it is hard to phrase questions which involve different representations in the path integral approach (unless of course one passes through the canonical approach and makes use of the Dirac transformation theory). Consider for example an harmonic oscillator. If one wants to ask for the probability for the particle which passes through the interval Δ_0 of the position space at time t_0 and the interval Δ_1 at time t_1 to end up in an interval Δ_2 at time t_2, we immediately know we should integrate over the set of paths which pass through the given intervals at the given times. Now suppose we change the question somewhat by replacing the condition at the intermediate time t_1 and ask instead that, at that time, the particle have an energy $\hbar\omega(n+\frac{1}{2})$ for a fixed n. Now, it is no longer obvious what paths to consider. More generally, paths or histories are trajectories in *some* configuration space (i.e. a Lagrangian sub-manifold of the phase space) and path integrals can easily cope with questions that refer to that fixed sub-space. In quantum theory, by contrast, the domain space of wave functions may have nothing to do with the classical phase space. For example, in the case of the harmonic oscillator, we could use the energy representation in which the quantum states are functions $\Psi(n) = \langle n \mid \Psi \rangle$ of an integer which (do not constitute a Lagrangian sub-manifold of the phase space and therefore) can not serve as a classical configuration space. Path integrals are not well-adapted to such representations. But it may well be that it is precisely such representations that are best suited to incorporate quantum dynamics. The loop representation, in particular, falls in this category.

The last problem is the one I already mentioned in the beginning of this sub-section. Path integrals provide transition amplitudes but not the kinematical structure; it has to be supplied from outside. In a real sense, therefore, the path integral and the canonical approaches generally complement each other. In quantum gravity, it is possible that ultimately the two will be used together so that the framework has both classical and quantum covariance. The dynamics coded in the quantum constraints in the canonical approach could perhaps be re-interpreted in terms of mathematically well-defined path integrals to demonstrate that there is a well-defined sense in which the theory enjoys space-time covariance. Such an interpretation would be especially helpful in the analysis of whether topology changes occur in quantum gravity and, if so, whether they have physically significant ramifications in the low energy regime. The full quantum covariance could be established by switching to the canonical framework.

It may also turn out, however, that in the gravitational case, canonical quantization can not be reconciled with path integrals in the full quantum theory. As the discussion in the next sub-section indicates, even the "spatial" 3-manifold could be a secondary construct. At the fundamental level, there may only be discrete structures and combinatorial operations. In such a scenario, the continuum picture and objects such as manifolds may arise as useful mathematical constructs only in semi-classical physics. And the theory would have the desired covariance only in these regimes. It may be that the predictions of the two methods agree in such regimes but in the fully Planckian domain, the two theories are quite different.

To summarize then, the canonical methods are well adapted to non-perturbative treatments of quantum gravity. However, there appears to be a fundamental tension between the quantum gravity scenarios that are natural to canonical approaches and space-time pictures that we are so accustomed to in the classical regime. Whether this tension is real or only apparent is not yet clear. Technical progress, particularly in the path integral approach, would be of great help to settle this issue.

2.2 An underlying vision

In this workshop, Thomas Thiemann discussed 2+1-dimensional gravity in some detail and Hermann Nicolai presented an extension (due to de Wit, Matschull and himself (1993)) of those results to give an elegant clarification of the issue of the "size" of the space of physical states in supergravity. I would now like to use 2+1 gravity for a different purpose: to illustrate my own expectations of non-perturbative quantum gravity in 3+1 dimensions. (For details on 2+1-dimensional gravity, in addition to these proceedings, see, e.g., Carlip (1990, 1993), and Ashtekar (1991), chapter 17.)

Let me begin with a brief historical detour. Conceptually –and, in certain respects, also technically– 2+1-dimensional general relativity is very similar to the 3+1-dimensional theory. There is no background structure; the theory is diffeomorphism invariant. One and the same object –the space-time metric $g_{\mu\nu}$– determines the geometry *and* encodes the gravitational field. In the canonical description, the Hamiltonian again vanishes in the spatially compact case and the dynamics is driven by first class constraints. Furthermore, in geometrodynamics, the general structure of the constraint algebra is the same as in the 3+1 theory (see, e.g., Beig, this volume); it is not a true Lie algebra. Consequently, for a number of years, it was believed that a non-perturbative, canonical quantization of this theory is as difficult as that of the 3+1-dimensional theory. In particular, not a single solution to the Wheeler-DeWitt equation was known. In the perturbative treatment, one finds that Newton's constant is again dimensionful (with dimension [L], thus a *positive* power of length in the $\hbar = c = 1$ units) and simple power counting arguments suggested that the theory is not renormalizable. Therefore, there appeared several papers in the literature saying that 2+1-dimensional general relativity is as intractable as the 3+1-dimensional theory.

This would have been somewhat puzzling since in 2+1-dimensions there are no local degrees of freedom; all solutions to the Einstein field equations are flat. Indeed, the conclusions of these papers turned out to be wrong. There are no divergences. Using connections as basic variables rather than metrics, the quantum theory can in fact be solved exactly in the spatially compact case for any genus. (Achucarro & Townsend (1986), Witten (1988). See also Ashtekar et al (1989) and Nelson & Regge (1989)). Since then, the theory has also been solved in the metric representation in the case when the spatial topology is that of a 2-torus (Moncrief 1989, Hosoya & Nakao (1990)) and work is in progress on the higher genus case. One of the lessons that one can draw from this work is that connections and holonomies are better suited to the mathematics of the quantum theory than metrics and light cones. It is *not* that the theory can not be solved in the metric picture; as I just remarked, this has already been achieved in the 2-torus topology. Rather, connections seem to be better adapted to deal with the quantum constraints and to write down Dirac observables. Variables which are most useful to the macro-physics are not the ones which make the micro-physics most transparent.

There are several equivalent ways of quantizing 2+1 gravity on a torus (see, e.g., Carlip 1993). The variety of representations that result provide us with a rich example of the Dirac transformation theory in action. Here, I will consider three and argue that each clarifies and emphasizes a different aspect of quantum gravity. At first sight, it appears that each representation provides its own, distinct picture of reality. And yet, thanks to the Dirac transformation theory, they are all equivalent.

First, there is the loop representation. Here, quantum states are functions of homotopy classes of loops on the spatial 2-manifold; the domain space of quantum states is thus *discrete*. In the case of a torus, each homotopy class is represented by a pair of integers (n_1, n_2) which tell us how many times the loop winds around the two generators of the homotopy group. Hence, in this case, the quantum states $\Psi(n_1, n_2)$ are functions just of two integers. The basic observables of the theory, $T^0(n_1, n_2)$ and $T^1(n_1, n_2)$, are also labelled by two integers and their action on states just shifts the arguments of the wave functions. Thus, the whole mathematical structure is *combinatorial*. There is no space, no time, no continuum. Hence there is no diffeomorphism group to implement and no issue of space-time covariance to worry about. I would like to regard this as the "fundamental description." Now, we learn from Thomas Thiemann's article in this volume that this description is completely equivalent to that in terms of connections on a 2-torus†. Put differently, suppose we knew nothing about general relativity but were provided with just the combinatorial description given above. Staring at this description, a clever young student could have realized that the description would become "nice and geometric" if one were to introduce a "fictitious, mathematical" 2-manifold, namely a 2-torus, and regard the arguments (n_1, n_2) of quantum states as labels for the homotopy classes of closed loop on this torus. Excitedly, she tells us that, once this is done, one can equally consider the quantum states as functions of two real numbers (a_1, a_2) which can be regarded as labels for gauge equivalent flat connections on the torus. After working in this picture for a while, we realize that the theory can be regarded as quantization of an interesting classical theory which features connections on a 3-dimensional manifold, obtained by taking the product of our fictitious torus with a fictitious real line. To our astonishment, we then notice that the theory has the peculiar property of being diffeomorphism invariant. Furthermore, we notice that it can be thought of as a theory of metrics of signature $-++$ and the connection which we discovered earlier is the spin connection of this metric. We can then go on and study the light cones and causal properties of this classical theory and be surprised that all this rich and unexpected structure arises in the classical limit. We discover that, in this classical approximation, the 3-manifold can be thought of as "space-time" in which things live. Intrigued by this notion of time and dynamics, we return to the quantum theory and realize that with a little bit of work, we can construct a third representation in which quantum states are functions of *three* variables subject to a differential equation which tells us how the state changes or "evolves" as we increase the third variable. We realize that variable could be interpreted as an "internal clock."

Thus, we have three different quantum pictures. The "fundamental" description is combinatorial. The second, in terms of connections, is a "timeless" description. We can interpret the 2-manifold as "space" and see that the spatial diffeomorphisms are unitarily implemented. But there is no time; we have a "frozen formalism" a la Bergmann and Komar. Finally, in the third description, we reintroduce time through an internal

† The two are related by the loop transform. A certain amount of caution is needed in the definition of the loop transform however because the space on which one integrates –the moduli space of flat connections– has non-compact connected components on which the Wilson loop functions are unbounded (Marolf 1993, Ashtekar & Loll (1993)). This important technical point was overlooked in the earlier treatment (Ashtekar et al 1989).

clock. Now the description resembles that of a parametrized particle. We are on familiar ground.

My hope is that we will arrive at a similar description in the 3+1 theory. There will be several equivalent representations, each illuminating its favorite facets of quantum gravity. Thanks to the Dirac transformation theory, the numerous expectations we have of the theory –which often seem even to contradict each other– will in fact be compatible. There is already some evidence in favor of this scenario. The loop and connection representations are qualitatively similar to those in the 2+1-dimensional case. (In fact, the emphasis on connections in the 2+1-dimensional theory was motivated, in the first place, by the connection dynamics formulation of the 3+1 theory.) In the loop representation, the diffeomorphism constraint leads us to knot classes and hence to combinatorial operations. The connection representation will lead us directly to a timeless, frozen formalism. And as we will see in section 4, in the weak field truncation, we can identify one of the components of the connection as an internal clock and recover quantum dynamics from the Hamiltonian constraint. Thus, we do have a number of pointers. Of course, whether this scenario is borne out in detail remains to be seen. In my view, the weakest point may be the last one; we may be able to isolate internal clocks only under suitable approximations.

3. Recent mathematical developments

It is obvious from the articles by Brügmann, Loll and Thiemann that the loop representation plays an important role in this approach to quantum gravity. It is therefore important to put this representation and results obtained therefrom on a sound mathematical footing. It is worth emphasizing that this is not just an abstract desire for rigor. In a real sense, in non-perturbative quantum gravity, we are entering a completely uncharted territory and the mathematical techniques required are quite new. In most of the work to date, one has proceeded by making convenient choices –e.g., of the regularity conditions on loop states, of factor-ordering and of limiting procedures used to regulate operators– as the need arose. These choices do seem "reasonable" and there *is* an overall coherence to the entire scheme. However, it is not clear that there aren't other choices that are equally plausible and that the results will not change if these choices were made. There are no uniqueness results. And none will come by unless one elevates the degree of mathematical precision. In a sense, this is analogous to the situation in interacting quantum field theory in Minkowski space, where there is again no a priori justification for using the Fock representation and the associated regularity conditions on states and operators. (In fact, in view of Haag's theorem (see, e.g., Streater

& Wightman (1964)), there are good mathematical reasons for *not* using it!). However, in that case, over the years one has acquired a great deal of theoretical experience and, more importantly, there is enormous experimental data which supports the strategy as a working hypothesis. In non-perturbative gravity, on the other hand, one lacks both and the issue of control over the assumptions and uniqueness results become much more important.

Fortunately, over the last two years several mathematical developments have occurred which have the potential of improving the situation dramatically. The finished and published results deal only with real $SU(n)$ or $U(n)$ connections. However, work is in progress to extend these results to the complex-valued connections that one encounters in gravity and it appears that the key results which enable one to define the loop transform and the loop representation will go over to that case as well†. For brevity and precision, in this section, I will restrict myself to $SU(2)$ connections and report on the results that have been already obtained.

To define the loop transform, we need an integration theory on the space $\mathcal{A}/\mathcal{G}$ of connections modulo gauge transformations. In the 2+1-dimensional case, this problem is easy to solve because the (appropriate components of the) moduli space of *flat* $SU(1,1)$ connections can be naturally given the structure of a finite dimensional symplectic manifold (see e.g. Ashtekar (1991), chapter 17); one can simply use the Liouville volume element to evaluate integrals. Thus, although the domain space of quantum states is non-linear, the integration theory is simple because the space is *finite* dimensional. In 3+1 space-time dimensions, the situation is again simple for the case of *linearized gravity* –the theory of free gravitons in Minkowski space. This theory can be cast in the language of connections (see, e.g., Ashtekar (1991), chapter 11). Integration theory is again

† The idea is to let the gauge group be $SU(2)$ –as is needed for the triads E_i^a to be real– but to allow complex-valued connections A_a^i (which take values in the Lie-algebra of $\mathbb{C}SU(2)$, complexified $SU(2)$. Thus, we are using the original Hamiltonian framework in which A_a^i is regarded as a complex-valued coordinate on the real phase space of general relativity. One then uses appropriate extensions of the techniques outlined below to develop calculus on the resulting $\mathcal{A}/\mathcal{G}$. In particular, one can give a precise meaning to the notion of holomorphic functionals used in the connection representation. We expect that the appropriate parts of the integration theory will also go through and enable us to define the loop transform rigorously, although here the work is still incomplete. If all our expectations are borne out, we would also have incorporated the reality conditions at the kinematical level, i.e. prior to the imposition of constraints.

well-developed; the domain space is now linear and one can simply use the Gaussian measure as in free field theories in Minkowski space. Thus, in this case, in spite of the presence of an infinite number of degrees of freedom, the integration theory is straightforward because of the underlying *linearity*. The problem is significantly more difficult if $\mathcal{A}/\mathcal{G}$ is *both* non-linear *and* infinite dimensional.

Let us now discuss this case. Let us begin by fixing an analytic 3-manifold Σ which will of course represent a Cauchy surface in space-times to be considered. We will consider $SU(2)$ connections on Σ. Since any $SU(2)$ bundle over a 3-manifold is trivial, we can represent any connection by a Lie-algebra valued 1-form A_a^i on Σ, where a is the spatial index and i, the internal†. Denote by $\mathcal{A}$ the space of smooth (say C^2) $SU(2)$-connections equipped with one of the standard (Sobolev) topologies (see, e.g. Mitter and Viallet (1981)). $\mathcal{A}$ has the structure of an affine space. However, what is of direct interest to us is the space $\mathcal{A}/\mathcal{G}$ obtained by taking the quotient of $\mathcal{A}$ by (C^3) local gauge transformations. In this quotient construction, the affine structure is lost; $\mathcal{A}/\mathcal{G}$ is a genuinely non-linear space with complicated topology.

To define the integration theory, we will adopt an algebraic approach. To see the main idea, consider integrals $\int f d\mu$ of continuous functions f on a compact, Hausdorff space S with respect to a regular (but not necessarily strictly positive) measure $d\mu$. The space of continuous functions has the structure of an Abelian $C^\star$-algebra with identity and the map $f \to < f > := \int f d\mu$ is a functional on the algebra satisfying the following two properties: i) positivity (which should really be called non-negativity), i.e. $< \bar{f} f > \ \geq 0$ for any element f in the algebra; and, ii) linearity, i.e. $< f + \lambda g > = < f > + \lambda < g >$ for all elements f and g of the algebra and complex numbers λ. Thus, given a measure, we acquire a positive linear functional on the given Abelian $C^\star$-algebra. Now, standard theorems ensure us that *every* regular measure on S arises in this manner. The idea now is to use this 1-1 correspondence to define measures on the space of connections modulo gauge transformations.

Let us begin by constructing an appropriate sub-algebra of the Abelian $C^\star$-algebra of continuous, bounded functions on $\mathcal{A}/\mathcal{G}$. This is to be the algebra of configuration variables. Therefore, the obvious strategy is to

† Results reported in this section for which explicit references are not provided are all taken from Ashtekar and Isham (1992) and Ashtekar and Lewandowski (1993a,b). The last two papers and those by Baez (1993a,b) contain significant generalizations which include allowing more general gauge groups, allowing the manifold Σ to be of arbitrary dimension and allowing the connections to live in non-trivial bundles.

use the Wilson loop functionals. Following the procedure and notation used by other speakers, given any closed loop α on the 3-manifold Σ, let us define the Wilson-loop functional T_α on $\mathcal{A}/\mathcal{G}$:

$$T_\alpha(A) := \tfrac{1}{2}\,\mathrm{Tr}\ \mathcal{P}\ \exp\ G \oint_\alpha A.dl, \tag{3.1}$$

where the trace is taken in the fundamental representation of $SU(2)$ and Newton's constant G appears because, in general relativity, it is GA_a^i that has the dimensions of a connection; in gauge theories, of course, this factor would be absent. For technical reasons, we will have to restrict ourselves to piecewise analytic loops α. (This is why we needed Σ to be analytic. Note that the loops need not be smooth; they can have kinks and self-intersections but only at a finite number of points.) As explained in detail by Brügmann and Loll, due to $SU(2)$ trace identities, the product of any two Wilson-loop functionals can be expressed as a sum of other Wilson loop functionals. Therefore, the vector space generated by finite complex-linear combinations of these functions has the structure of a $\star$-algebra (where the $\star$-operation is just complex-conjugation). The functionals T_α are all bounded (between -1 and 1). Hence, the sup-norm (over $\mathcal{A}/\mathcal{G}$) is well-defined and we can take the completion to obtain a $C^\star$-algebra. We will call it the *holonomy $C^\star$-algebra* and denote it by $\mathcal{HA}$. Elements of $\mathcal{HA}$ are to be thought of as the configuration variables of the theory.

Since $\mathcal{HA}$ is an Abelian $C^\star$-algebra with identity, we can apply a standard theorem due to Gel'fand and Naimark to conclude that $\mathcal{HA}$ is isomorphic with the $C^\star$-algebra of all continuous functions on a compact, Hausdorff space $sp(\mathcal{HA})$, the spectrum of the given $C^\star$-algebra $\mathcal{HA}$. The spectrum itself can be recovered from the algebra $\mathcal{HA}$ directly; its points are homomorphisms from $\mathcal{HA}$ to the $\star$-algebra of complex numbers. Now, we know that the elements of $\mathcal{HA}$ suffice to separate points of $\mathcal{A}/\mathcal{G}$; given any two elements of $\mathcal{A}/\mathcal{G}$ there is at least one loop α such that T_α takes different values on the two elements. Therefore it follows from the Gel'fand-Naimark theorem that $\mathcal{A}/\mathcal{G}$ is *densely embedded in $sp(\mathcal{HA})$*. To emphasize this point, from now on, we will denote the spectrum by $\overline{\mathcal{A}/\mathcal{G}}$ and regard it as a completion of $\mathcal{A}/\mathcal{G}$ (in the Gel'fand topology). Integration theory will be defined on $\overline{\mathcal{A}/\mathcal{G}}$ rather than $\mathcal{A}/\mathcal{G}$. This enlargement is in accordance with the common occurrence in quantum field theory: while the classical configuration (or phase) space may contain only smooth fields (typically taken to belong to be the Schwartz space), the domain space of quantum states is a completion of this space in an appropriate topology (the space of distributions). From the remarks I made above, it is clear that regular measures on $\overline{\mathcal{A}/\mathcal{G}}$ will correspond to positive linear functions on the holonomy algebra $\mathcal{HA}$. This is the general strategy we will follow.

However, a key difficulty with the use of the Gel'fand theory is that one generally has relatively little control on the structure of the spectrum. A reasonable degree of control is essential since (at the kinematical level) the quantum states in the connection representation are to be functions $\Psi(\bar{A})$ on $\overline{\mathcal{A}/\mathcal{G}}$. In the present case, we are fortunate: a simple and complete characterization of the spectrum is available. To present it, I first need to introduce a key definition. Fix a base point x_o in the 3-manifold Σ and regard two (piecewise analytic) closed loops α and α' to be equivalent if the holonomy of any connection in $\mathcal{A}$, evaluated at x_o, around α is the same as that around α'. We will call each equivalence class (a holonomically equivalent loop or) a *hoop* and denote the hoop to which a loop α belongs by $\tilde{\alpha}$. For example, α and α' define the same hoop if they differ by a reparametrization or by a line segment which is immediately re-traced. (For piecewise analytic loops and $SU(n)$ connections, one can show that these two are the most general operations; two loops define the same hoop if and only if they are related by a combination of reparametrizations and retracings.) The set of hoops has, naturally, the structure of a group. We will call it the *hoop group* and denote it by $\mathcal{HG}$. In terms of this group, we can now present a simple characterization of the Gel'fand spectrum $\overline{\mathcal{A}/\mathcal{G}}$:

> *Every homomorphism $\hat{H}$ from the hoop group $\mathcal{HG}$ to the gauge group $SU(2)$ defines an element $\bar{A}$ of the spectrum $\overline{\mathcal{A}/\mathcal{G}}$ and every $\bar{A}$ in the spectrum defines a homomorphism $\hat{H}$ such that $\bar{A}(\tilde{\alpha}) = \frac{1}{2}\mathrm{Tr}\ \hat{H}(\tilde{\alpha})$. This is a 1-1 correspondence modulo the trivial ambiguity that homomorphisms $\hat{H}$ and $g^{-1}\cdot\hat{H}\cdot g$ define the same element $\bar{A}$ of the spectrum.*

Clearly, every regular connection A in $\mathcal{A}$ defines the desired homomorphism simply through the holonomy operation: $\hat{H}(\tilde{\alpha}) := \mathcal{P}\ \exp G \oint_\alpha A.dl$, where α is any loop in the hoop $\tilde{\alpha}$. However, there are many homomorphisms which do not arise from smooth connections. This leads to "generalized connections" –i.e. elements in $\overline{\mathcal{A}/\mathcal{G}} - \mathcal{A}/\mathcal{G}$. In particular, there exist $\bar{A}$ in $\overline{\mathcal{A}/\mathcal{G}}$ which have support at a single point and are thus "distributional." Note that this characterization of the spectrum $\overline{\mathcal{A}/\mathcal{G}}$ is *completely algebraic*; there is no continuity assumption on the homomorphisms. This property makes the characterization very useful in practice.

As noted above, positive linear functionals on $\mathcal{HA}$ are in 1-1 correspondence with regular measures on (the compact Hausdorff space) $\overline{\mathcal{A}/\mathcal{G}}$. It turns out that the positive linear functions, in turn, are determined completely by certain "generating functionals" $\Gamma(\alpha)$ on the space $\mathcal{L}_{x_o}$ of loops based at x_o:

> *There is a 1-1 correspondence between positive linear functionals on $\mathcal{HA}$ (and hence regular measures on $\overline{\mathcal{A}/\mathcal{G}}$) and functional $\Gamma(\alpha)$*

on $\mathcal{L}_{x_o}$ satisfying:
i) $\sum_i a_i T_{\alpha_i} = 0 \Rightarrow \sum_i a_i \Gamma(\alpha_i) = 0$; *and,*
ii) $\sum_{i,j} \bar{a}_i a_j (\Gamma(\alpha_i \circ \alpha_j) + \Gamma(\alpha_i \circ \alpha_j^{-1}) \geq 0$.
for all loops α_i and complex numbers a_i.

The first condition implies that the functional Γ is well-defined on hoops. Hence we could have taken it to be a functional on $\mathcal{HG}$ from the beginning. Thus, we see that there is a nice "non-linear duality" between the spectrum $\overline{\mathcal{A}/\mathcal{G}}$ and the hoop group $\mathcal{HG}$: Elements of $\overline{\mathcal{A}/\mathcal{G}}$ are homomorphisms from $\mathcal{HG}$ to $SU(2)$ and regular measures on $\overline{\mathcal{A}/\mathcal{G}}$ correspond to certain functionals on $\mathcal{HG}$. Finally, if one is interested in measures on $\overline{\mathcal{A}/\mathcal{G}}$ which are invariant under the (induced) action of diffeomorphisms on Σ, one is led to seek functionals $\Gamma(\alpha)$ which depend not on the individual loops α but rather on the (generalized) knot class to which α belongs. (The qualification "generalized" refers to the fact that here we are allowing the loops to have kinks, overlaps and self-intersections. Until recently, knot theorists considered only smoothly embedded loops.) Thus, there is an interesting –and potentially powerful– interplay between knot theory and representations of the holonomy algebra $\mathcal{HA}$ in which the diffeomorphism group of Σ is unitarily implemented.

Finally, we can make the integration theory more explicit as follows. Consider a subgroup S_n of the hoop group $\mathcal{HG}$ which is generated by n independent hoops. We can introduce the following equivalence relation on $\overline{\mathcal{A}/\mathcal{G}}$: $\bar{A} \approx \bar{A}'$ if and only if their action on all elements of S_n coincides, i.e., if and only if $\bar{A}(\tilde{\alpha}) = g^{-1} \cdot \bar{A}'(\tilde{\alpha}) \cdot g$ for all $\tilde{\alpha} \in S_n$ and some (hoop independent) $g \in SU(2)$. Thus, intuitively, to be equivalent, $\bar{A}$ and $\bar{A}'$ have to agree at points on the set of loops in S_n; their behavior elsewhere on Σ does not matter. *It turns out that the quotient space is isomorphic to* $[SU(2)]^n/Ad$ (where the adjoint action of $SU(2)$ on $[SU(2)]^n$ is given by $(g_1, ..., g_n) \rightarrow (g^{-1} \cdot g_1 \cdot g, ..., g^{-1} \cdot g_n \cdot g)$ where g is an arbitrary element of $SU(2)$ and $(g_1, ..., g_n)$ is an arbitrary element of $[SU(2)]^n$). This is the key result that enables us to make the space of integrable functions and the integration procedure itself more transparent. It says that the quotients obtained using two different sub-groups of the hoop group are isomorphic if and only if the two sub-groups are generated the same number of independent hoops. The details of the subgroups do not matter. There is a certain "universality" to the structure of the quotient.

We want to be able to integrate the generalized Wilson loop functionals $T_{\tilde{\alpha}}(\bar{A}) := \bar{A}(\tilde{\alpha})$ on $\overline{\mathcal{A}/\mathcal{G}}$. Using the quotient construction given in the last paragraph, we will first present a convenient characterization of the space of all integrable functions on $\overline{\mathcal{A}/\mathcal{G}}$. Let us begin with a definition. A function f on $\overline{\mathcal{A}/\mathcal{G}}$ will be said to be *cylindrical* if it is the pull-back to $\overline{\mathcal{A}/\mathcal{G}}$ of a smooth function $\tilde{f}$ on $[SU(2)]^n/Ad$ for some sub-group S_n of the hoop group. These will be the integrable functions and they include the generalized Wilson loops. Next, we define integrals of these functions f on $\overline{\mathcal{A}/\mathcal{G}}$ explicitly using the integrals of $\tilde{f}$ on $[SU(2)]^n/Ad$, provided of course we equip $[SU(2)]^n/Ad$ with suitable measures $d\mu_n$ for each n. We can then first define a positive linear functional Γ' on the space of cylindrical functions f via:

$$\Gamma'(f) := \int_{[SU(2)]^n/Ad} \tilde{f} d\mu_n \,. \tag{3.2}$$

For the functional to be well-defined, of course, the family of measures $d\mu_n$ on $[SU(2)]^n/Ad$ must satisfy certain consistency conditions. It turns out that these requirements can be met. The resulting functionals $\Gamma'(f)$ define regular measures $d\mu'$ on $\overline{\mathcal{A}/\mathcal{G}}$ such that $\Gamma'(f) = \int f d\mu'$.

A particularly natural choice (and, not surprisingly, the first to be discovered) is to let $d\mu_n$ be simply induced on $[SU(2)]^n/Ad$ by the Haar-measure on $SU(2)$. With this choice, we have the following results:

i)The consistency conditions are satisfied; the left side of (3.2) is well-defined for all cylindrical functions f on $\overline{\mathcal{A}/\mathcal{G}}$;
ii) The generalized holonomies $T_{\tilde{\alpha}}$ are cylindrical functionals on $\overline{\mathcal{A}/\mathcal{G}}$ and $\Gamma(\alpha) := \Gamma'(T_{\tilde{\alpha}})$ defined via (3.2) serves as a generating functional for a faithful, cyclic representation of the honomony $C^\star$-algebra $\mathcal{HA}$ which ensures that $d\mu$ is a regular, strictly positive measure on $\overline{\mathcal{A}/\mathcal{G}}$;
iii) The measure $d\mu$ is invariant under the induced action of the diffeomorphism group on Σ.

(The knot invariant defined by $d\mu$ is a genuinely generalized one; roughly, it counts the number of self-overlaps in any given loop.)

This measure is in some ways analogous to the Gaussian measure on linear vector spaces. Both can be obtained by a "cylindrical construction." The Gaussian measure uses the natural metric on $\mathbb{R}^n$ while the above measure uses the natural (induced) Haar measure on $[SU(2)]^n/Ad$. They are both regular and strictly positive. This leads us to ask if other properties of the Gaussian measure are shared. For instance, we know that the Gaussian measure is concentrated on distributions; although the smooth fields are dense in the space of distributions in an appropriate topology, they are contained in a set whose total measure is zero. Is the situation similar here? The answer turns out to be affirmative. The classical configuration space $\mathcal{A}/\mathcal{G}$ with which we began is dense in the domain space $\overline{\mathcal{A}/\mathcal{G}}$ of quantum states in the Gel'fand topology. However, $\mathcal{A}/\mathcal{G}$ is contained in a set whose total measure is zero. The measure is again concentrated on "generalized" connections in $\overline{\mathcal{A}/\mathcal{G}}$ (Marolf & Mourão (1993)).

In a certain sense, just as the Gaussian measures on linear spaces originate in the harmonic oscillator, the new measure on $\overline{\mathcal{A}/\mathcal{G}}$ originates in a (generalized) rotor (whose configuration space is the $SU(2)$ group-manifold). However, the measure is, so to say, "genuinely" tailored to the underlying non-linearity. It is *not* obtained by "perturbing" the Gaussian measure.

With the measure $d\mu$ at hand, we can consider the Hilbert space $L^2(\overline{\mathcal{A}/\mathcal{G}}, d\mu)$ and introduce operators on it. This is *not* the Hilbert space of physical states of quantum gravity since we have not imposed constraints. It is a fiducial, *kinematical* space which is to enable us to regularize various operators (in particular, the quantum constraint operators). The configuration operators are associated with the generalized Wilson loop functionals: $\hat{T}_{\tilde{\alpha}} \circ \Psi(\bar{A}) = \bar{A}(\tilde{\alpha})\Psi(\bar{A})$. One can show that they are bounded, self-adjoint operators on the Hilbert space. There are also "momentum operators" –associated with closed, 2-dimensional ribbons or strips in the 3-manifold Σ– which are gauge invariant and linear in the electric field. One can show that these are also self-adjoint (but unbounded). Finally, since $d\mu$ is invariant under the induced action of the diffeomorphism group of Σ, this group acts unitarily.

Next, we can now make the Rovelli-Smolin loop transform (Rovelli & Smolin (1990), see also Gambini & Trias (1986), Brügmann, Loll, this volume) rigorous. Since each generalized Wilson loop function $T_{\tilde{\alpha}}(\bar{A})$ is cylindrical and belongs to the Hilbert space $L^2(\overline{\mathcal{A}/\mathcal{G}}, d\mu)$, the loop transform can be defined simply as a scalar product:

$$\psi(\tilde{\alpha}) := \int_{\overline{\mathcal{A}/\mathcal{G}}} \overline{T_{\tilde{\alpha}}}(\bar{A})\Psi(\bar{A})\, d\mu\,. \tag{3.3}$$

As desired, it enables us to pass from functionals $\Psi(\bar{A})$ of generalized connections to functionals $\psi(\tilde{\alpha})$ of hoops†. With the rigorously defined transform at hand, we can take over various operators from the connection to the loop representation. For the Wilson loop operators $\hat{T}^0_{\tilde{\alpha}}$ as well as the strip operators –which are the smeared versions of $\hat{T}^a_{\tilde{\alpha}}(s)$– we recover the same answers as have been used in all the calculations in the loop representation. For the operator $\hat{T}^{ab}(s,t)$, which is used in the regularization of the metric and of the Hamiltonian constraint operators, work is still in progress.

---

† Since the left side is a function of hoops, we should, strictly, use the terms hoop representation and the hoop transform. However, in various calculations, it is often convenient to lift these functionals from the hoop group $\mathcal{HG}$ to the space of loops $\mathcal{L}_{x_o}$. Therefore, as in most of the literature on the subject, we will not keep a careful distinction between loops and hoops in what follows.

Finally, let us consider the constraints. Our task here is to represent the quantum constraints as well-defined operators on the Hilbert space $L^2(\overline{\mathcal{A}/\mathcal{G}}, d\mu)$, take them over to the loop side and solve them. For the diffeomorphism constraint, most of this task has been carried out. The operator is well-defined and self-adjoint on the connection states and can be taken over to the loop side unambiguously. The result is the expected one: the operator acts on the loop states $\psi(\alpha)$ by displacing the loop α in the argument via the corresponding diffeomorphism. Its functional form is the one used in the literature. It is therefore intuitively clear that the solutions to this constraint are functions of (generalized) knot classes. The question is: What type of functions are we to allow? As explained in Petr Hajiček's article, typically, zero is a continuous eigenvalue of the constraint operators whence elements of the kernel are not normalizable. Thus, the availability of the Hilbert space by itself does not provide us with the regularity conditions that must be satisfied by the *physical* states. For this, as Hajiček pointed out, we need a rigged Hilbert space construction. To fix the ideas, let me recall the situation in the case of a relativistic particle, where the classical constraint is $P^\alpha P_\alpha + \mu^2 = 0$. In this case, the kinematical Hilbert space can be taken to be $L^2(\mathbb{R}^4)$. This space is needed to translate the classical constraint function to a well-defined operator (whose kernel can then be found). The operator, of course, is $-\hbar^2\eta^{\alpha\beta}\partial_\alpha\partial_\beta + \mu^2$. It is self-adjoint on the kinematic Hilbert space $L^2(\mathbb{R}^4)$. However, no (non-zero) element in its kernel is normalizable. These elements belong to the rigged Hilbert space; in the momentum space, they are distributions with support on the mass shell. To find out the regularity conditions on the (generalized) knot invariants, we need to carry out a similar construction and isolate the appropriate rigged Hilbert space. Work is in progress on this problem.

For the Hamiltonian constraint, there is still no progress to report on the mathematical side. This is not surprising in the light of the fact that work on the operators $\hat{T}^{ab}(s,t)$ is still in progress. But the general line of attack is clear. At the heuristic level, there have been several distinct approaches to the problem of constructing this operator in the loop representation but they have all led to equivalent results (Brügmann & Pullin (1993)). In an optimistic view, this could be taken as a signal that the operator *does* exist rigorously.

To summarize then, there has been a considerable amount of rigorous work in the last two years and the goal of giving a precise meaning to the loop transform and the constraint operators as well as that of extracting the regularity conditions on the physical states now seems attainable in the near future. The main open problems are: i) extending the theory to incorporate complex connections; ii) writing the Hamiltonian constraint as a well-defined operator on $L^2(\overline{\mathcal{A}/\mathcal{G}}, d\mu)$; and, iii) constructing the appropriate rigged Hilbert spaces to obtain regularity conditions on the physical states. Work is in progress on all these problems.

Thus, it appears that the integration theory based on the measure $d\mu$ will provide the mathematical basis for this non-perturbative approach to quantum gravity. However, these techniques may be used also in other theories of connections which are diffeomorphism invariant and perhaps even in Yang-Mills theory which is not diffeomorphism invariant. We saw that the full domain space of quantum theory, $\overline{\mathcal{A}/\mathcal{G}}$, can be thought of as the space of homomorphisms from the full hoop group $\mathcal{HG}$ to the gauge group $SU(2)$. Given a finitely generated sub-group S_n of the hoop group, we can consider the space of homomorphisms from it to $SU(2)$. This provides the space $[SU(2)]^n/Ad$ which is precisely the domain space of quantum states of a lattice gauge theory where the lattice is not rectangular but tailored to the given subgroup S_n of the gauge group. Thus, what we have is a set of "floating lattices," each associated with a finitely generated subgroup of the hoop group. The space $\overline{\mathcal{A}/\mathcal{G}}$ can be rigorously recovered as a projective limit of the configuration spaces of lattice theories (Marolf & Mourão 1993). This construction is potentially quite powerful; it may enable one to take continuum limits of operators of lattice theories in a completely new fashion. The limit is obtained not by taking the lattice separation to zero but by enlarging lattices to probe the continuum connections better and better, i.e., by considering larger and larger subgroups of the hoop group. This is a good illustration of the synergetic exchange of ideas between general relativity and gauge theories that is made possible by this approach to quantum gravity.

4. Emergence of low energy physics

One of the important features that any non-pertrubative approach to quantum gravity must have is that, in a suitable limit, it should reproduce the laboratory scale physics correctly. Does the theory admit states which can be interpreted as smooth geometries on large scales? Is there at least an approximate notion of time which is compatible with a space-time picture? Can one regard non-gravitational physical fields as evolving unitarily with respect to this time? Is there an approximate Hamiltonian governing this evolution? One often takes for granted that the answers to such questions would be obviously "yes." However, in genuinely non-perturbative treatments, this is by no means clear a priori; one may be working in a sector of the quantum theory which does not admit the correct or unambiguous classical limit. For example, the sector may correspond to a confined phase which has no classical analog or the limit may yield a wrong number even for the macroscopic dimensions of space-time!

In this section, I will present two examples of such results that have been obtained within the present non-perturbative framework which were not discussed by other speakers. The first involves the existence of quantum states which approximate classical geometries at macroscopic scales (Ashtekar et al, 1992). I will begin by showing that certain operators representing *geometrical observables* can be regulated in a way that respects the diffeomorphism invariance of the underlying theory. What is more, these regulated operators are finite *without any renormalization.* Using these operators, one can ask if there exist loop states which approximate smooth geometry at large scales. Not only is the answer in the affirmative but, furthermore, *these states exhibit a discrete structure of a definite type at the Planck scale.* (For further details, see, e.g., Rovelli & Smolin (1990), Ashtekar (1992) and Smolin (1993).) The second example involves the issue of time in a certain weak field truncation of the theory. Here, I will only summarize overall the situation, emphasizing those points which will be needed in the next section (For details, see, Ashtekar (1991), chapter 12 and, for more recent developments, Rovelli and Smolin (1993).)

4.1 Non-perturbative regularization

Let us begin with the issue of regularization. As noted by Giulini (this volume), in the present framework, the spatial metric is constructed from products of "electric fields" E^a_i. It is thus a "composite" field given by $q^{ab}(x) = E^{ai}(x)E^b_i(x)$. (This field is, as emphasized by other speakers, a density of weight 2. To keep the notation simple, however, I will drop the tildes that are generally used to denote the density weights.) In the quantum theory, therefore, this operator must be regulated. The obvious possibility is point splitting. One might set $q^{ab}(x) = \lim_{y\to x} E^{ai}(x)E^b_i(y)$. However, the procedure violates gauge invariance since the internal indices at two *different* points have been contracted. There is, however, a suitable modification that will ensure gauge invariance. Consider the field $T^{aa'}[\alpha](y',y)$, labelled by a closed loop α and points y and y' thereon, defined in the classical theory by:

$$T^{aa'}[\alpha](y,y') := \frac{1}{2}\,\mathrm{Tr}\left[(\mathcal{P}\exp G\int_{y'}^{y} A_b dl^b)E^a(y')(\mathcal{P}\exp G\int_{y}^{y'} A_c dl^c)E^{a'}(y)\right]. \tag{4.1}$$

In the limit as α shrinks to zero, $T^{aa'}[\alpha](y,y')$ tends to $-4q^{aa'}$.

Now, in quantum theory, one can define the action of the operator $\hat{T}^{aa'}[\alpha](y,y')$ directly on the loop states $\psi(\beta)$. The explicit form will not be needed here. We only note that using the bra-ket notation, $\psi(\beta) = \langle\beta\,|\,\psi\rangle$, the action can be specified easily. Indeed, $\langle\beta\,|\circ\hat{T}^{aa'}[\alpha](y,y')$ is rather simple: if a loop β does not intersect α at y or y', the operator simply annihilates the bra $\langle\beta \mid$ while if an intersection does occur, it breaks and re-routes the loop β, each routing being assigned a specific weight. One may therefore try to define a quantum operator $\hat{q}^{aa'}$ as a limit of $\hat{T}^{aa'}[\alpha]$ as α shrinks to zero. The resulting operator does exist after suitable regularization *and* renormalization. However (because of the density weights involved) the operator necessarily carries a memory of the background metric used in regularization. Thus, the idea of defining the metric operator again fails. In fact one can give general qualitative arguments to say that there are no *local*, operators which carry the metric information *and* which are independent of background fields (used in the regularization). Thus, in quantum theory, the absence of background fields introduces new difficulties. That such difficulties would arise was recognized quite early by Chris Isham and John Klauder.

There do exist, however, *non-local* operators which can be regulated in a way that respects diffeomorphism invariance.

As the first example, consider the function $Q(\omega)$ –representing the smeared 3-metric– on the classical phase space, defined by

$$Q(\omega) := \int d^3x \ \ (q^{ab}\omega_a\omega_b)^{\frac{1}{2}}, \tag{4.2}$$

where ω_a is any smooth 1-form of compact support. (Note that the integral is well-defined without the need of a background volume element because q^{ab} is a density of weight 2.) It is important to emphasize that, in spite of the notation, $Q(\omega)$ is *not* obtained by smearing a *distribution* with a test field; because of the square-root, $Q(\omega)$ is *not* linear in ω. We can, nonetheless define the corresponding quantum operator as follows. Let us choose on Σ test fields $f_\epsilon(x,y)$ (which are densities of weight one in x and) which satisfy:

$$\lim_{\epsilon\to 0}\int_\Sigma d^3x f_\epsilon(x,y) g(x) = g(y) \tag{4.3}$$

for all smooth functions of compact support $g(x)$. If Σ is topologically $\mathbb{R}^3$, for example, we can construct these test fields as follows:

$$f_\epsilon(x,y) = \frac{\sqrt{h(x)}}{\pi^{\frac{3}{2}}\epsilon^3}\exp -\frac{\mid \vec{x}-\vec{y}\mid^2}{2\epsilon^2}, \tag{4.4}$$

where $\vec{x}$ are the Cartesian coordinates labeling the point x and $h(x)$ is a "background" scalar density of weight 2. Next, let us define

$$q_\epsilon^{aa'}(x) = -\frac{1}{4}\int_\Sigma d^3y \int_\Sigma d^3y' f_\epsilon(x,y) f_\epsilon(x,y') T^{aa'}(y,y'). \tag{4.5}$$

As ϵ tends to zero, the right side tends to q^{ab} because the test fields force both the points y and y' to approach x, and hence the loop passing through y, y', used in the definition of $T^{aa'}(y, y')$, to zero. It is now tempting to try to define a local metric operator $\hat{q}^{aa'}$ corresponding to $q^{aa'}$ by replacing $T^{aa'}(y, y')$ in (4.5) by its quantum analog and then taking the limit. One finds that the limit does exist provided we first renormalize $\hat{q}_\epsilon^{aa'}$ by an appropriate power of ϵ. However, as before, the answer depends on the background structure (such as the density $h(x)$) used to construct the test fields $f_\epsilon(x, y)$. If, however, one tries to construct the quantum analog of the *non-local* classical variable $Q(\omega)$, this problem disappears. To see this, let us first express $Q(\omega)$ using (4.5) as:

$$Q(\omega) = \lim_{\epsilon \to 0} \int_\Sigma d^3x (q_\epsilon^{aa'} \omega_a \omega_{a'})^{\frac{1}{2}}. \tag{4.6}$$

The required quantum operator $\hat{Q}(\omega)$ on the loop states can now be obtained by replacing $T^{aa'}(y, y')$ by the operator $\hat{T}^{aa'}(y, y')$. A careful calculation shows that: i) the resulting operator *is* well-defined on loop states; ii) no renormalization is necessary, i.e., the limit is automatically *finite*; and, iii) the final answer carries no imprint of the background structure (such as the density $h(x)$ or, more generally, the specific choice of the test fields $f_\epsilon(x, y)$) used in regularization. To write out its explicit expression, let me restrict myself to smooth loops α without any self-intersection. Then, the action is given simply by:

$$\langle \alpha \mid \circ \hat{Q}(\omega) = l_P^2 \oint_\alpha ds |\dot{\alpha}^a \omega_a| \cdot \langle \alpha \mid, \tag{4.7}$$

where $l_P = \sqrt{G\hbar}$ is the Planck length, s, a parameter along the loop and $\dot{\alpha}^a$ the tangent vector to the loop. In this calculation, the operation of taking the square-root is straightforward because the relevant operators are diagonal in the loop representation. This is analogous to the fact that, in the position representation of non-relativistic quantum mechanics, we can set $< x| \circ \sqrt{\hat{X}^2} =< x| \cdot |x|$ without recourse to the detailed spectral theory. The G in l_P of (4.7) comes from the fact that GA_a^i has the usual dimensions of a connection while $\hbar$ comes from the fact that $\hat{E}_i^a$ is $\hbar$ times a functional derivative. The final result is that, on non-intersecting loops, the operator acts simply by multiplication: the loop representation is well-suited to find states in which the 3-geometry –rather than its time evolution– is sharp.

The second class of operators corresponds to the area of 2-surfaces. Note first that, given a smooth 2-surface S in Σ, its area $\mathcal{A}_S$ is a function on the classical phase space. We first express it using the classical loop variables. Let us divide the surface S into a large number N of area elements $S_I, I = 1, 2...N$, and set $\mathcal{A}_I^{\mathrm{appr}}$ to be

$$\mathcal{A}_I^{\mathrm{appr}} = -\frac{1}{4}\left[\int_{S_I} d^2S^{bc}(x)\eta_{abc}\int_{S_I} d^2S^{b'c'}(x')\eta_{a'b'c'}\,T^{aa'}(x,x')\right]^{\frac{1}{2}}, \quad (4.8)$$

where η_{abc} is the (metric independent) Levi-Civita density of weight -1. It is easy to show that $\mathcal{A}_I^{\mathrm{appr}}$ approximates the area function (on the phase space) defined by the surface elements S_I, the approximation becoming better as S_I –and hence loops with points x and x' used in the definition of $T^{aa'}$– shrink. Therefore, the total area $\mathcal{A}_S$ associated with S is given by

$$\mathcal{A}_S = \lim_{N\to\infty}\sum_{I=1}^{N}\mathcal{A}_I^{\mathrm{appr}}. \quad (4.9)$$

To obtain the quantum operator $\hat{\mathcal{A}}_S$, we simply replace $T^{aa'}$ in (4.8) by the quantum loop operator $\hat{T}^{aa'}$. This somewhat indirect procedure is necessary because, as indicated above, there is no well-defined operator-valued distribution that represents the metric or its area element *at a point.* Again, the operator $\hat{\mathcal{A}}_S$ turns out to be finite. Its action, evaluated on a nonintersecting loop α (for simplicity), is given by:

$$\langle\alpha| \circ \hat{\mathcal{A}}_S = \frac{l_P^2}{2}\, I(S,\alpha)\cdot\langle\alpha|, \quad (4.10)$$

where $I(S,\alpha)$ is simply the *unoriented* intersection number between the 2-surface S and the loop α. (One obtains the *un*oriented intersection number here and the absolute sign in the integrand of (4.7) because of the square-root operation involved in the definition of these operators.) Thus, in essence, "a loop α contributes half a Planck unit of area to any surface it intersects."

The fact that the area operator also acts simply by multiplication on non-intersecting loops lends further support to the idea that the loop representation is well-suited to "diagonalize" operators describing the 3-geometry. Indeed, we can immediately construct a large set of simultaneous eigenbras of the smeared metric and the area operators. There is one, $\langle\alpha|$, associated to every nonintersecting loop α. Note that the corresponding eigenvalues of area are *quantized* in integral multiples of $l_P^2/2$. There are also eigenstates associated with intersecting loops which, however, I will not go into since the discussion quickly becomes rather involved technically.

4.2 Weaves

With these operators on hand, we can now turn to the construction of quantum loop states that approximate the classical metric h_{ab} on Σ on a scale large compared to the Planck length. The basic idea is to weave the classical metric out of quantum loops by spacing them so that, on average, precisely one line crosses *any* surface element whose area, *as measured by the given* h_{ab}, is one Planck unit. Such loop states will be called *weaves*. Note that these states are not uniquely picked out since our requirement is rather weak. Indeed, given a weave approximating a specific classical metric, one can obtain others, approximating the same classical metric.

Let us begin with a concrete example of such a state which will approximate a *flat* metric h_{ab}. To construct this state, we proceed as follows. Using this metric, let us introduce a random distribution of points on $\Sigma = \mathbb{R}^3$ with density n (so that in any given volume V there are $nV(1+\mathcal{O}(1/\sqrt{nV}))$ points). Center a circle of radius $a = (1/n)^{\frac{1}{3}}$ at each of these points, with a random orientation. We assume that $a << L$, so that there is a large number of (non-intersecting but, generically, *linked*) loops in a macroscopic volume L^3. Denote the collection of these circles by Δ_a. As noted in section 4, due to trace identities, products of Wilson loop functionals $T_{\tilde{\alpha}}$ can be expressed as linear combinations of Wilson loop functionals. As a consequence, it turns out that the bras defined by multi-loops are equivalent to linear combinations of single loop bras. Therefore, for each choice of the parameter a, there is a well-defined bra $\langle\Delta_a|$. This is our candidate weave state.

Let us consider the observable $\hat{Q}[\omega]$. To see if $\langle\Delta_a|$ reproduces the geometry determined by the classical metric h_{ab} on a scale $L >> l_p$, let us introduce a 1-form ω_a which is *slowly varying on the scale* L and compare the value $Q[\omega](h)$ of the classical $Q[\omega]$ evaluated at the metric h_{ab}, with the action of the quantum operator $\hat{Q}[\omega]$ on $\langle\Delta_a|$. A detailed calculation yields:

$$\langle\Delta_a|\circ\hat{Q}[\omega] = \left[\frac{\pi}{2}\,(\frac{l_p}{a})^2\,Q[w](h) + \mathcal{O}(\frac{a}{L})\right]\cdot\langle\Delta_a|. \qquad (4.11)$$

Thus, $\langle\Delta_a|$ is an eigenstate of $\hat{Q}[\omega]$ and the corresponding eigenvalue is closely related to $Q[\omega](h)$. However, even to the leading order, the two are unequal *unless* the parameter a –the average distance between the centers of loops– *equals* $\sqrt{\pi/2}\,l_p$. More precisely, (4.11) can be interpreted as follows. Let us write the leading coefficient on the right side of this equation as $(1/4)(2\pi a/l_p)(nl_p^3)$. Since this has to be unity for the weave to reproduce the classical value (to leading order), we see that Δ_a should contain, on average, one fourth Planck length of curve per Planck volume, where lengths and volumes are measured using h_{ab}.

The situation is the same for the area operators $\hat{A}_S$. Let S be a 2-surface whose extrinsic curvature varies slowly on a scale $L >> l_P$. One can evaluate the action of the area operator on $\langle\Delta_a|$ and compare the eigenvalue obtained with the value of the area assigned to S by the given flat metric h_{ab}. Again, the eigenvalue can be re-expressed as a sum of two terms, the leading term which has the desired form, except for an overall coefficient which depends on the mean separation a of loops constituting Δ_a, and a correction term which is of the order of $\mathcal{O}(\frac{a}{L})$. We require that the coefficient be so adjusted that the leading term agrees with the classical result. This occurs, again, precisely when $a = \sqrt{\pi/2}l_p$. It is interesting to note that the details of the calculations which enable one to express the eigenvalues in terms of the mean separation are rather different for the two observables. In spite of this, the final constraint on the mean separation is *precisely* the same.

Let us explore the meaning and implications of these results.

1) It is generally accepted that, to obtain classical behavior from quantum theory, one needs two things: i) an appropriate coarse graining, and, ii) special states. In our procedure, the slowly varying test fields ω_a and surfaces S with slowly varying extrinsic curvature enable us to perform the appropriate coarse graining while weaves –with the precisely tuned mean separation a– are the special states. There is, however, something rather startling: The restriction on the mean separation a –i.e., on the *short distance* behavior of the multi-loop Δ_a– came from the requirement that $\langle\Delta_a|$ should approximate the classical metric h_{ab} on *large scales* L!
2) In the limit $a \to \infty$, the eigenvalues of the two operators on $\langle\Delta_a|$ go to zero. This is not too surprising. Roughly, in a state represented by any loop α, one expects the quantum geometry to be excited just at the points of the loops. If the loops are *very* far away from each other as measured by the fiducial h_{ab}, there would be macroscopic regions devoid of excitations where the quantum geometry would seem to correspond to a zero "metric".
3) The result of the opposite limit, however, *is* surprising. One might have naively expected that the best approximation to the classical metric would occur in the continuum limit in which the separation a between loops goes to zero. However, the explicit calculation outlined above shows that this is not the case: as a tends to zero, the leading terms in the eigenvalues of $\hat{Q}[\omega]$ and $\mathcal{A}_S$ actually diverge! (One's first impulse from lattice gauge theories may be to say that the limit is divergent simply because we are not rescaling, i.e., renormalizing the operator appropriately. Note, however, that, in contrast to the calculations one performs in lattice theories, here, we *already* have

a well defined operator in the continuum. We are only probing the properties of its eigenvectors and eigenvalues, whence there is nothing to renormalize.) It is, however, easy to see the reason underlying this behavior. Intuitively, the factors of the Planck length in (4.7) and (4.10) force each loop in the weave to contribute a Planck unit to the eigenvalue of the two geometrical observables. In the limit $a \to 0$, the number of loops in any fixed volume (relative to the fiducial h_{ab}) grows unboundedly and the eigenvalue diverges.

4) It is important to note the structure of the argument. In non-perturbative quantum gravity, there is no background space-time. Hence, terms such as "slowly varying" or "microscopic" or "macroscopic" have, a priori, no physical meaning. One must do some extra work, introduce some extra structure to make them meaningful. The required structure should come from the very questions one wants to ask. Here, the questions had to do with approximating a classical geometry. Therefore, we could *begin* with a classical metric h_{ab}. We used it repeatedly in the construction: to introduce the length scale L, to speak of "slowly varying" fields ω_a and surfaces S, and to construct the weave itself. The final result is then a consistency argument: If we construct the weave according to the given prescription, then we find that it approximates h_{ab} on macroscopic scales L, provided we choose the mean separation a to be $\sqrt{\pi/2}l_p$, where all lengths are measured relative to the same h_{ab}.

5) Note that there is a considerable non-uniqueness in the construction. As we noted already, a given 3-geometry can lead to distinct weave states; our construction only serves to make the existence of such states explicit. For example, there is no reason to fix the radius r of the individual loops to be a. For the calculation to work, we only need to ensure that the loops are large enough so that they are generically linked and small enough so that the values of the slowly varying fields on each loop can be regarded as constants plus error terms which we can afford to keep in the final expression. Thus, it is easy to obtain a 2-parameter family of weave states, parametrized by r and a. The condition that the leading order terms reproduce the classical values determined by h_{ab} then gives a relation between r, a and l_P which again implies discreteness. Clearly, one can further enlarge this freedom considerably: There are a lot of eigenbras of the smeared-metric and the area operators whose eigenvalues approximate the classical values determined by h_{ab} up to terms of the order $\mathcal{O}(\frac{l_P}{L})$ since this approximation ignores Planck scale quantum fluctuations.

6) Finally, I would like to emphasize that, at a conceptual level, the important point is that the eigenvalues of $\hat{Q}[\omega]$ and $\mathcal{A}[S]$ can be *discrete.*

Let me conclude the discussion on weaves with two remarks. First, it is not difficult to extend the above construction to obtain weave states for curved metrics g_{ab} which are slowly varying with respect to a flat metric h_{ab}. Given such a metric, one can find a slowly varying tensor field $t_a{}^b$, such that the metric g_{ab} can be expressed as $t_a{}^c t_b{}^d h_{cd}$. Then, given a weave of the type $\langle\Delta|$ considered above approximating h_{ab}, we can "deform" each circle in the multi-loop Δ using $t_a{}^b$ to obtain a new weave $\langle\Delta|_t$ which approximates g_{ab} in the same sense as $\langle\Delta|$ approximates h_{ab}. (See, also, Zegwaard (1993, 1994) for the weave corresponding to the Schwarzschild black-hole.) The second remark is that since the weaves are eigenbras of the operators that capture the 3-geometry, they do not approximate 4-geometries. To obtain a state that can approximate Minkowski space-time, for example, one has to consider a loop state that resembles a "coherent state" peaked at the weave Δ_a. In that state, neither the 3-geometry nor the time-derivative thereof would be sharp; but they would have minimum spreads allowed by the uncertainty principle. This issue has been examined in detail by Iwasaki and Rovelli (1993).

Since these results are both unexpected and interesting, it is important to probe their origin. We see no analogous results in familiar theories. For example, the eigenvalues of the fluxes of electric or magnetic fields are not quantized in QED nor do the linearized analogs of our geometric operators admit discrete eigenvalues in spin-2 gravity. Why then did we find qualitatively different results? The technical answer is simply that the familiar results refer to the *Fock representation* for photons and gravitons while we are using a completely different representation here. Thus, the results *are* tied to our specific choice of the representation. Why do we not use Fock or Fock-like states? It is not because we insist on working with loops rather than space-time fields such as connections. Indeed, one *can* translate the Fock representation of gravitons and photons to the loop picture. (See, e.g., Ashtekar et al (1991) and Ashtekar & Rovelli, (1992).) And then, as in the Fock space, the discrete structures of the type we found in this section simply disappear. However, to construct these loop representations, one must use a flat background metric and essentially every step in the construction violates diffeomorphism invariance. Indeed, there is simply no way to construct "familiar, Fock-like" representations without spoiling the diffeomorphism invariance. Thus, the results we found are, in a sense, a direct consequence of our desire to carry out a genuinely non-perturbative quantization without introducing any background structure. However, we do not have a uniqueness theorem singling out the measure $d\mu$ which was used to define the loop transform and hence to construct the loop representation used here. There do exist other diffeomorphism invariant measures which will lead to other loop representations. The measure we have used is the most natural and the simplest among the known ones. Whether the results presented here depend sensitively on the choice of the measure is not known. Therefore, it would be highly desirable to have a uniqueness theorem which tells us that, on physical grounds, we should restrict ourselves to a specific (class of) measure(s).

My overall viewpoint is that one should simultaneously proceed along two lines: i) one should take these results as an indication that we are on the right track and push heuristic calculations within this general framework; and, ii) one should try to put the available heuristic results on rigorous mathematical footing to avoid the danger of "wandering off" in unsound directions.

4.3 The issue of time

The results reported in the last two sub-sections were kinematical; constraints were not yet imposed. This is because our questions themselves referred to 3-metrics which fail to be Dirac observables; in the classical analysis, 3-metrics are well-defined only on the full (and hence kinematical) phase space rather than on the reduced phase space (the space of physical or "true" degrees of freedom). On the other hand, as Hajiček explained in his article, the issue of time is *not* kinematical; it is intimately intertwined with the scalar or the Hamiltonian constraint. Therefore, we must now bring in this constraint. We will show that in a certain truncation of the full quantum theory, the issue of time can be resolved satisfactorily in the asymptotically flat context.

It turns out that, for technical reasons sketched below, (in pure gravity) it is the connection representation that seems best-suited to handle this issue. Thus, the question we wish to now ask is the following: can we single out a component of the connection, ${}^{T}\!A$ which can serve as "internal" time? More precisely, can we re-interpret the scalar –or Hamiltonian– constraint as telling us how the "true" or the dynamical degrees of freedom ${}^{D}\!A$ evolve with respect to the internal clock ${}^{T}\!A$? The questions themselves are not new. They have been asked in the context of geometrodynamics since the sixties. It turned out, however, that one can not isolate time in this manner in the metric representation of geometrodynamics even in the weak field truncation (Kuchař (1970)). Why does one then hope that the situation would be any better in connection dynamics?

Let us begin by analyzing the source of the difficulty. Consider, first, a parametrized, non-relativistic system. Let the phase space coordinates be $(q^1, ...q^n; p_1, ...p_n)$ and let the constraint be:

$$p_1 + H(q^1, q^2, ..., q^n; p_2, ...p_n) = 0. \qquad (4.12)$$

Now, if in the quantum theory, we used the q-representation, the quantum constraint has the same form as the time-dependent Schrödinger equation,

$i\hbar\partial\Psi(q^i)/\partial q^1 = \hat{H}\circ\Psi(q^i)$, where q^1 plays the role of time. On the other hand, the momentum representation is equally viable mathematically. The quantum constraint would now take the form $\hat{p}^1\Psi(p^i) = -\hat{H}\circ\Psi(p^i)$; it does *not* resemble the time-dependent Schrödinger equation. In particular, it is now hard to isolate time from among the configuration variables on which the wave function depends. Thus, although two representation may be mathematically equivalent, one may be better suited to deal with the issue of time. Since the Dirac transformation theory enable us to pass from one to another, the issue here is that of convenience. However, since so many of the issues related to the interpretation of the framework hinge on the availability of a (possibly approximate) time variable, a representation which is well-suited to extract an internal clock has central importance. Let us now return to gravity. Recall from Beig's lectures that, in the asymptotically flat case, the Hamiltonian of geometrodynamics can be written as $H_N(q,p) = C_N(q,p) + \mathcal{E}(q)$, where N is any lapse which tends to 1 at infinity at an appropriate rate and $\mathcal{E}(q)$ is the ADM energy integral which depends only on the 3-metric q_{ab} but not on the conjugate momentum p^{ab}. Hence, the scalar constraint corresponding to this lapse can be written in the form:

$$\mathcal{E}(q) - H_N(q,p) = 0, \tag{4.13}$$

which is identical in form to (4.12), with $-\mathcal{E}(q)$ playing the role of p_1. The similarity suggests that it is the variable that is "conjugate" to $-\mathcal{E}(q)$ that should play the role of time. This idea is attractive also because $\mathcal{E}(q)$ has the interpretation of the total energy of the system. Furthermore, the structure of (4.13) is unaffected by the presence of matter sources; it is universal. However, the similarity also suggests that the metric representation is not likely to be well-suited to extract time from among the gravitational degrees of freedom. The connection representation would be better suited since $\mathcal{E}(q)$ would be a differential operator in the A-representation. We will see below that the expectation is borne out in the weak field truncation of the theory. Note finally that this argument also explains why it is unlikely that the gravitational loop variables would be well-suited to extract time: as we saw in the last two sub-sections, the loop representation is diagonal in the operators carrying information of the 3-geometry†. With these general remarks out of the way, we can now explain the precise sense in which the scalar constraint equation can be interpreted as a time dependent Schrödinger equation.

---

† If we couple gravity to matter, however, the picture can change. It is then possible that one of the matter variables can play the role of an internal clock. This is an attractive strategy especially in the spatially compact case.

We will now assume that the underlying 3-manifold Σ is topologically $\mathbb{R}^3$. Let us begin by introducing a background point in the phase space, $({}^oA_a^i = 0, {}^oE_i^a)$, where ${}^oE_i^a$ is a flat triad and expand out the fields $\hat{A}_a^i$ and $\hat{E}_i^a$ that appear in the constraint operators, in the powers of the deviations $\hat{A}_a^i - 0$ and $\hat{E}_i^a - {}^oE_i^a$. Since we now have access to a background triad, ${}^oE_i^a$, it will be convenient to convert the internal indices to vector indices on all dynamical fields. Thus, quantum states can now be regarded as functionals of $A_{ab} := A_a^i\, {}^oE_{ib}$. Next, using the flat metric ${}^oq_{ab}$ obtained from ${}^oE_i^a$, we can decompose the symmetric part of A_{ab} into its transverse-traceless, longitudinal and trace parts. Let us now impose the quantum constraints order by order in the deviation. The first order equations imply that the wave functions must depend only on the symmetric, transverse-traceless parts. That is, all other parts of $\hat{A}_{ab}$ are at least of second order. Next, imposing the scalar constraint to the second order yields:

$$-\tfrac{2}{G}\,(\Delta\ {}^T\hat{E})\ \circ\ \Psi({}^DA, {}^TA) = G\ ({}^{TT}\hat{A}_{ab})^\star\ \ ({}^{TT}\hat{A}^{ab})\ \cdot\ \Psi({}^DA, {}^TA), \quad (4.14)$$

where Δ is the Laplacian with respect to ${}^oq_{ab}$, ${}^T\hat{E}$ is the trace part of $\hat{E}^{ab}$, ${}^{TT}\hat{A}_{ab}$ is the (symmetric) transverse traceless part of $\hat{A}_{ab}$, and where, as before, ${}^TA$ and ${}^DA$ are the trace and the dynamical (i.e. all but trace) parts of A_{ab}. It is now natural to introduce the variable $\tau(x)$ which is conjugate to the operator on the left side of this equation: $\tau(x) = (-G/2)\Delta^{-1}.{}^TA(x)$, to write the equation more explicitly as:

$$\hbar\ \frac{\delta\Psi({}^DA, {}^TA)}{\delta\tau(x)} = G\ ({}^{TT}\hat{A}_{ab})^\star\ \ ({}^{TT}\hat{A}^{ab})\ \cdot\ \Psi({}^DA, {}^TA). \quad (4.15)$$

Eq (4.15) tells us how to determine the value of the physical state $\Psi({}^DA, {}^TA)$ *everywhere* on the configuration space from its value on one ${}^TA =$ *const.* hypersurface. In this sense, the equation can be thought of as a quantum constraint. It tells us that the physical states are not freely specifiable; their functional form is constrained. To obtain the "evolution" interpretation, it suffices to use just "one component" of this functional differential equation. The idea is simply to integrate the equation on Σ which corresponds to taking its moment with a lapse $N = 1$. We can simplify the left side of the resulting equation using the fact that the states are all *holomorphic* functionals of A_{ab} and the right side by going to the momentum space. The result is:

$$i\hbar\Big(\int d^3x \frac{\delta}{\delta \mathrm{Im}\tau(x)}\Big)\circ\Psi({}^DA, {}^TA)$$
$$= \hbar\Big(\int d^3k|k|\ {}^{TT}A_{ab}(k)\ \frac{\delta}{\delta {}^{TT}A_{ab}(k)}\Big)\circ\Psi({}^DA, {}^TA). \quad (4.16)$$

Thus, the imaginary part, $\mathrm{Im}\tau(x)$, of $\tau(x)$ plays the role of time. The operator on the right side is precisely the Hamiltonian of the truncated theory. Thus, in the second order truncation, the scalar constraint reproduces the time-dependent Schrödinger equation. This feature remains intact if one includes matter fields as sources.

As I have emphasized, in non-perturbative quantum gravity, we do not have access to a classical space-time. Even in the above truncation procedure, we worked only with the 3-manifold Σ and expanded operators around some background fields on Σ. Thus, we did not have access to a classically defined time variable. Rather, we were able to isolate, from among the mathematical variables contained in A_a^i, a preferred time t which serves as an "internal clock" with respect to which the wave function evolves. Put differently, by identifying time in the components of A_a^i, we have derived the Schrödinger equation of weak-field gravity without having a direct access to a space-time metric or even a 4-manifold. After having obtained the result, we can look back and see that we could have obtained the same result in a space-time picture. That is, the true dynamics takes place in the infinite dimensional configuration space of connections. However, *in suitable approximations* that dynamics can be re-interpreted as taking place in a suitably constructed 4-dimensional space-time. There may, however, be instances when the approximation would break down and no space-time picture is adequate. However, if any non-perturbative description is to be viable at all, it better be the case that it reduces to Minkowskian physics with Schrödinger evolution for quantum fields in an appropriate limit. As Kuchař (1970) pointed out almost 25 years ago, this is hard to achieve directly in geometrodynamics. Connection dynamics, as we have seen, is better suited for this task.

Finally, there is another approach to the extraction of time: coupling to matter fields. Recently, Rovelli and Smolin (1993) have made significant progress by coupling gravity to a scalar and other matter fields and using the scalar field to define the internal clock. The remarkable aspect of the development is that the resulting *true* quantum Hamiltonian is a manageable operator in the loop representation. The precise domain in which time can be so extracted –and hence, the nature of the implicit approximation scheme– is, however, not clear yet. Nonetheless, the fact that concrete calculations can be done in the full, non-truncated theory, is already quite exciting. The problem of time is an old one and much effort has gone into trying to find a clean, exact solution to that problem. These approximation methods and concrete calculations may well provide the type of insight that has been missing.

5. Outlook

As articles in this volume illustrate, in recent years there has been notable progress on some of the difficult issues in quantum gravity. In particular, we have seen that connection dynamics offers a unified mathematical framework for all four fundamental interactions. Not only do the loop space methods enable us to go back and forth successfully between gauge theories and gravity but they have also made contact with to the fertile area of knot theory. On the mathematical side, the subject has achieved a surprising degree of maturity over the last two years and one can look forward to further contributions from the mathematical physics community -the constructive field theorists, the $C^{\star}$-algebra community, topologists and knot theorists- in the coming years.

As Brügmann discussed in some detail, the program is of course far from being complete. But the situation in 2+1-dimensional gravity suggests that the various difficulties we face may be primarily technical. With enough work, in the near future we should be able to establish that either the diverse goals of the program can in fact be met or that the program encounters *specific* unsurmountable obstacles. The second type of result would also represent concrete progress. After all, the program is driven by a "radical conservatism" —it is based on the well established principles of general relativity and quantum theory and does not begin by guessing what the Planck scale structure ought to be. Therefore, if it runs into specific problems, it is likely that these limitations themselves will suggest the required modifications.

What directions is the program likely to take in the near future? It is perhaps of interest to list a few open problems whose solution will considerably improve the current state of the field. This is *not* an exhaustive list. My primary goal is to illustrate, with concrete examples, promising directions in the hope that these examples will stimulate young researchers as well as experts in various areas to contribute to the field:

1) *Mathematical issues:* As we saw in section 3, the level of mathematical precision in the Rovelli-Smolin loop transform has improved considerably. However, very little is known regarding the inverse transform. Can we characterize the loop functionals in the image of the transform in a simple way without having to refer to the connection representation? Can we define a measure on the hoop group and define the inverse transform? Does the equivalent of the Plancherel theorem – which makes the Fourier transform so powerful– hold? A promising direction is being pursued by Gambini and his collaborators where the hoop group is replaced by a group of "smoothened out" hoops (Di Bortolo et al, 1993). This group has the structure of a Lie group and offers a new approach to the problem of regularization of vari-

ous operators. This may well be the framework needed to make the inverse transform well-defined.

2) *Euclideanization:* In constructive quantum field theory, the developments which culminated in the Osterwalder-Schrader system of Euclidean axioms caused a burst of activity. Suddenly, a variety of powerful techniques became available. The key result was the demonstration of the equivalence of the Osterwalder-Schrader system to the Wightman system: if we have a theory satisfying the first set, there exists a theory satisfying the second, even though it may be hard in practice to construct it explicitly. Thus, it is still the Lorentzian theory that is of physical interest; it is just that the mathematical problem of constructing such a theory can be reduced to a more tractable problem in the Euclidean framework. Does an analogous Euclidean system exist for general relativity? We know that naive attempts at trying to get at the physical, Lorentzian theory via the obvious Euclidean constructions can not succeed (Mena-Marugan, 1993). Are there more subtle constructions? The discovery of such a procedure would be a key contribution since many of the technically hard problems simplify enormously in the Euclidean domain.

3) *Midisuperspaces:* The midi-superspace of solutions to the (3+1-dimensional) Einstein equation with one (space-like) Killing field is an especially fertile area for future work. Mathematically, this system is equivalent to 2+1-dimensional general relativity coupled to a nonlinear sigma-model. If the Killing field in question is hypersurface orthogonal, the matter field reduces to a single scalar field satisfying the wave equation. (If, in addition, the norm of the Killing field is constant, the matter field disappears and we have vacuum 2+1 gravity, which we already know how to quantize.) Thus, in these midisuperspaces, the problem reduces to that of quantizing 2+1 gravity with matter. It is a genuine field theory with an infinite number of degrees of freedom. However, it has two key simplifications. First, the reality conditions now become significantly simpler: through an "internal Wick-rotation" along the Killing direction, one is led to consider *real* $SU(1,1)$ connections in place of the complex $SU(2)$ ones (Ashtekar & Varadarajan, 1992a). Second, in the asymptotically flat case, the total Hamiltonian of the system is (not only non-negative but also) *bounded from above*, indicating strongly that the quantum theory may be finite (Ashtekar & Varadarajan, 1992b). This midisuperspace thus has just the right blend of technical simplifications and physical generality to serve as a valuable stepping stone to the full theory.

4) *Solutions to the Hamiltonian constraint:* The "smoothened loop" techniques introduced by Gambini and his collaborators have led to interesting new solutions to the Hamiltonian constraints which are related to some well-known knot invariants (Brügmann et al, 1992a,b). Furthermore, it is clear that these are just the simplest applications of the framework. These methods are powerful and have opened up a new line of attack to the problem of solving the Hamiltonian constraint. Obtaining explicit solutions and understanding their physical meaning is important because that would strengthen our intuition considerably. However, it is also highly desirable to get a handle on the *structure* of the space of solutions. To define the inner-product on physical states, for example, it is not necessary to have the solutions explicitly. What we need is an understanding of the various mathematical structures that naturally exist on this space. So far, this issue has attracted very little attention. Finally, recently Barbero-Gonzalez (1993) has recast general relativity in 3 and 4 dimensions as a theory of *two* connections in which the triad field is secondary and does not even enter the basic equations. This is an exciting development with a lot of potential and may, in particular, open up new avenues to the issue of solving the quantum constraints.
5) *Approximation methods* To address physical issues, we need to develop the approximation methods further. For example, we saw in section 4.3 that by a truncation procedure around Minkowskian initial data, one can recover the familiar laboratory physics from the scalar constraint. An obvious question then is what would happen if we truncated around a black-hole initial data. Can one automatically recover the Hawking effect? There are indications that one would. The analysis of "embroidery around weaves" due to Iwasaki and Rovelli (1993) is likely to play an important role in this procedure. The final result will be a derivation of the Hawking effect from the loop representation of full quantum gravity (i.e., from "above") rather than from quantum field theory in curved space-times (i.e., from "below"). A second problem involves matter couplings. Rovelli and Smolin (1993) have considered the coupling to a scalar field while Matschull and Nicoli (1993a, b) and Morales-Técotl and Rovelli (1994) have discussed fermion coupling. As I mentioned in section 4.3, Rovelli and Smolin use the scalar field to define an approximate time variable and thus recover a time dependent Schrödinger equation with a true Hamiltonian. For spinor field sources, they are developing an approximation technique to analyze the action of this Hamiltonian. A great deal of physical understanding is expected to come from such analyses of matter couplings.

Different people, of course, have different expectations of quantum gravity. To conclude, I would like to ask your indulgence for a moment or two as I express my personal views and prejudices.

The aim of the canonical program, as I see it, is to find out if a consistent quantum version of general relativity coupled to matter fields can exist non-perturbatively. Thus, one focusess on certain issues and, at least at the first go, ignores others. The focus is on quantum geometry, on diffeomorphism invariance, on internal clocks, on mathematical methods needed in any non-perturbative treatment of gravity, and, although these were not discussed at this workshop, on new approximation methods. The program is certainly far from being complete. However, it has not run into an unsurmountable obstacle either. The basic spirit, I believe, is the same as the one that drove physicists when they were groping in the dark during the development of quantum mechanics. One pushes the promising ideas and the plausible techniques as hard as one can. Either they keep working or break down at some point. If they break down, one focuses on where and why they fail. Occasionally, the breakdown suggests brand new strategies and changes the direction of the effort substantially. We know that, during the development of quantum mechanics, most ideas did not work and, at any given time, almost everyone was barking up the wrong tree. And yet this strategy of pushing the ideas ruthlessly as far as they can go always produced interesting results. In the same spirit, the hope now is that we will learn something interesting and perhaps even something deep. The evolution from metrics to connections to loops to weaves and knots represents progress along these lines.

Even if the program succeeds and produces a consistent, non-perturbative theory, there is no guarantee, of course, that this theory would be the correct one. Indeed, to make concrete progress, some issues were put aside at first and, in the final analysis, these may well turn out to be so central to the problem that they have to be incorporated right at the start. For example, it may well be, as the majority of the particle physics community holds, that real progress would not occur unless one has, from the very beginning, a principle which unifies the dynamics of all interactions. It may be inappropriate to seek insights into the quantum nature of geometry without incorporating *all* excitations of the superstring. It may also be, as several relativists believe, that real progress would not occur unless the approach is geared to tackle the measurement problems of quantum mechanics from the start; unless the mathematical structure underlying quantum mechanics is made to absorb some fundamental non-linearity at the outset. Most people working on this canonical approach are sympathetic to these ideas in a general way. Indeed, work is in progress on both of these frontiers. However, the general sentiment is that the approach need not be *based* on such premises. Once the subject has evolved sufficiently, new ideas will come up to tackle such issues. There is often a tendency to underestimate the value of having a consistent, non-perturbative theory irrespective of whether it ultimately turns out to be the correct one physically. There appears to be an overriding sentiment that a quantum theory of gravity *must* solve all sorts of problems including those that are not, at least in any obvious way, intrinsic to gravity. And when one lists all these issues and all the associated problems, one is often so struck by the enormity of the task that a sense of hopelessness seems to take over. For a long time now, I think, there has been an undercurrent of pessimism at least in some parts of the relativity community: One does what one can but secretly (or, sometimes, openly!) believes that the task is *way* beyond us. This may well be the case. But I think it is also not obvious that this is really the case. Perhaps we should not try to solve all problems at once. We can suitably restrict our goals and pursue these "modest" programs with full enthusiasm and hope of success. I think we need a more outgoing attitude here, more aggressive spirits and more optimistic hearts!

Acknowledgments

I benefited a great deal from the talks as well as informal discussions with the participants of this workshop. My special thanks go to Helmut Friedrich for organizing this stimulating meeting and for his patience with all administrative matters. This work was supported in part by the NSF grant PHY93-96246 and by the Eberly research fund of the Pennsylvania State University.

References

Achucarro, A. & Townsend, P. (1986) Phys. Lett. **B180**, 85-89.

Agishtein, A. & Migdal, A. (1992) Mod. Phys. Lett. **7**, 1039-1061.

Amati, D. Ciafaloni, M. & Veneziano, G. (1990) Nucl. Phys. **B347**, 550-590.

Aspinwall, P. S. (1993) Mirror symmetry, talk at the 1993 Rutherford-Appleton Meeting.

Ashtekar, A. (1991) *Non-perturbative canonical gravity*, World Scientific, Singapore.

Ashtekar, A. (1992) Mathematical problems of non-perturbative quantum general relativity, SU-GP-92/11-2; to appear in the proceedings of the 1992 Les Houches summer school, B. Julia (ed), Elsevier.

Ashtekar, A., Husain, V., Rovelli, C., Samuel, J. & Smolin, L. (1989) Class. Quan. Grav. **6**, L185-L193.

Ashtekar, A., Rovelli, C. & Smolin, L. (1991) Phys. Rev. **D44**, 1740-1755.

Ashtekar, A. & Isham, C. (1992) Class. Quan. Grav. **9**, 1069-1100.
Ashtekar, A. & Rovelli, C. (1992) Class. Quan. Grav. **91** 1121-1150.
Ashtekar, A., Rovelli, C. & Smolin L. (1992) Phys. Rev. Lett. **69** 237-240.
Ashtekar, A. & Varadarajan, M. (1992a,b) in preparation.
Ashtekar, A. & Lewandowski, J. (1993a) Representation theory of analytic holonomy C^* algebras; to appear in *Knots and quantum gravity*, J. Baez (ed), Oxford University press.
Ashtekar A. & Lewandowski, J. (1993b) Differential calculus on $\mathcal{A}/\mathcal{G}$, in preparation.
Ashtekar, A. & Loll, R. (1994) New loop representations for 2+1-dimensional gravity.
Baez, J. (1993a) Diffeomorphism-invariant generalized measures on the space of connections modulo gauge transformations, hep-th/9305045; to appear in the Proceedings of conference on quantum topology, L. Crane and D. Yetter (eds).
Baez J. (1993b) Link invariants, functional integration and holonomy algebras,
hep-th/9301063.
Barbero-Gonzalez, J. F. (1993) General relativity as a theory of two connections, CGPG-93/9-5; Int. J. Mod. Phys. **D**(in press).
Brügmann, B., Gambini, R. & Pullin, J. (1992a) Phys. Rev. Lett. **68**, 431-434.
Brügmann, B., Gambini, R. & Pullin, J. (1992b) In: *XXth International conference on differential geometric methods in physics*, S. Cotta & A. Rocha (eds), World Scientific, Singapore.
Brügmann, B. & Marinari, E. (1993) Phys. Rev. Lett. **70** 1908-1911.
Brügmann, B. & Pullin, J. (1993) Nucl. Phys. **390** 399-438.
Carlip, S. (1990) Phys. Rev. **D42**, 2647-2654.
Carlip, S. (1993) Six ways to quantize (2+1)-dimensional gravity, gr-qc/9305020.
de Wit, B., Matschull, H.J. & Nicolai, H. (1993) Phys. Lett. **B318**, 115.
Di Bortolo, R. Gambini & Griego, J. (1992) Comm. Math. Phys. (to appear)
Gambini, R. & Trias A. (1986) Nucl. Phys. **B278**, 436-448.
Gross, D. & Mende, P (1988) Nucl. Phys. **B303**, 407-454.
Heisenberg, W. (1963) Interview by T.S. Kuhn of February 25th.
Hosoya, A. & Nakao, K. (1990) Class. Quan. Grav. **7** 163-176.
Iwasaki, J. & Rovelli, C. (1993) Int. J. Mod. Phys. **D1** 533-558.
Kuchař (1970) J. Math. Phys. **11** 3322-3334.
Loll, R. (1992) Nucl. Phys. **B368**, 121-142.

Marolf, D. (1993) Loop representations of 2+1 gravity on a torus, SUGP-93/3/1.
Marolf D. & Mourão J. (1993) Carrier space of the Ashtekar-Lewandowski measure, in preparation.
Mena-Marugán, G. A. (1993) Reality conditions for Lorentzian and Euclidean gravity in the Ashtekar formulation, CGPG-93/9-3; Int. J. Mod. Phys. **D** (in press).
Matschull, H.J. & Nicolai, H. (1993) DESY 93-073; Nucl Phys. **B** (in press)
Mitter, P.K. & Viallet C. (1981) Commun. Math. Phys. **79** 43-58.
Moncrief, V. (1989) J. Math. Phys. **30** 2907-2914.
Nelson, J. & Regge, T. (1989) **A4**, 2021-2030.
Nicolai, H. & Matschull, H.J. (1993) J. Geo. & Phys. **11** 15-62.
Pais, A. (1987) In: *Paul Adrian Maurice Dirac*, B. N. Kursunoglu & E. P. Wigner (eds), Cambridge University Press, Cambridge, page 96.
Morales-Técotl, H.A. & Rovelli, C. (1994) UNiversity of Pittsburgh pre-print.
Pullin, J. (1993) Knot theory and quantum gravity in loop space: A premier, hep-th/9301028, to appear in the proceedings of the Vth Mexican school on particles and fields, J.L. Lucio (ed), Word Scientific, Singapore.
Rovelli, C. & Smolin, L. (1990) Nucl. Phys. **B331**, 80-152.
Rovelli, C. & Smolin, L. (1993) The physical Hamiltonian in nonperturbative quantum gravity, CGPC-93/8-2.
Smolin, L. (1993) In: *General relativity and gravitation 1992*, R.J. Gleiser et al (eds), Institute of Physics, Bristol, pages 229-261.
Streater, R. & Wightman, A. (1964) *PCT, spin and statistics, and all that*, Benjamin, New York.
Witten, E. (1988) Nucl. Phys. **B311**, 46-78.
Zegwaard, J. (1993) Phys. Lett. **B378** 217-220.
Zegwaard, J. (1994) *The loop representation of canonical quantum gravity and its interpretation* Ph.D. thesis, University of Utrecht.

Springer-Verlag and the Environment

We at Springer-Verlag firmly believe that an international science publisher has a special obligation to the environment, and our corporate policies consistently reflect this conviction.

We also expect our business partners – paper mills, printers, packaging manufacturers, etc. – to commit themselves to using environmentally friendly materials and production processes.

The paper in this book is made from low- or no-chlorine pulp and is acid free, in conformance with international standards for paper permanency.

Lecture Notes in Physics

For information about Vols. 1–397
please contact your bookseller or Springer-Verlag

Vol. 398: T. M. M. Verheggen (Ed.), Numerical Methods for the Simulation of Multi-Phase and Complex Flow. Proceedings, 1990. VI, 153 pages. 1992.

Vol. 399: Z. Švestka, B. V. Jackson, M. E. Machedo (Eds.), Eruptive Solar Flares. Proceedings, 1991. XIV, 409 pages. 1992.

Vol. 400: M. Dienes, M. Month, S. Turner (Eds.), Frontiers of Particle Beams: Intensity Limitations. Proceedings, 1990. IX, 610 pages. 1992.

Vol. 401: U. Heber, C. S. Jeffery (Eds.), The Atmospheres of Early-Type Stars. Proceedings, 1991. XIX, 450 pages. 1992.

Vol. 402: L. Boi, D. Flament, J.-M. Salanskis (Eds.), 1830-1930: A Century of Geometry. VIII, 304 pages. 1992.

Vol. 403: E. Balslev (Ed.), Schrödinger Operators. Proceedings, 1991. VIII, 264 pages. 1992.

Vol. 404: R. Schmidt, H. O. Lutz, R. Dreizler (Eds.), Nuclear Physics Concepts in the Study of Atomic Cluster Physics. Proceedings, 1991. XVIII, 363 pages. 1992.

Vol. 405: W. Hollik, R. Rückl, J. Wess (Eds.), Phenomenological Aspects of Supersymmetry. VII, 329 pages. 1992.

Vol. 406: R. Kayser, T. Schramm, L. Nieser (Eds.), Gravitational Lenses. Proceedings, 1991. XXII, 399 pages. 1992.

Vol. 407: P. L. Smith, W. L. Wiese (Eds.), Atomic and Molecular Data for Space Astronomy. VII, 158 pages. 1992.

Vol. 408: V. J. Martínez, M. Portilla, D. Sàez (Eds.), New Insights into the Universe. Proceedings, 1991. XI, 298 pages. 1992.

Vol. 409: H. Gausterer, C. B. Lang (Eds.), Computational Methods in Field Theory. Proceedings, 1992. XII, 274 pages. 1992.

Vol. 410: J. Ehlers, G. Schäfer (Eds.), Relativistic Gravity Research. Proceedings, VIII, 409 pages. 1992.

Vol. 411: W. Dieter Heiss (Ed.), Chaos and Quantum Chaos. Proceedings, XIV, 330 pages. 1992.

Vol. 412: A. W. Clegg, G. E. Nedoluha (Eds.), Astrophysical Masers. Proceedings, 1992. XX, 480 pages. 1993.

Vol. 413: Aa. Sandqvist, T. P. Ray (Eds.); Central Activity in Galaxies. From Observational Data to Astrophysical Diagnostics. XIII, 235 pages. 1993.

Vol. 414: M. Napolitano, F. Sabetta (Eds.), Thirteenth International Conference on Numerical Methods in Fluid Dynamics. Proceedings, 1992. XIV, 541 pages. 1993.

Vol. 415: L. Garrido (Ed.), Complex Fluids. Proceedings, 1992. XIII, 413 pages. 1993.

Vol. 416: B. Baschek, G. Klare, J. Lequeux (Eds.), New Aspects of Magellanic Cloud Research. Proceedings, 1992. XIII, 494 pages. 1993.

Vol. 417: K. Goeke P. Kroll, H.-R. Petry (Eds.), Quark Cluster Dynamics. Proceedings, 1992. XI, 297 pages. 1993.

Vol. 418: J. van Paradijs, H. M. Maitzen (Eds.), Galactic High-Energy Astrophysics. XIII, 293 pages. 1993.

Vol. 419: K. H. Ploog, L. Tapfer (Eds.), Physics and Technology of Semiconductor Quantum Devices. Proceedings, 1992. VIII, 212 pages. 1993.

Vol. 420: F. Ehlotzky (Ed.), Fundamentals of Quantum Optics III. Proceedings, 1993. XII, 346 pages. 1993.

Vol. 421: H.-J. Röser, K. Meisenheimer (Eds.), Jets in Extragalactic Radio Sources. XX, 301 pages. 1993.

Vol. 422: L. Päivärinta, E. Somersalo (Eds.), Inverse Problems in Mathematical Physics. Proceedings, 1992. XVIII, 256 pages. 1993.

Vol. 423: F. J. Chinea, L. M. González-Romero (Eds.), Rotating Objects and Relativistic Physics. Proceedings, 1992. XII, 304 pages. 1993.

Vol. 424: G. F. Helminck (Ed.), Geometric and Quantum Aspects of Integrable Systems. Proceedings, 1992. IX, 224 pages. 1993.

Vol. 425: M. Dienes, M. Month, B. Strasser, S. Turner (Eds.), Frontiers of Particle Beams: Factories with $e^+ e^-$ Rings. Proceedings, 1992. IX, 414 pages. 1994.

Vol. 426: L. Mathelitsch, W. Plessas (Eds.), Substructures of Matter as Revealed with Electroweak Probes. Proceedings, 1993. XIV, 441 pages. 1994

Vol. 427: H. V. von Geramb (Ed.), Quantum Inversion Theory and Applications. Proceedings, 1993. VIII, 481 pages. 1994.

Vol. 428: U. G. Jørgensen (Ed.), Molecules in the Stellar Environment. Proceedings, 1993. VIII, 440 pages. 1994.

Vol. 429: J. L. Sanz, E. Martínez-González, L. Cayón (Eds.), Present and Future of the Cosmic Microwave Background. Proceedings, 1993. VIII, 233 pages. 1994.

Vol. 430: V. G. Gurzadyan, D. Pfenniger (Eds.), Ergodic Concepts in Stellar Dynamics. Proceedings, 1993. XVI, 302 pages. 1994.

Vol. 431: T. P. Ray, S. Beckwith (Eds.), Star Formation and Techniques in Infrared and mm-Wave Astronomy. XIV, 314 pages. 1994.

Vol. 432: G. Belvedere, M. Rodonò, G. M. Simnett (Eds.), Advances in Solar Physics. Proceedings, 1993. XVII, 335 pages. 1994.

Vol. 433: G. Contopoulos, N. Spyrou, L. Vlahos (Eds.), Galactic Dynamics and N-Body Simulations. XIV, 417 pages. 1994.

Vol. 434: J. Ehlers, H. Friedrich (Eds.), Canonical Gravity: From Classical to Quantum. Proceedings, 1993. X, 267 pages. 1994.

New Series m: Monographs

Vol. m 1: H. Hora, Plasmas at High Temperature and Density. VIII, 442 pages. 1991.

Vol. m 2: P. Busch, P. J. Lahti, P. Mittelstaedt, The Quantum Theory of Measurement. XIII, 165 pages. 1991.

Vol. m 3: A. Heck, J. M. Perdang (Eds.), Applying Fractals in Astronomy. IX, 210 pages. 1991.

Vol. m 4: R. K. Zeytounian, Mécanique des fluides fondamentale. XV, 615 pages, 1991.

Vol. m 5: R. K. Zeytounian, Meteorological Fluid Dynamics. XI, 346 pages. 1991.

Vol. m 6: N. M. J. Woodhouse, Special Relativity. VIII, 86 pages. 1992.

Vol. m 7: G. Morandi, The Role of Topology in Classical and Quantum Physics. XIII, 239 pages. 1992.

Vol. m 8: D. Funaro, Polynomial Approximation of Differential Equations. X, 305 pages. 1992.

Vol. m 9: M. Namiki, Stochastic Quantization. X, 217 pages. 1992.

Vol. m 10: J. Hoppe, Lectures on Integrable Systems. VII, 111 pages. 1992.

Vol. m 11: A. D. Yaghjian, Relativistic Dynamics of a Charged Sphere. XII, 115 pages. 1992.

Vol. m 12: G. Esposito, Quantum Gravity, Quantum Cosmology and Lorentzian Geometries. Second Corrected and Enlarged Edition. XVIII, 349 pages. 1994.

Vol. m 13: M. Klein, A. Knauf, Classical Planar Scattering by Coulombic Potentials. V, 142 pages. 1992.

Vol. m 14: A. Lerda, Anyons. XI, 138 pages. 1992.

Vol. m 15: N. Peters, B. Rogg (Eds.), Reduced Kinetic Mechanisms for Applications in Combustion Systems. X, 360 pages. 1993.

Vol. m 16: P. Christe, M. Henkel, Introduction to Conformal Invariance and Its Applications to Critical Phenomena. XV, 260 pages. 1993.

Vol. m 17: M. Schoen, Computer Simulation of Condensed Phases in Complex Geometries. X, 136 pages. 1993.

Vol. m 18: H. Carmichael, An Open Systems Approach to Quantum Optics. X, 179 pages. 1993.

Vol. m 19: S. D. Bogan, M. K. Hinders, Interface Effects in Elastic Wave Scattering. XII, 182 pages. 1994.

Vol. m 20: E. Abdalla, M. C. B. Abdalla, D. Dalmazi, A. Zadra, 2D-Gravity in Non-Critical Strings. IX, 319 pages. 1994.

Vol. m 21: G. P. Berman, E. N. Bulgakov, D. D. Holm, Crossover-Time in Quantum Boson and Spin Systems. XI, 268 pages. 1994.

Vol. m 22: M.-O. Hongler, Chaotic and Stochastic Behaviour in Automatic Production Lines. V, 85 pages. 1994.

Vol. m 23: V. S. Viswanath, G. Müller, The Recursion Method. X, 259 pages. 1994.

Vol. m 24: A. Ern, V. Giovangigli, Multicomponent Transport Algorithms. XIV, 427 pages. 1994.

MIX
Papier aus verantwortungsvollen Quellen
Paper from responsible sources
FSC® C105338

If you have any concerns about our products,
you can contact us on
ProductSafety@springernature.com

In case Publisher is established outside the EU,
the EU authorized representative is:
Springer Nature Customer Service Center GmbH
Europaplatz 3, 69115 Heidelberg, Germany

Printed by Libri Plureos GmbH
in Hamburg, Germany